AF802883

Introduction to
Quality Engineering

First Edition

Mehmet Ali Arslan
Gebze Technical University

MAA

Library of Congress Control Number: 2019933815

Introduction to Quality Engineering

First Edition, 2019

Copyright © 2019, by its Author

All rights reserved. No part of this book may be reproduced, in any form or by any means, without permission in writing from the Author.

The author and publisher make no warranty of any kind, expressed or implied, with regard to any teachings, knowledge, information, procedure, program etc. contained/presented in this book. The author and publisher shall not be liable in any event for incidental or consequential damages in connection with, or arising out of, the furnishing, performance, or use of these programs, files, knowledge etc. contained in this book. This book is not attempting to render quality, statistical or other professional services. If such services are required, the assistance of an appropriate professional should be sought.

ISBN: 978-0-9992005-6-8

MAA – Self Published Book, New York
Editor/Cover Designer: Mehmet Ali Arslan

Printed in Europe

The author/publisher of this book can be contacted at **dr.arslam@gmail.com**
Please also visit **www.maabook.com**

To my mother, father and children

Contents

Preface

Engineering the *robustness* into design manifests itself as *quality* under true management skills. The *quality engineering* is a *wise* combination of interdisciplinary engineering and management; both utilized in establishing the overall product/process quality to acquire customer satisfaction. The quality engineering mainly deals with uncertainty, therefore it struggles with many non-deterministic events from start to finish in all aspects of production/process with complex human behaviors included. Why is quality? Because companies always value increasing quality/productivity improvements so that they can be part of the economic competition in global markets. Every person of the company (business managers, engineers, quality people, technicians, workers, interns, etc.) is from different backgrounds with experiences so that everyone is responsible for the company's success.

In this context, the teachings of quality engineering based on statistical and management concepts can offer an unprecedented contribution to the company's growth. How? Because the *events*, affecting product/process quality, take place in non-deterministic manners and statistics can offer a pertinent methodology to model/understand and analyze such events. But companies also require and value management leadership. Effective management is always sought in any organization in establishing quality insights, dealing with customers and nicely/respectfully treating the company employees. Good management can contribute to directing/coordinating employees in the implementation of quality tools in the company.

The first edition of *Introduction to Quality Engineering* is truly a comprehensible book, emphasizing coverage on the main subjects of quality including statistics, engineering, and management, all joined for benefit of the company. This book is specially written for the use of all levels of students from a wide range of backgrounds and designed to equip them with hands-on knowledge and experience in understanding and implementing the quality. Therefore, the basics of quality introduced in this book require entry-level of algebra and beginning level of computer skills, thus making this book straightforward to adapt for any student. In the literature, one can find many books written on quality and its applications. While no single book can meet all expectations of students, yet there still exists challenging/increasing need in teaching/experiencing quality and its hands-on applications with a quick and efficient way of learning and application.

The focus of this book is to introduce the students to quality engineering with quick turn-around learning time and get them started using most important quality tools/methods on the solution of their own. These goals are supported/enhanced rather clearly and efficiently by using both traditional hand calculations and computer-aided statistical software, i.e. *Excel* and Minitab. All key applications are demonstrated in the carefully chosen example problems. Supplementary exercises are further provided for students to test and forge his or her practice skills at the end of the chapters.

Finally, this book can be suitably used as a main or an auxiliary introduction book for any level of quality engineering course for students from all educational levels and disciplines with its easy to use and practical content. Thank you for adopting this book, and I hope you benefit from the first edition.

Mehmet Ali Arslan-2019

Chapter 1

Quality Basics

For many years, it has been an on-going struggle for companies to survive in a hectic business environment. There is no doubt that being *profitable* is the key and directly linked to continuing profitable sales(economics) which require a *"good"* amount of market share with elusive positive reputation coming from satisfied and happy customers. Thus, we can assert that; the *market share* and *profitability* are two major indicators for any company's success as for being competitive in the local/global markets. Preference of a product, from customer's perspective, is a practical optimization problem, which mainly depends on both (preferably low/reasonable) *product cost* and (high) *product quality*. It a must for companies to minimize their waste, the lead time and rework to acquire both. Moreover, *quality* may be regarded as more important than the immediate *low-cost* in short term, when considering long term costs due to warranty repair claims, bad reputation/returns, stock market share losses etc. In this book, our primary focus will be on the main concept of "*Quality*".

Quality is a meaningful and powerful term that everybody somehow loves to adopt and see/touch in their lives. *Quality* is desired, welcomed, loved and so it is sought/wanted by everyone. Nevertheless, it might be perceived or understood differently. Therefore, in the eye of the beholder, it has many aspects when being evaluated. You might say how come? Well, there are almost eight billion people on earth each from a different culture, region, geography, expectations, experiences, climate, land, with different needs, prejudice etc. so that their understanding of "*quality*" very well may differ quite dramatically! Although quality is a reality! at the end, it is a perception. If the perception of the people who vote for the certain product's quality, greatly converges to a common judgment, then we can talk about true assessment of quality.

Finally, we have no choice but to accept the *quality* as a key which opens the doors of preference or choice in the eye of the customer. For any product or process, as illustrated in Figure 1.1, we can clearly see the dependency between the market share in the numerator (i.e. maximizing) and the reduced cost in the denominator (i.e. minimizing), which are all obtained with improved quality in design and manufacturing stages. Finally, we can maximize the profit (also be called *S/N* ratio as our optimization problem's objective).

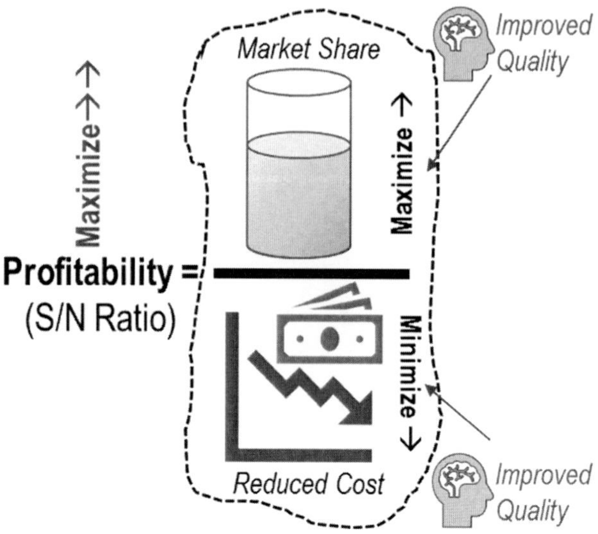

FIGURE 1.1 Profitability vs. quality

Many companies with big market-share put their product or process quality at most priority to assure customer satisfaction and loyalty, later makes customers prefer buying the same company's product(s) again and again. It is said that the best advertisement is a happy customer. To do so, quality must be an "*attribute*" of a product at the required or higher level possible at a projected/affordable cost which is deemed/comprehended "*okay*" by the customer.

So companies need "*smart*" designs that they never let them down! Such designs are called "*Robust Design*". ***Robust:*** **1540s, from Middle French *robuste* (14c.) and directly from Latin *robustus* "strong and hardy" literally "*as strong as oak*"… (Online Etymology Dictionary).**

1.1 What is Robust Design?

Dr. Genichi Taguchi, in the 1950's was working for the Japanese Company, is the initiator and developer of *robust design* methodology. His critical insight establishes the recognition that even small variations in a system design parameter(s) can significantly affect the overall product performance. Taguchi realized that traditional approach using tolerances and specification limits can be rather misleading and restrictive, yielding poor quality products. Sometimes instead of tightening design tolerances, finding an alternate solution using "*smart re-design*" so as to an alternative but insensitive solution to variations is obtained for product/process of interest.

Robust design is a methodology, where the *quality* is both designed and integrated into the product or process with "*wisdom*". Of course, wisdom needs guidance and utilizes the mathematical science of statistics. Statistics is a tremendous asset, yet it is not very well understood and appreciated by most of the people! As you may notice any product's quality measure has its basics in statistics. Therefore, we are going to bear most quality characteristics onto statistics in the progress of this book chapters step by a step; so that all robust design tools are clearly understood and put into application in real-life by all levels of students from different backgrounds.

In the industry, the concept of *robust design* starts in the early product design phase, next applied over the product's manufacturing cycle with precision craftsmanship and ends up with the final-use period of the product in customer's hands. In essence, the performance (perceived quality) of the product may exhibit varying characteristics. Why? Because, every product/process suffers from variations i.e. time, temperature, pressure, humidity, voltage changes etc. In brief, if any design is less sensitive to the unwanted inner or outer effects (also called "*noise*"), the better design for product quality is acquired! That is easy to say but it might be highly demanding to do!

Apparently, we need to eliminate/minimize all effects of *noise* on the product affecting negatively its working performance so that we can accomplish a *robust design;* thus we could improve the quality of products. To come up with a robust design, we must try to eliminate or minimize the noise first! Somehow if you may not be able to do that for some instances, then you should try to come up with a re-design so that your product is less sensitive to the noise. The *Robust design* presents itself by being insensitive to the noises which cause variations on the product performance. In a more simple definition *robustly designed products never lets you down!* when the other similar products are doing so! Why? Because your product is better designed/produced to handle or tolerate variations! which kills your product performance!

- For example, if your car does not start at -25 °C, your car is not designed to undertake low ambient temperature or your fuel line is frozen under these conditions. You cannot drive your car at +40 °C with air condition on since your radiator and fan or radiator fluid are not designed/suitable well enough to do so! Eventually, your car's engine overheats and if you do not cool it down properly it may even burn down.

- You may be a good driver for a couple hours driving, but say over four hours of driving, you might be a disaster to the others and yourself. Excessive driving time, being tired and loosing motivation for driving etc. have destructed your skills and made you a dangerous/hazardous driver. There need to be security measures taken to keep you and traffic safe!

- Or, you live in a rural area and your TV is brand new and equipped with the latest technology. But there are fluctuations in electric voltage in your grid of electric network. You do not always get 120/240 Volts, as nominal Voltage for your TV. Since your TV is not designed to compensate voltage fluctuations, you may never get neat pictures and broadcasting. Eventually, if you do not buy a regulator for voltage drifts, you may get your brand new TV input voltage unit damaged. Clearly, your TV manufacturer has to consider this kind of variation and design the TV which could operate at disastrous voltage drifts so that no warranty claims are filed upon due to this reason.

- Or, you have purchased a brand new hot-water heater, it does not have enough heating capability(power) and you do not get hot water in cold winter days. You would not want that.

Then, you might ask yourself; Why are customers so selective? Because they are not earning their salary so easily! That's why, they want to spend their money on the "*goods*" those never let them down! and make them satisfied and happy.

After previous discussions, we can state the relationship between quality and robust design; the *quality* may be regarded as the final consequence/outcome of *robust design*, just like the preferred/outstanding performance (state of capability, product attribute) of product or process. *Quality* is the out-looking face or character of the product which is designed, preferably using *robust design* techniques.

FIGURE 1.2 Chain of relationship: robust design, quality, and customer satisfaction

We can draw the very basic *flow chart* in serial manner; i.e. the chain of relationship for all of the above three arguments (i.e. quality, robust design, and customer satisfaction) are depicted briefly in Figure 1.2; That is,

- engineers work in teams on the pre-determined product that the customer is looking for; car, TV, cellphone etc. i.e. customer needs
- by applying robust design techniques on this product's relevant key characteristics, we plan and implement robust design strategies
- after that, we manufacture the product with "*wisdom*" as planned in previous step. Adopt "*design for manufacturing*" techniques with required *precision* machining to meet robust design requirements
- get the product right on the target specs and make high quality products so that it would be heavily/vigorously demanded by customers
- finally, the product is in the market and these products are sold with good amount of profit, and all production and marketing efforts are turned into cash for company to survive!
- over the period of product use (i.e. economical life of a car is ten years), product quality is realized/appreciated by customers. This makes customers satisfied and happy!
- nevertheless, there is no good advertisement than happy customers!

The robust design methodologies have a wide variety of application areas in many sectors, i.e. aeronautical, transportation, electronics, textile, franchising etc., to increase quality, to lower cost, to maximize profits by attracting the interest of the customers by providing them top *quality* products or services.

Traditional tools such as DOE (Design of Experiments), especially Taguchi methods etc. may be used in a robust design, where we do need input/output arrays (i.e. data) to evaluate the certain characteristics of products or processes. In brief, DOE requires;

- the collection or generation system/product data with wisdom to use in statistical assessments. This information is so valuable and provides an initial step towards handling the desired level of any chosen outcome.

- naming, defining and measuring this key outcome = reflection of quality, as perceived by the customers.

1.2 What is Quality?

We do not need a fancy or condense wording to define the *quality*. Let's try our straight forward people's definition. *Quality* is a relative term which can be viewed from various aspects, in the eye of the customer, if a product (or service) that you got makes you happy for its intention of use for certain time, and you keep saying *"I did a good job by purchasing this product"* and you keep thinking "*you got your money's worth*".

Then, you are a happy customer with the product or service you got. You are satisfied! Why? Because you felt/experienced/enjoyed/lived/had the "*Quality*".

Quality is defined in American National Standards Institute/American Society of Quality Control Standard A3-1987 as;

"The totality of features and characteristics of a product or service that bear on its ability to satisfy a given need"

Satisfaction of the needs always draws the bottom line for *quality*. But it is a fact that satisfaction concept also depends on the individual's expectations. To be more scientific, *Quality* needs to be put/expressed in *mathematical* form for appropriate/dependable/valid problem handling with satisfaction "*integrated*" in it. In such situations, Statistics can provide us the strong fundamentals for developing a valid/proven assessment of *quality* with a certain confidence.

Low cost, *reliable*, consistent and never letting you down performance of a product (or its expected features) or service are always important when you are talking about the *quality* and customer satisfaction.

In his well-known book, J. M. Juran defines these two fundamental quality components as;

i. *a product's features*; consider a battery that delivers 12V 60 *Ah* output for normal starting of an automobile engine ignition

ii. *a product's conformance* to these features; in any excessive driving or climate conditions, design voltage and amperage are going to be delivered by a battery for ignition.

Apparently, *product features* what makes the product attractive to the customer. Further, in the lifetime usage and operating range of the product, *conformance* to these features is vital to make sure, the product is capable of delivering on/near target performance consistently even under variation. Both quality components are the elements of robust design and customer satisfaction.

Customer satisfaction is not so easy! and can easily be considered as a multi-objective optimization problem. How? Let's assume you need to purchase a cell-phone cover, the question is, what do you expect from this cover? This clearly also depends on the gender of the user, i.e. male or female,

On top of that, you may want,

- a reasonable **low** cost cover,
- a **nice** looking cover,
- a reliable, **high** strength cover, in case you drop your phone, it would be safe and you do not need to worry about it. Your phone is protected under all conditions, hot, cold weather and on all platforms. Beside, you are very well aware of limitations, nothing is perfect!
- A **high** consistency in *conformance*, i.e. the same/similar expected performance and beautiful appearance, buttons are well operating, camera zoom, speaker, audio spaces are well designed with your phone etc. Easy phone in/out by hand without looseness or blockage.

As you would notice, there are at least four objectives need to be taken care of when making an assessment/decision on the quality of the cell-phone cover before spending any money. Of course the solution will differ for groups/gender/background etc. of buyers as pointed out.

1.3 Organization of the Book

This book is designed to be easily accessible and quite handy, with its rich and useful, content, applied to related quality engineering problems. The order of the chapters is crafted in a way to teach and motivate the students to use and apply relevant ideas and tools in their own problems with hands-on knowledge.

The book begins with a brief introduction of quality and robust design concepts with their comprehensible definitions. It also emphasizes the importance of both market share and profitability for survival and their strong ties with quality as presented in this Chapter 1, where customer satisfaction is also highlighted as a crucial outcome/consequence of quality. In Chapter 2, an introduction to quality engineering and its application characteristics are presented, especially dimensions of quality, cost of quality and a framework for quality improvement are introduced with examples.

In Chapters 3, core subjects in descriptive statistics are introduced for establishing fundamental statistics background for use in the context of quality engineering, in the consequent chapters. In Chapter 4, probability distributions are briefly introduced and normal distribution - *as being the most important and widely used probability distribution in statistics* - is introduced in detail. In Chapters 5, statistical inference on single and two samples/populations are introduced for various scenarios. This chapter also provides a valuable insight for use in the following Chapter 6 and 7 which are designed to introduce several analysis techniques of both simple linear and multiple regression analysis, respectively to explore/model causal/empiric relationship model between input and output variable(s). Outcoming regression model's validity and expected capability of the regression model is further emphasized/evaluated by using several well-known statistical metrics. In this context, *response surface methodology* is also outlined and all practical steps in constructing such an efficient response surface function are explained in detail.

Chapter 8 steps into the introduction to designed experiments methods and their popular use in parameter design. Chapter 9 introduces and discusses the use of both full-factorial-design and Taguchi design experiments. Chapter 9 also compares full factorial design vs. Taguchi design experiments in terms of their effectiveness and the number of experiments/efforts needed to execute each design.

Chapter 10 introduces and emphasizes the usefulness of Poka-Yoke methodology to prevent human mistakes etc. in production/manufacturing. The importance of Poka-Yoke as fault/mistake prevention methodology is highlighted. Chapter 11 presents production control charts and Chapter 12 further introduces the process capability to use in product quality assessment and monitoring to make sure stable/in-control production thus products.

Chapter 13 focuses on the quality loss function i.e. especially goal post and Taguchi loss function methodologies and their comparisons to quantify the amount of quality loss in product quality characteristic evaluations. Finally, the book is concluded by highlighting the total quality management's importance in quality engineering with its conceptual/desired interactions and related methodology between management and employees in Chapter 14.

1.4 Exercises

1) Describe quality using your own terms.

2) What is a robust design? Choose an example design and manufacturing problem, discuss what might be your precautions to make your example problem more robust.

3) Discuss what features of the product of your choice must possess for you to say the product is good? I got my money's worth!

4) Investigate descriptive and inferential statistics with examples.

5) Investigate how to obtain probability density function(p.d.f.) for any given distribution.

Chapter 2

Engineering the Quality

The *quality engineering* has very comprehensive meaning which accommodates various methods/tools i.e. concept of management, auditing, quality tools/approaches, product and process control/monitoring and design, applied statistics, cost management, human-oriented philosophies, computer-aided systems, and planning etc. to meet the customer needs with quality products/processes.

Considering the business point of view, companies must invest and do more research and development (R&D) studies on their products or processes to be competitive. According to *"Capital IQ data, Thomson Reuters Eikon data, Strategy & analysis"*, total R&D spending by the top 20 (global) companies was $214.5 billion in 2018, i.e. up to 25% of their revenues to improve characteristic features and quality of their products/processes/services, so that they can offer/sell them (remember, quality is one of the key attribute of product to get preferred by the customers) and stay alive in the very sensitive global market conditions. I might add,

"For businesses staying on the top requires - being smart, bold and wealth spent"

In our time, globalization is getting more and more popular, i.e. information travels fast and "*goods*" are transported easily too. Local companies and businesses feel the pressure in such a highly competitive environment. Keep in mind that, as far as local companies concerned, many local people and their families do depend on their company's success for bright/stable tomorrows with job security. Therefore, all companies must do their best to be successful in keeping the wealth in-house, of course by making their household happy, first. This is perfectly understandable.

How this success is maintained? Well, in this context, two main indicators quickly catch up our attention, i.e. *market share* and *profitability*. Because, they have been two major "*players/end results*" in businesses for looking at feature with great expectations. The stronger the players, the better the team/company plays. Based on this argument, we can state that the following three main factors/attributes drive above two players (i.e. market share and profitability*)* for both controlling and improving the companies competitiveness;

i. ***Unit product cost***; the "*possibly lower*" prices are quite tempting from customers perspective, but sometimes quality may be desired/sought more than a low price itself! How? There is always a *trade-off* in many decisions customers make! i.e. cost and quality must negotiate/sacrifice at some middle point of agreement.

ii. ***Product quality***; ever wanted attribute of the product that quality products never (or almost never to be realistic) lets you down! perform its intended jobs, and makes you feel you get your money's worth to the last pennies. Happy customer means better sales in the future.

iii. ***Lead time***; from concept and raw material inception to the final sale/dispatch of the finished product to the customer in which wisely planned efficient works/efforts spent are turned/transferred into maximum gain(profit) for the products sold.

If we combine these three objectives into one multi-objective, our engineering (optimization) problem can be stated as follows;

As a company, you would like to have great products with overwhelmingly higher quality when compared to the similar products at the market, while minimizing lead time and lowering unit product cost! In other words, maximize quality, minimize lead time and unit product cost all, concurrently! On top of that, you need to satisfy multiple product/quality requirements/constraints which are studied as in "*dimensions of quality*".

In practice, quality engineering is very vast subject and has multidisciplinary engineering, statistical and managemental aspects. In the context of the book, we will be focusing more on first two aspects where analytical tools are also presented for improving quality of products. In the following, in a metaphorical sense, "*degree of freedom of quality*" is introduced in terms of characteristic "*dimensions of quality*", next!

2.1 The Dimensions of Quality

Quality is a multi-objective outcome of the several dimensions/factors where each dimension contributes to quality to the certain degree, and the contribution rank/order for each dimension can very well vary depending on the type of the product or process. Garvin(1987) and Moen(1991) et. al. have collected/summarized these dimensions as presented in the following;

- *Performance*, the main degree of accomplishment for operating task as output
- *Features*, an auxiliary supporting distinctive characteristic of the product
- *Time*, 4th dimension to move things in a usable manner. Time is money! Time is limited! And time is what we need to spend to convert raw material into the final product. Time must be spent very carefully efficiently and in a most productive manner. I have witnessed to so many people they think they are working hard to solve their problem for a week, but there are some people get the same job done in a couple of hours. Working hard does not mean working productive! It is a relative term!
- *Reliability,* functioning capability of the product without failure under intended conditions for stated/declared period of time.
- *Durability*, a capability of remaining functional for a relatively long lifelong usage of a product, with minor maintenance or repair.
- *Uniformity*, identical/consistent (i.e. very low variation) outcome of all products being manufactured, or services rendered.
- *Consistency,* unmatchable adherence to the correspondence/uniformity among the behaviors for the same operations. You always get what you expected!
- *Serviceability*, the capacity of being able to resolve (customer) complaints/ problems thus not making adversaries but satisfied customers.
- *Aesthetics,* a reflection of beauty to the senses
- *Personal interface,* interactive affections between individuals, i.e. truthful, punctual, kind, funny, serious etc.
- *Harmlessness,* a state of environment friendly, feeling secure, healthy and peaceful now and tomorrow
- *Perceived Quality,* indirect/untold affection/inferences/sense about a product makes someone feel good about it, during its use. Sometimes, perceived quality may be misleading for some people, as well. Good cars may not sell while low quality ones do sell. Social events or understanding may cause to this, or someone may object our above argument about on good/low quality car brand, and assert that his/her car preference is good and valid!

- *Advertisement,* bringing the products to the user's attention and causing familiarity with product/brand or its features may be good thing if done rightfully; so that it increases the sales, the good budget supports quality improvements for the companies. Then, product quality increases product sales. Remember, many good quality products may vanish if their sales do not support the investor, therefore quality products should be advertised for customer's own benefit, as well.

I added the latest *Advertisement* dimension which is a major industry; using radio, TV, newspaper and the increasing internet time. It greatly uses/affects people expectations/ perception/likes/dislikes and make them aware and familiar with the product. You might have a great product but if nobody is aware even such a product does exist. Then, inferior products are purchased and used. This is no good from the quality perspective.

It is also a responsibility that the good word about good quality products must spread. If good quality products are preferred, it is good for society, for the environment, for companies, because completion among companies, most of the time, may trigger making of even better quality products/goods. Nevertheless, through advertisement sometimes inferior quality products may also be advertised and people may be brainwashed, this is certainly not what we meant above!

2.2 Cost of Quality

Dr. Joseph M. Juran in 1951 mentions this concept in his "Quality Control Handbook". Later this concept had found its popularity by Philip B. Crosby in his book "Quality is Free" (Beecroft, 2000). Unfortunately, most costs of quality measures remained less obvious i.e. buried in the process, as hidden(ghost) contributors.

Resembling the iceberg, we see only the tip, below the water surface is an imaginational, i.e. not easy to capture mystery. While the real costs of poor quality in operations/delivery of a product lie below the surface of the water, many businesses could only measure the tip of the iceberg (more obvious ones).

Most of the time, quality will add more investment cost to the production. But, that is unavoidable, and when considering the wide-span duration of the production cycle and product usage, this cost will be compensated in the mid/long term. Mainly, we can itemize these costs into four items as depicted in Figure 2.1; i.e. first two as *"good"* and next two as *"bad"* for quality.

FIGURE 2.1 Break-down for cost of *quality*

- ✓ *Prevention cost,* the limited amount of resources spent in the earlier phase of the design, can create a much greater impact on the product quality which could never be acquired when done at the later stages of manufacturing, even much greater amount of money is spent.

  Most activities designed/arranged in a way to prevent poor quality, i.e. quality planning, supplier selection, preliminary/detailed design reviews, new product reviews, error checking procedures, Poke-Yoke, in house capability/quality reviews, in house trainings (multidisciplinary team projects, work security training, process reviews, six-sigma training) etc.

- ✓ *Appraisal cost,* includes costs as part of the quality control process to make sure offered products/services to meet customer expectations and conformance to the regulatory requirements. These costs could include expenses for field tests, inspections, measuring and auditing etc. We must prevent defective products from reaching out to the customers.

  Creativity is encouraged for companies and its ok to incur appraisal costs to inspect/detect faulty products before they got in customers hands, i.e. materials and services consumed, real-time in-process tests, final inspection and tests, on-site field tests, several audits for product, service offered, equipment calibration, final product/process adjustments, feedback etc.

- ❖ *Internal failure costs,* associated with defective products spotted in the house before shipping to the customer, i.e. any waste, rework, any re-*times* as re-design, re-manufacturing, re-order, re-analysis, re-test, downtimes etc. It is good to spot defective products internally.

- ❖ *External failure costs*, associated with defective products spotted after shipping to the customer, i.e. warranties, recalls, bad reputation, complaints, customer product abuse, accidents/delay in transportation cost, insurance costs, fines etc.

This is certainly the worst case for reputation and fixing external failures do cost and it hurts!

If you sum up all of these costs, you find the total quality cost, but many of the costs of quality remain elusive since they are not easy to identify by traditional measurement systems, remember iceberg.

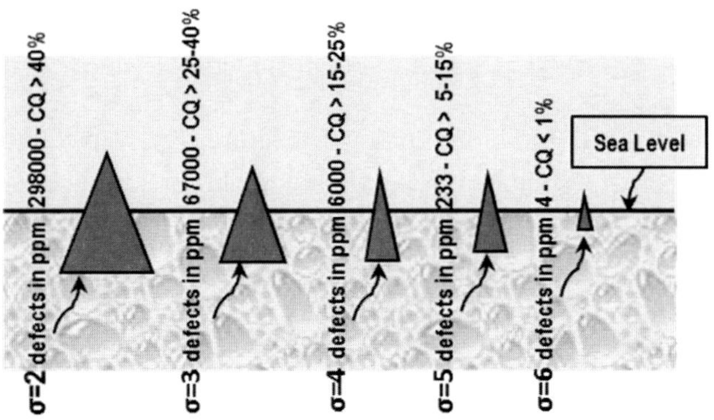

FIGURE 2.2 Iceberg of CQ - Cost of quality for Sigma levels

Sigma level determines the performance/success of a company in creating defective parts in *ppm* (i.e. σ=2, *298000 parts per million* is defective, which means almost 30% of the parts are defective!) based upon this waste, cost of quality is given in the literature. As this relationship is clearly stated, we can go ahead and illustrate these facts on a representative iceberg masses as presented in Figure 2.2.

For example, when σ=4, the company loses between 15-25% of its annual revenue due to quality costs. An improvement in quality to σ=5 would increase net profits quite dramatically.

Obviously, identifying and reducing these costs will significantly lower business costs. The amount invested in quality cost may vary from as low as 5% to as high as 40% of sales. Quality costs do differ from industry to industry or company to company. It is a matter of vision.

This brings up the vital concept of "*measuring all costs*" to great importance. When you cannot measure anything coming in or going out, you cannot account for that and never act upon that thing for book keeping, recording or improving. The power of true *book-keeping* is indisputable crucial for any company for monitoring, steering the wheels, quality/cost improvement and taking the present and future decisions.

2.3 A Framework for Quality Improvement

Although quality improvement must be a continuous job/responsibility, product design stage is the just about the right time to build quality into the products. Because at this stage, designers have more flexibility/freedom for design changes/improvements to make at the cheapest cost. Engineers know that the design itself is a *repetitive* process and most of the time to obtain a usable/feasible product design, all classical product design development steps must be repeated several times.

Companies carry out *wisely* planned *activities of quality* in production/process to minimize variation and come up with quality and productivity thus higher profits, with minimum waste, are earned.

In 1989, Taguchi et al. divides these activities into five major stages. Now, let's briefly look at product quality improvement opportunities when considering an example production cycle. The first step to initiate the production is to conduct *product planning* from there we move on;

1. *Product planning:* to meet the customer needs in functionality/quality and fulfill expectations, products are thoroughly planned using well-established methodologies and principles before products are being launched.

2. *Product design:* based upon previous product planning inputs, the design of robust products are launched to hit target values of product characteristics. For product, system design, parameter design, and tolerance design can be performed at this stage as introduced later.

3. *Process design:* stage plans all the activities required for production in accordance with the product design stage. System design, parameter design, and tolerance design can also be performed here in the process design stage.

4. *Production/manufacturing*: using all previous three stages, products are just produced as per the specifications, consistently. This stage may require monitoring and adjusting product characteristics for quality by adopting statistical process control. Quality inspections can be performed after this stage as pictured in Figure 2.3.

5. *(Product maintenance)/Service after purchase:* for keeping the product fully functional we may need to offer maintenance/service for certain periods. Further, we may repair or replace the defective products in case they reach to the customer to compensate their losses and unhappiness.

There is one important point to consider that, there may always be forward or backward communication or feedback between these stages, i.e. since design through after sales are directly linked to the remaining. Therefore, depending on the emerging needs, we need to consider interactions between these stages and re-iterate the flow when required as depicted in Figure 2.3.

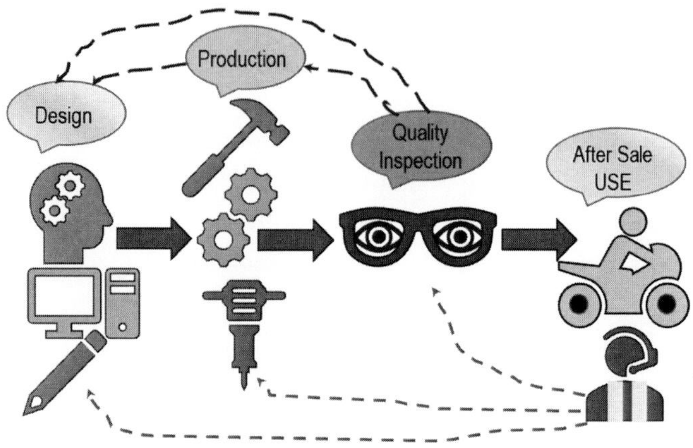

FIGURE 2.3 Activities of quality with interactions

Roughly all of these steps constitutes production cycle, depending on the type of the product, this cycle may take years. For example, according to recent news, an airliner latest engine development reportedly took almost twenty years of extensive research and development, manufacturing, testing time invested. Some cell phone companies also spend more than a year to modify their existing cell phone with new features so that they either maintain or increase their sales.

Under these explanations, let's briefly examine the type of efforts at different instants of overall production cycle. The quality improvement can be done in;

- Preferably *off-line* quality control studies in the product design/development phase are designed and developed even before the product prototype is ready or when being developed, hey this is a continuous job.
- Further, *on-line* quality control applications (sometimes *on-line* sampling but *off-line* inspection is performed when throughput is so many and it is impossible to test each of them in real-time) during manufacturing are enforced to make sure manufacturing process is in-control the way we want (stable, repeatable), thus to reduce defective parts.
- After sale use of the product by providing vital maintenance schedule and any required services to the products, so it serves to the customer as planned and promised. Companies always need/love loyal happy customers.

But one should keep in mind that once the product design is frozen and manufacturing starts off, companies hands are tied, and dramatic changes to the product design is either not possible or devastating in terms of cost. Once the product starts appearing from production line, at the expense of more severe cost, quality can also be improved at the production stage. *Statistical Process Control* (SPC) can be used to reduce non-conforming product characteristics. Even at the quality control stage some limited adjustments, most of the time, with limited correction to the product can be done. This issue is further explained in the next section.

2.3.1 Designing Quality

As this fact introduced in the earlier section, due to unmatchable freedom of change/ modification/re-design in the design phase, product design has to be done with quality attribute characteristics as designed/integrated into the design phase.

For example, as a well-known tech-company, you are going to design mold with all internal cavities/runners/sprues to produce TV back cover from plastic. You need to consider all design details for the cover with required thicknesses for "*low*" plastic usage and weight. Meanwhile the required strength/rigidity must be constructed too. Further, many small window openings for cooling purposes need to be placed where high heat generation occurs due to electronic parts heating, during TV operation. Further, the 3D cover part needs to be designed with required indentations to cover the electronic components in the back of the TV. Further, snap fits/screw holes should also be designed and placed properly for easy assembly.

Next, cover plastic material selection has to be done properly for the required strength, heat transfer capabilities and increased plastic molding capabilities. Clean manufacturing technology must also be used at all stages. Once the process planning is finished for both the design and manufacturing stages, a big injection mold needs to be designed, CNC machined and assembled.

Next, you would use your mold, in the high pressure plastic injection machine, where TV covers are injected plastic under high pressures into TV mold. In this operation, molten plastic is injected from several routes/runners for complete fill up of TV cover cavities without voids, bubbles, misruns, briefly defects. Finally, quality inspections are performed.

Assume in the product quality inspections, it was realized that design engineers made a mistake in cover design and you are at a point that huge plastic cover mold is manufactured and many TV cover parts are injected with deficiencies, eventually all you got is so many defective TV covers as waste.

How you would fix the problem? You may try to modify the TV cover mold? Sometime it is possible to re-modify the TV cover mold, but most of the time not! Then, you need to go all the way back to the design phase and make all corrective activities, following second time TV cover mold manufacturing, injection molding procedure as mentioned above again! Huge cost in money, time and resources are wasted! Hey, design and manufacturing is a serious job.

What if this design problem was discovered in early TV cover design phase, I mean, before any manufacturing actives made. Then, it is an easy fix without any cost to the company. This is exactly what we mean "*designing quality into the product in the early design phase*". It provides you freedom which means "*none or little cost*". Once the things are *"frozen",* we mean all required design and manufacturing are complete, then you become constrained in freedom of functionality as well as quality. Worst of all , what if you sold all of these defective TVs and customer complaints and warranty returns happen. It is a nightmare! How would you fix this?

Quality loss occurs when there are deviations in products operation characteristics from the target reference values. Taguchi establishes a very famous methodology/philosophy for designing quality into products. He suggests that the *product design* process should be performed in three-phase as outlined below;

- ***System design:*** concentrates on the development of system skeleton to operate under a set of nominal conditions. System design requires technical knowledge/experience to support the whole system using science and engineering knowledge to produce a basic prototype design.

- ***Parameter design*:** after the system architecture has been chosen, the next step is parameter design. Finding the optimum controllable system parameters (i.e. their levels) such that the product's functional characteristics (as an objective) are optimized while having minimal sensitivity to inner/outer noises. This is briefly an effective parameter tuning job. Parameter design is the very effective impact of Taguchi's approach for achieving high quality without an increase in cost.

- ***Tolerance design:*** is utilized by tightening tolerances around the chosen target values of controllable system parameters after/following *parameter design* if the resulting/expected improvement to reduce variation has been insufficient. Of course, this often results in an increase in production/manufacturing costs.

Of these three-phases in product design processes, the Taguchi method primarily focuses on *parameter design.*

2.3.2 Taguchi Quality Loss

Quality loss paradigm is used to ensure that the manufactured products/process fall within specification limits which are set in the design phase as explained in the previous section, i.e. on TV cover design and its manufacturing example.

All products meeting design specification limits are considered as good quality products. But this assessment is not complete/easy as it sounds, yet deviations may partially invalidate our rules for quality loss, therefore conforming/non-conforming product decisions must rely upon "*a sound*" quality loss functions. Taguchi quality vs. traditional loss concepts is introduced in detail in Chapter 9.

In Figure 2.4, the flow chart of very well-known Taguchi quality loss methodology is introduced in terms of the flow of the process. Where first customer needs are explored and identified.

Next, multidisciplinary teams of specialist get together and identify all important design factors and their levels, according to their design objective characteristics, i.e. min. noise, max. cooling power, nominal shaft diameter etc.

Next, experimental design is arranged using orthogonal arrays (see Chapter 8) while considering interactions among the design factors.

Once the problem set up is complete, then calculate/run the Taguchi design, analyze these results and make sure they make sense. After these analyses, we obtain a combination of design factor levels and their impact level in "*rank*", on the selected design (objective) characteristic.

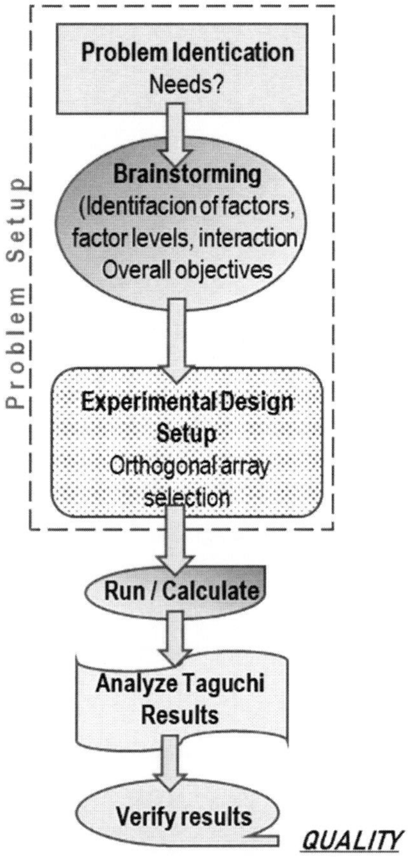

FIGURE 2.4 Taguchi design/analysis flow chart

Finally, use these optimum design factor levels in real experiment/simulation and verify that your optimum results are consistent with the real system output. If not, revise all design and repeat these steps until you reach reasonably good results, those would yield quality product characteristics.

2.3.3 Robust Design and Signal-to-Noise ratio

Robust design is a concept that promotes designing robust systems/processes in broad-sense, which are less sensitive to the noises while performing greatly. In this context, *Signal* can be considered an output (i.e. quality measure, objective) of the system, *noise* is the unwanted inner/outer effects that the signal is exposed to. Since it affects our system's true functioning capability/performance in an unwanted manner, we do not like the noise!

Signal-to-Noise ratio (*S/N*) basically integrates signal objective with noise factors in one objective function, thus making our optimization problem simpler as *unconstrained* where we would like to always maximize the *S/N* ratio, as also depicted in Figure 1.1.

This means we want high signal as outcome in the numerator, but low noise in the denominator, both will serve making *S/N* ratio maximum as objective. *S/N* formulations differs for various objective types, i.e. maximization, minimization, and nominal cases. Detailed information on the *S/N* ratio will be presented in Chapter 9.

2.4 Product Quality Improvement During Manufacturing

Any quality improvement opportunity must be realized and utilized during manufacturing/production as well. The main idea here is to minimize *waste* by reducing *variation*. Here *waste* term has a very comprehensive meaning in this context! "*Waste is defined as anything does not add value to the product*" in Japanese *Muda* definition. Just in time *(JIT)* manufacturing systems are based on *3M* philosophy; *Muda:* waste, *Mura:* unevenness, *Muri:* overburden.

To serve to the purpose, start with designing/integrating quality into products during the design phase. Next, prepare a process plan and based upon this plan, manufacture the products with the desired level of quality features (i.e. precision, operation range etc.). Of course this requires a stable and in-control process even under all noise factors. This is our challenge!

The steps for quality improvement during design/manufacturing/production and quality control check processes are simply demonstrated as shown in Figure 2.5.

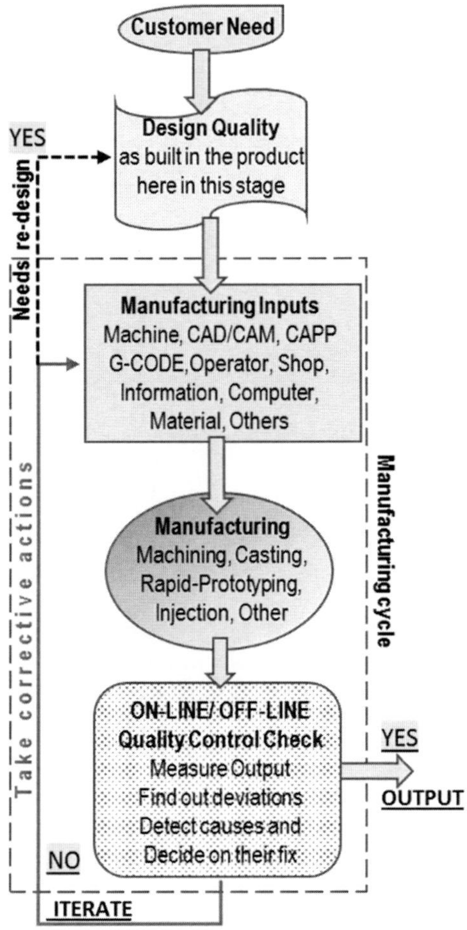

FIGURE 2.5 Quality improvement process in design/manufacturing cycle

Statistical Process Control is commonly used in monitoring process stability and improving process capability. Also note that in conjunction with design of experiments (DOE), SPC can also be used to reduce the process variability.

In 1991, Montgomery introduced many effective process improvement tools, as presented next!

Histogram: A histogram is a very useful distribution(shape) plot of the numerical data which is also used to estimate of the probability distribution of a continuous variable. Histograms are a type of bar plot for data that group the data into bins, either in vertical or horizontal directions, more information is introduced in Chapter 3 in detail.

Check Sheet: A check sheet is a typically a blank form that is prepared for quick, easy and efficient recording/collecting of the desired real-time data at the location where the data is generated. Collected/recorded data can be either quantitative or qualitative. It can be many of form and format depending on the needs.

An example of check sheet is presented in Figure 2.6, which shows certain losses to the publishing house, due to various reasons (i.e. a total of five main reasons) in the second week of January for an example book title defects/returns etc.

TOMORROW'S DREAM Publishing Co. Weekly Check Sheet						
Book T i t l e: Do Good						
Operator N a m e: Adam K.		**B a t c h:** 1000 per order			**D a t e:** Jan. 7-11,2018	
Defect/Reason	Monday	Tuesday	Wednesday	Thursday	Friday	Total
Scratch	3	1	2	1	1	8
Crushed Spine	2			3		5
Blank Page		1				1
Customer Return					3	3
Shipping damage	1		1			2
Total	6	2	3	4	4	19

FIGURE 2.6 Check sheet for an example book title

Pareto chart: It states that 80% of the effects (i.e. errors/problems/sales/benefits) are coming from 20% of the causes (i.e. variables/participants/factors). Considering this is generally true for workplaces, where a small fraction of the hard-working people get most of the job done. They take responsibility and courage for most of the time but the remaining people get the credit from their job outcome.

TABLE 2.1 Pareto chart for cell-phone user complaints

	A	B
1	**Cell-Phone Complaints**	
2		
3	Complaint	Count
4	Overpriced	30
5	Noisy	44
6	Poor reception	71
7	Breaking easily	4
8	Low battery-life	722
9	Slow respond	13
10	Freezing	12
11	Starts slowly	111
12	**Total Complaints=**	**1007**
13		
14		

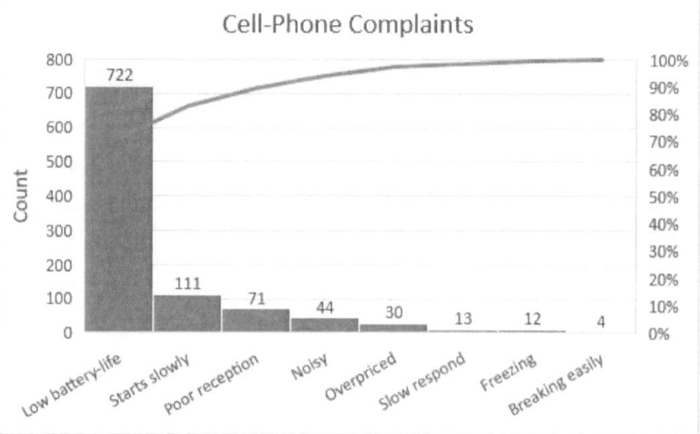

Let's assume as per an example, in the initial market reviews, the new model cell phones are evaluated by n=1007 sample users from different backgrounds, ages, and gender. The resulting data as presented in Table 2.1 is collected. We would like to analyze these data using a Pareto chart. We go ahead and draw the Pareto chart in *Excel* easily.

By examining this plot, we can realize that "*Low battery-life*" and "*Starts slowly*" complaints are two major ones, having (722+111)/1007= 83% of the all complaints. Accordingly, we can direct and focus our R&D efforts towards the elimination of these two complaints with the highest priority.

Cause-and-effect diagram: due to its similarity is also known as the *fishbone* diagram. It is a graphics tool that helps identify, sort, and display possible causes of a problem or a quality characteristic. When you have a serious problem, before pondering on the solution, it's recommended to explore all of the factors causing the problem. You do not want to get stuck to the limited opinions, then you might very well fail in assessing/finding the feasible solution. In this case, you need to repeat the process, but at the same time, you do not want to go back and forth empty handed! Time is an invaluable asset.

An example of a cause-and-effect diagram is presented in Figure 2.7 for a shop, manufacturing coil springs for automotive companies. But recently they suffer from serious problems in poor quality coil spring feedbacks and returns. A special team of people get together and draw this diagram. By examining the cause-and-effect diagram, several causes and their effects on the product are identified and concluded that these issues have to be dealt with care. A shop may require an investment in the workforce and its training, new advanced CNC machines etc.

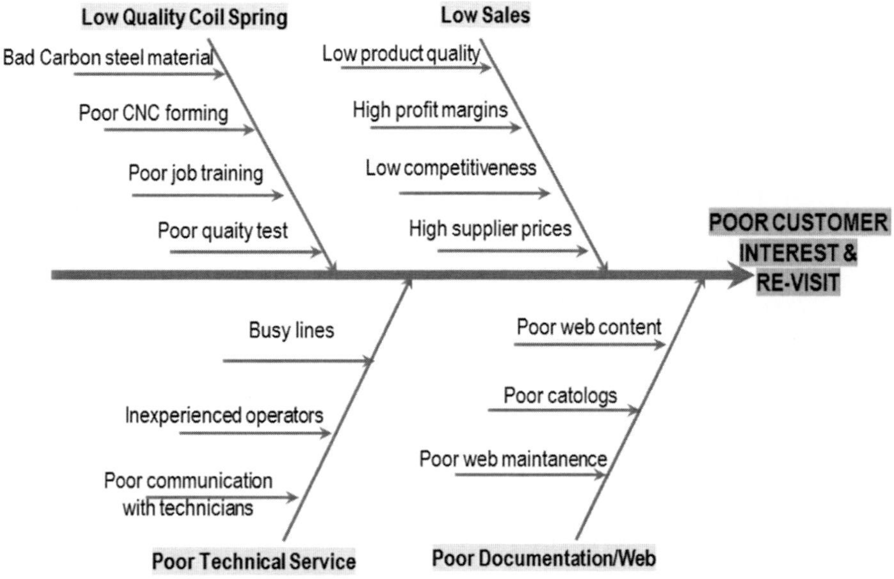

FIGURE 2.7 Cause-and-effect diagram for poor quality coil springs

Defect Concentration Diagram: is a simple visual representation as a diagram or map which shows all detected defects/problem areas with all relevant views displayed; for example if we re-visit book return case, all relevant factors(i.e. scratch, crashed spine, shipping damage, unclear cover color etc.) can be shown on 2D representative book covers with appropriate markings so that most of the important criterions are taken into account in the solution of book quality as shown in Figure 2.8.

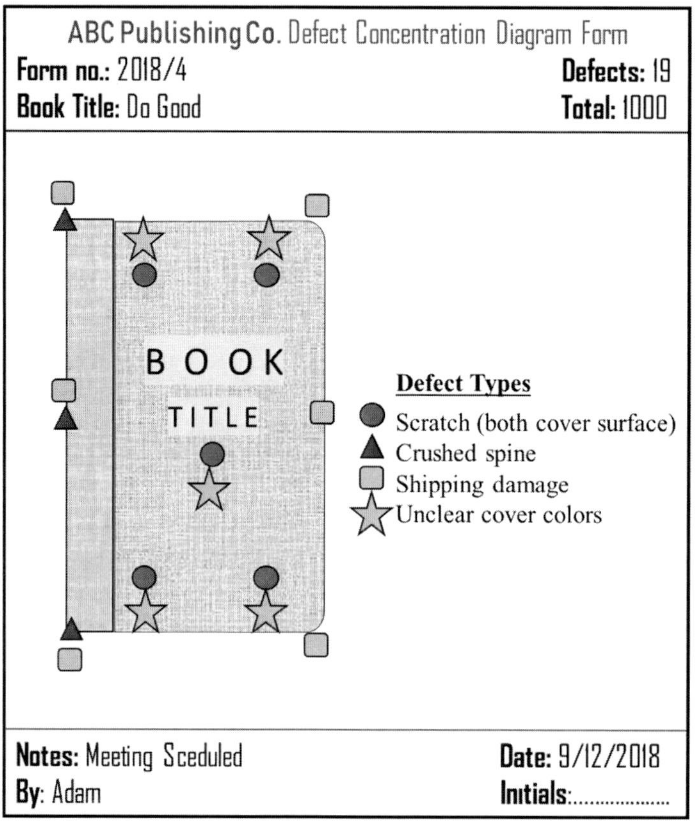

FIGURE 2.8 Defect concentration diagram for a sample book

Scatter Diagram: is also called a *scatter plot, scatter graph* which is very popular and useful *dot/line* plot using cartesian coordinates to display the pertinent relationship between two variables for a set of data. It is introduced in Chapter 3 in detail.

Control Chart: is a graph used to study/monitor how a product/process characteristics data vary over time. Data are sorted and plotted in increasing time order. A control chart always has a central line for the average, an upper line for the upper control limit and a lower line for the lower control limit are used for assessing safe specification boundaries. More are introduced in Chapter 11 in detail.

2.5 Poka-Yoke

The *Lean manufacturing process* is a continuing combination of efforts focusing on the waste (remember Japanese *Muda: waste* concept) minimization within manufacturing systems while simultaneously maximizing productivity. Therefore fault/defect detection should always be first done at early stages or during manufacturing/assembly as the processes contribute towards obtaining finished products.

Poka-Yoke (a Japanese term meaning *"mistake proofing"*) is an effective mechanism in a "*Lean manufacturing process*" that helps avoiding mistakes of all kind in manufacturing, assembly, operation and even in customer service etc. In 1961, reportedly invented by Shigeo Shingo, an engineer working for Toyota. Most of the time, different purpose sensors and special tools i.e. tooling-setup and molds are used to detect and avoid operator errors in the workplace. This concept is introduced in Chapter 10 in detail.

2.6 Failure Mode and Effect Analysis (FMEA)

Reportedly, developed in the 1940s by the U.S. military, failure modes and effects analysis (*FMEA*) is one of the earliest important reliability improvement methods. Today it is still used as a *structured* and *qualitative* reasoning approach to lower the *possibility of failure*. *FMEA* is a comprehensive step-by-step approach for identifying all possible failures in a design-stage/manufacturing/assembly/product/process or service.

If *FMEA* is broken down into two components, "*Failure mode*" is to determine the ways/modes, in which product/process might fail. Note that failures are any errors or defects affecting the customer, and can be in form of either potential or actual. *"Effects analysis"* deals with studying the consequences of those failures.

FMEA enhances engineering by applying the *knowledge* and *experience* of multidisciplinary experts (which are composed of true-experienced people, such as process owners, experienced designer/engineers/technicians/operators/practitioners) in order to determine/foresee the potential failure modes by assessing their risk/probability of failures.

Failures are prioritized according to their,

- severity, the seriousness of their consequences (priority is assigned)
- frequently occurrences
- easier detection

FMEA takes actions to eliminate/reduce failures, starting with the highest-priority ones. *FMEA* also documents/records current knowledge and actions about the risks of failures, for use in continuous improvement. *FMEA* is preferably used during the early design stage to prevent failures. Later it's used for control, before and during ongoing operation of the process.

In short, *FMEA* is an ever-continuing method which begins during the earliest conceptual stages of design and continues throughout the life of the product or service. FMEA can be used when:

- ✓ a new product/process is being initiated, after *quality function deployment* (QFD).

  QFD is simply a process and set of tools used to effectively define customer requirements and translate them into detailed engineering specifications and plans to produce the products that fulfill those requirements. QFD is used to translate the voice of customers into measurable design targets and drive them from the assembly level down through the sub-assembly, component and production process levels. We can use FMEA when;

- ➢ changes are made to the product operating conditions.
- ➢ a change is made to either the product or process design.
- ➢ when the product design change affected the process or vice-versa.
- ➢ the new regulations are in use.

FMEA has several benefits in assisting the engineer to

- ➢ improve product/process reliability and quality
- ➢ increase customer satisfaction
- ➢ identify and eliminate the potential product/process failure modes
- ➢ prioritize product/process deficiencies
- ➢ capture/back up engineering/organization knowledge

2.6.1 Main Steps of FMEA

1) ***Identify potential failures and effects***; for the problem of your interest, analyze functional requirements and their possible effects with multidisciplinary experts. And, identify all possible failures modes. Such as, blade root cracking, increased high temperature effects, excessive tool wear and vibration etc.

2) ***Assess the problem severity;*** by assigning ratings from 1(lowest) to 10 (highest);
   1: No effect
   2: Very minor, almost unnoticeable
   3: Minor, partial system effect, an average customer can notice
   4-6 Moderate, most of the customers are not happy
   7-8: High, customers are dissatisfied
   9-10: Very high, the hazardous product can cause injury/death, customers are angry

3) ***Assess the frequency of occurrence;*** by assigning ratings from 1(lowest) to 10 (highest);

   1: No effect
   2-3: Low, a few failure incidents
   4-6 Moderate, occasional failures
   7-8: High, repeating failures
   9-10: Very high, hazardous failure is almost certain, customer's trust is at stake and customers are angry

4) ***Detect the failure;*** once the problems are fixed, we go ahead and test the product to witness the effectiveness of our precautions/fixes by assigning ratings from 1(lowest) to 10 (highest);

   The chance of ***fault will be caught by testing***
   1: certain
   2: almost certain
   3: highly probable
   4-6 Moderately probable
   7-8: Lowly probable
   9-10: un-detected/ no-way

Risk priority number **(RPN) = Severity * Occurrence * Detection**

RPN should be assessed for the entire product/process and documented in the FMEA. Apparently, the highest RPNs should get the highest priority for corrective measures taken for fixing i.e. change the design first, improve inspection testing and manufacturing capabilities etc.

Keep in mind that, after taking corrective measures, you need to go ahead and re-calculate the RPN and document the results in the FMEA to see the effectiveness of your actions.

2.6.2 Tire Installation FMEA Example

In the guidance of previous remarks, let's present here an example of a very simplified FMEA study for four-tire installation practices at an automobile assembly station. For this purpose professional team members got together and come up with an FMEA table with a total of five potential failure modes.

Next, team members score/rate *severity, occurrence* and *detection probabilities* (all of the participant scores can be averaged, if necessary). Table 2.2 brings up and displays the result of this scoring on example FMEA format.

TABLE 2.2 FMEA for tire installation practice

Failure mode and effects Analysis (FMEA)			**Date:** 9/15/2017 ***Rev:*** *1.1*	
Process Name: Automobile Tire Installation Practices **Process Number:** TIREINS-002			**Meeting room:** East-1	
Failure Mode	(A) SEVERITY Rate 1-10 10: Most severe	(B) OCCURRENCE Rate 1-10 10: Highest Probability	(C) DETECTION Rate 1-10 10: Lowest Probability	RPN Factor = A*B*C
1) Select wrong tire size	5	1	8	40
2) Wrong bolt tightening torque	9	3	5	**135**
3) Damaging threads while tightening	5	2	1	10
4) Faulty tire installation to the rims	8	2	2	32
5) Wrong tire air pressure fills	4	3	4	48

After examining Table 2.2, we clearly see that out of five potential failure modes, those have been identified by our professional team members, the *failure mode* i.e. number two, has an RPN value of 135 and therefore it gets the highest priority, for corrective measures assigned for the process improvement.

In such events, establishing sound/realistic guidelines for rating the values for severity, occurrence, and detection will contribute to RPN and make it as realistic/useful as possible.

2.7 Automated Inspection

Automated inspection is designed/utilized to make sure product/process conformance to the specifications set priori in the design for manufacturing. This process also detects machine/machine-tool malfunctions which may cause deviations in product conformance. It can be either *on-line* for all products manufactured or *off-line* for some *sample* amount product data those are selected *randomly* so that we can have an idea about the population data (all product characteristics) emerging from inspection line.

There are many inspection equipment for specific tasks. Company makes an appropriate selection of such equipment depending on the investment, annual return and product type, quality etc. factors. Such as, thickness measurements, diameter control using gauges, data systems, hardness measuring, surface defect detecting systems, coordinate measuring machine (CMM), optical inspection systems, machine vision systems, high speed cameras, robots, ultrasound systems, X-ray for printed circuit boards (PCB), etc.

Such an inspection station is depicted in Figure 2.9, where DVDs are automatically checked for certain quality characteristics (i.e. cracks, bad sectors etc.) and while good DVDs pass the inspection test, faulty ones fails and go to waste bin.

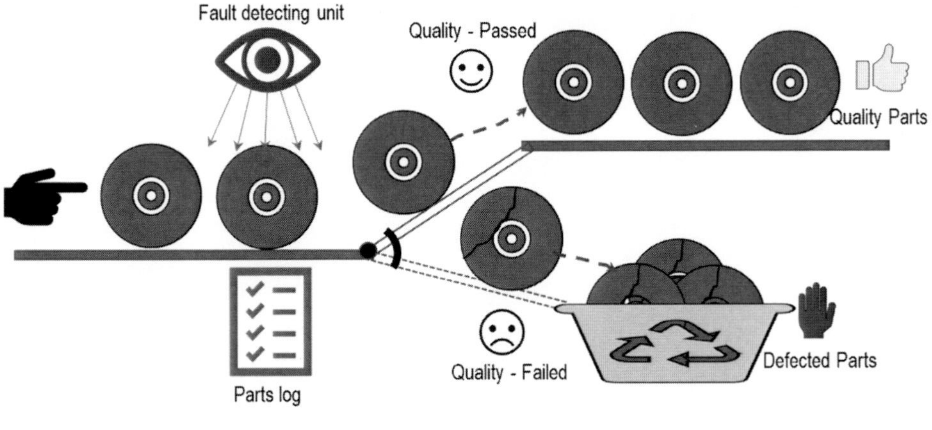

FIGURE 2.9 Automated inspection of DVDs

2.8 Total Quality Management (TQM)

TQM includes both management(top level) and employee(bottom level) practices in the organization, focuses on continuous quality improvement of products and services, empowered to ensure that the organization consistently meets customer satisfaction. TQM is introduced in Chapter 14.

2.9 Exercises

1) Describe *quality engineering* using your own terms.

2) Find an example FMAE analysis from the literature. Show the steps introduced/ explained in the book, on this example.

3) Give an example Pareto diagram for this purpose conduct a research within your reach and try to verify if 80/20% ratio is satisfied.

4) For a process or product of your choice, draw a defect concentration diagram.

5) Provide an example to on-line automated inspection systems that you are interested in? When do we need on-line inspection systems? What is the main reason to build on-line inspection systems?

6) What is a Robust design? Describe by using your own terms?

7) Investigate how CMM systems work? When do you need CMM?

Chapter 3

Basics of Statistics

Making an interpretation and inference on any data requires both describing the data by means of *graphical* display and statistical *numerical* evaluations, respectively. We will focus on data display part first, i.e. how to *summarize* and *display* the (quantitative) data which are obtained from the experiments or measurements. It is quite well established and routine procedure that, we always start our interpretation with studying the data by visualizing its behavioral characteristics. Several graphics/plots for this purpose are possible and introduced in this section. Since *Excel* and the other statistical software can be used to construct these plots, we are going to spend more focus on their interpretation rather than plotting.

As far as *numerical* methods are involved, the calculation of the basic data statistics are performed, next. Finally, we end up with both visual and mathematical i.e. *descriptive statistics* which would be studied i.e. combined in this chapter. As per brief introduction, consider having some data for *pressure, load, strength, part price, weight, conductivity, chemical composition, etch rate, diffusivity, hardening* etc. Often measurements/readings are taken for *n* sample from a very large data set, called as *population*.

Think about an example, where bolts are manufactured in a lot size of *100n*. Since we cannot measure the strength of every bolt, we can go ahead and evaluate *n samples* out of *100n* bolts. We would like to generalize the overall quality of the bolts in lots of *100n*, which have been manufactured by only using *n sample* of bolt strength measurements data. This is only possible if the *sample*(selected portion/part of the *population*) is a true representative of the *population.* Remember that, random data sampling can make such generalization possible!

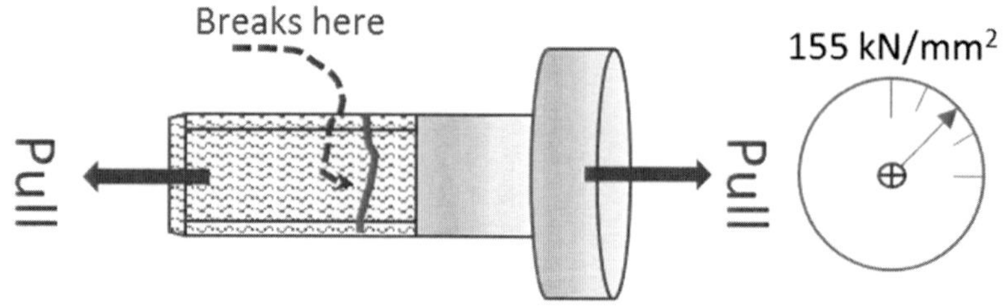

FIGURE 3.1 Bolt pull-test and a strength gauge for Example 3.1

Example 3.1: Bolt Strength

For this purpose, consider a similar example given above, where an M16x2 bolt (i.e. 16 mm diameter and 2 mm in pitch) is being tested for tensile strength at pull-test equipment as shown in Figure 3.1, where out of 2000 batch of bolts, $n=20$ sample of bolts are selected randomly, and their tensile strengths are tested and these strength values are recorded in kN/mm^2 for *samples,* as presented in Table 3.1. Note, this example will be studied in the next section for the demonstration of different plotting techniques.

TABLE 3.1 Example 3.1 tensile test data

Sample (n)	1	2	3	4	5	6	7	8	9	10
Strength [kN/mm^2]	155	175	162	179	166	169	176	176	166	172
	11	12	13	14	15	16	17	18	19	20
	181	186	166	174	189	183	182	169	168	173

3.1 Data Plot Display Methods

In the investigation and understanding of the data behavior, as mentioned before, the first step would be visualizing the data and observing its *distribution* and *trend*. In essence, as a starting point for a trained individual, a visual examination of data plots may reveal quite useful information about the data characteristics. For example, two initial features of the data; the *mean(average) location* and the *scatter/variability* of data, can be easily observed.

Also, note that all of the following plots are plotted using *Minitab* software. The use of such software packages is very straight forward and popular. Therefore we will not be giving any further information about how to make these plots, rather we will introduce these plots/diagrams in the following section.

3.1.1 The Dot Diagram

One of the simplest plot to start with is *the dot diagram* which is a quite handy 1D (one-dimensional) plot for displaying a small number of data i.e. say, up to $n=30$ observations of *sample* data. When you have nicely behaving data (easy visual interpretation), this sample data count n may go up even further.

The dot diagram is a convenient way to visualize some of the key data characteristics of such data. A plot of the *dot diagram* for Example 3.1 is presented in Figure 3.2, where note that when data values repeat (the same), the dots representing the measurement are placed on top of another, indicating the frequency of data at this location. As one sees from the following *dot diagram*, the range, between the lowest and highest data points is quite large! (Note that the following plots are drawn in *Minitab* statistics package).

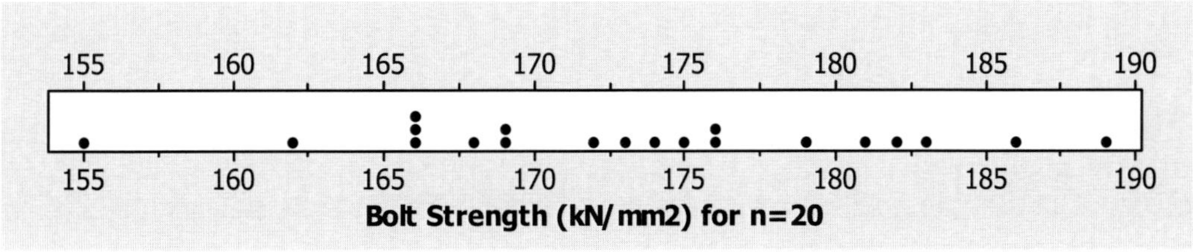

FIGURE 3.2 *The dot diagram* for Example 3.1 data

3.1.2 Scatter Plot

Scatter plot is one step relatively *better* data plot display method than the *dot* diagram. *Scatter plot* displays the data in 2D (two-dimensional), as visually being more comprehensive (i.e. ascending, descending gradients/correlation), i.e. a scatter plots are better to spot trends and correlations between two variables i.e. x *vs.* y, for studying the basic statistics.

Scatter Plot can usually be used to display the larger number of data than that of *the dot diagram*. In this plot, x represents the sample data count, while y representing observations/measurements in bolt strength, an example *scatter plot,* where each dot is connected with lines for Example 3.1, is presented in Figure 3.3, where the *mean* and scatter of the data can be observed/judged easily.

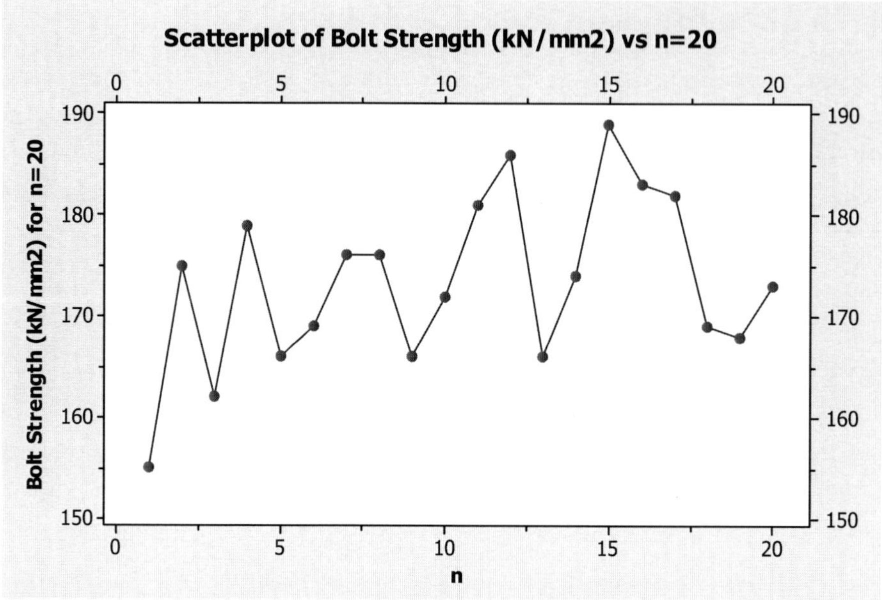

FIGURE 3.3 Scatterplot for Example 1.1 data

3.1.3 Box Plot

As a third alternative, *Box plot* offers much more comprehensive visual information with descriptive data statistics numbers i.e. *minimum, first quartile, median/mean marker, third quartile,* and *maximum. Box plot* is very useful for observing outliers (i.e. data points with *uncommon* overall behaviors), overall spread or spread about the *median. Box plots* are also frequently used for comparing two or more distributions. This plot comprises of quartiles in rectangular lower, median and upper boundaries plus vertical line at both sides i.e. connected to first and third quartiles. The box plot looks like a simple pump or bicycle suspension.

Initially, an ordered/sorted set of data is divided into four identical parts, the division lines/points are drawn and called *quartiles* as,

- A rectangular box, divided with *median, q2:second quartile* which has approximately 50% of the observations below its value
- Lower boundary(bottom box line) represents *q1:first quartile* which has approximately 25% of the observations below its value
- While upper boundary(upper box line) represents *q3:third quartile* which has approximately 75% of the observations below its value

- The quartiles may be different in box length, depending on data/median, and the *Inter-Quartile Range* or *IQR is defined as q3-q1*
- Vertical *lower line (i.e.* whisker) sticks from first the quartile downwards and vertical *upper line (i.e.* whisker) sticks up from the third quartile. The whiskers extend as far out as the minimum and maximum of the data, to a maximum of 1.5 times the IQR range.
  Beyond these limits, we might have outliers. *Upper and lower limits for outliers, Q3+1.5*IQR, Q1-1.5*IQR, respectively.* More, extreme points i.e. potential outliers will be displayed as *dots* beyond the whiskers if any!
- An example box plot for Example 3.1 is presented in Figure 3.4 with construction details explained below. Please note that for this example, there are no outliers, but dummy ones are shown here anyway to give full drawing insight into the structure of box plots to the readers.
- Note, (just like in Example 3.1), when the data are normal, we would expect to see just the box and whisker with no *dots*(outliers) outside.

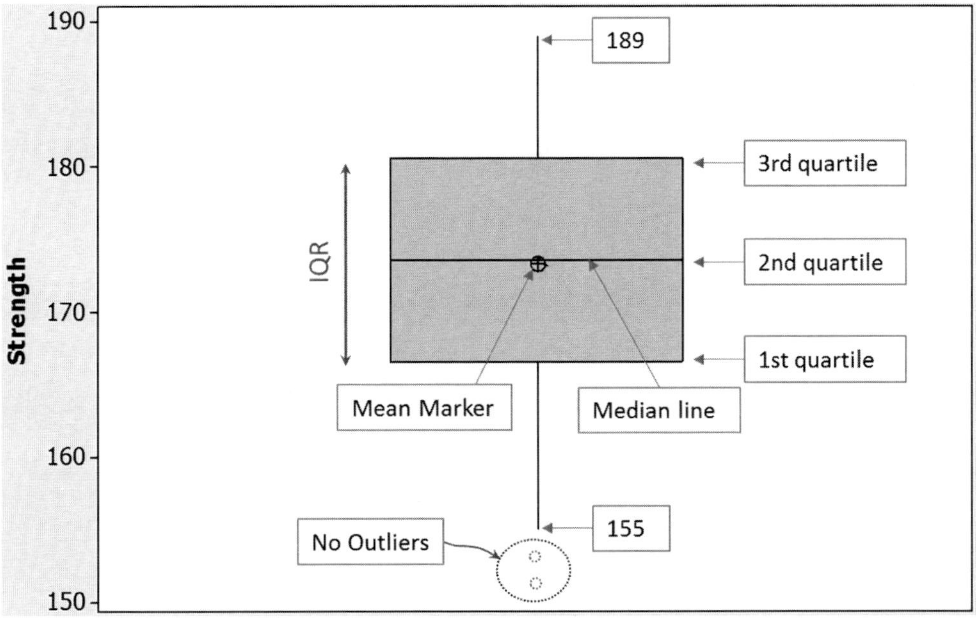

FIGURE 3.4 Scatter plot for Example 3.1 data

In Example 3.1, remember that there are *n=20* observations. *Minitab* calculates the first and third quartiles as the $(n+1)/4$ and $3(n+1)/4$ and also interpolates in accordance with these values when required, in order to find *q1* and *q3* quartiles.

As far as Example 3.1 is concerned, we can calculate the descriptive statistics to draw the box plot as follows,

$$\text{Median} = \frac{173 + 174}{2} = 173.50$$

$(n + 1)/4 = (20 + 1)/4$ =5.25 so using an descending ordered values of data 5 and 6, by interpolation, we get,

$$q_1 = 166 + 0.25\frac{168 - 166}{1} = 166.50$$

for $3(n + 1)/4$=15.75, by interpolation, we also get,

$$q_3 = 179 + 0.75\frac{181 - 179}{1} = 180.50$$

IQR = *q3-q1* = 180.50 - 166.50 = 14

Also, Minimum=155, Maximum=189, Mode=166, Range=189 - 155 = 34, Number of mode count = 3

Outlier limits, Q3+1.5*IQR = 180.5+1.5*14= 201.5>189 and Q1-1.5*IQR=166.5-1.5*14=145.5<155. No outliers!

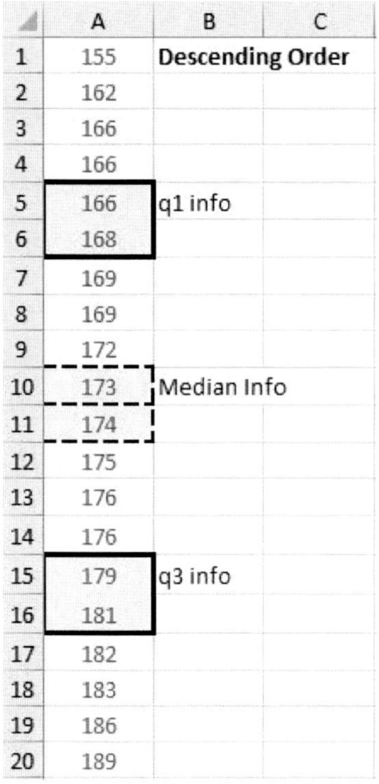

	A	B	C
1	155	Descending Order	
2	162		
3	166		
4	166		
5	166	q1 info	
6	168		
7	169		
8	169		
9	172		
10	173	Median Info	
11	174		
12	175		
13	176		
14	176		
15	179	q3 info	
16	181		
17	182		
18	183		
19	186		
20	189		

FIGURE 3.5 Median and *q1* and *q3* data

Descriptive Statistics data are also presented in Table 3.2, the description and calculation of these data are going to be introduced later in this chapter. Here, this table is given for your reference only so that you may compare this table data to box plot calculations to get more insight about box plot drawing and its interpretation. Overall, as you may notice in Figure 3.4, we have well-behaved and balanced data, hence the distribution.

TABLE 3.2 Descriptive box plot statistics for Example 1.1

Variable	Minimum	Q1	Median	Q3	Maximum	Range	IQR	Mode	# for Mode
Strength	155	166.50	173.50	180.50	189	34	14	166	3

3.1.4 Histogram Plot

A histogram is a versatile graph that is used to visualize the shape and spread of continuous data. In other words, the histogram can also be defined as a visual display of the *frequency distribution* of sample data. *A histogram* plot comprises of allocation of sample data values into pre-determined width intervals, called *bins*. Each bar on the histogram formed with the certain cluster of the data, falling within a pre-determined *bin interval.*

The histogram can provide a reasonably good shape indicator of the sample data distribution, when the sample data size, *n* is large enough. Yet, a histogram is a good estimate/starting point of the probability distribution function *(i.e. pdf)* of a continuous variable, and very useful/valuable information that can be obtained from this plot about the distribution characteristics of the data, i.e.

- 80% of students got less than *grade* ***C*** in the class of "*differential equations*" in Spring, 2015
- top 5% of students get grade ***A***
- bottom %30 of student gets grade ***F***

If the data are *symmetric*, where the *mean* and *median* of data are equal (the *median* is denoted as $\tilde{x}$). Otherwise, the distribution may be skewed to the left or right, indicating the data distribution trends.

In an ascending order (i.e. sorted data) of observation values, the median $\tilde{x}$ is the central observation value when the number of observations is *odd*. Otherwise, the average of *two central* observations gives the median. Besides, the sample *mode* is the most frequently observed observation value. A Histogram plot for Example 3.1 is shown in Figure 3.6 and as drawn in *Minitab.*

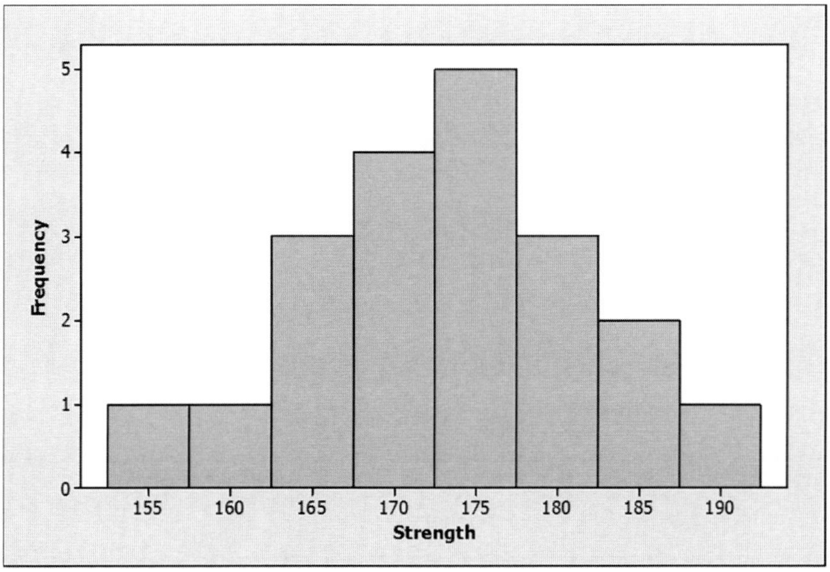

FIGURE 3.6 Histogram plot for Example 3.1 data

3.1.5 Normal Probability Plot

Histograms may not be a good choice to find out the distributions when the sample sizes are small. In such situations, we may adopt a graphical method for probability plotting to see, if the sample data conform to an assumed hypothesized distribution by visual examination. Also called *normal probability plot* is a graph that is used to evaluate the fit of a probability distribution to the data.

A good distribution fit is the one where the observations are near the fitted line, i.e. we place a thick pen on the data, if it covers the data points, we assume our distribution assumption to our data is *valid.* The normal probability plot can be useful in identifying *symmetric* and *skewed* distributions. The normal probability plot shown in Figure 3.7 has the following features.

- Each value is plotted along a fitted distribution line (middle solid line).
- In default, the approximate 95% confidence intervals (curved solid lines) may be shown
- It may display some statistics to help us evaluate the distribution fit to our data, i.e. mean, StDev. etc.

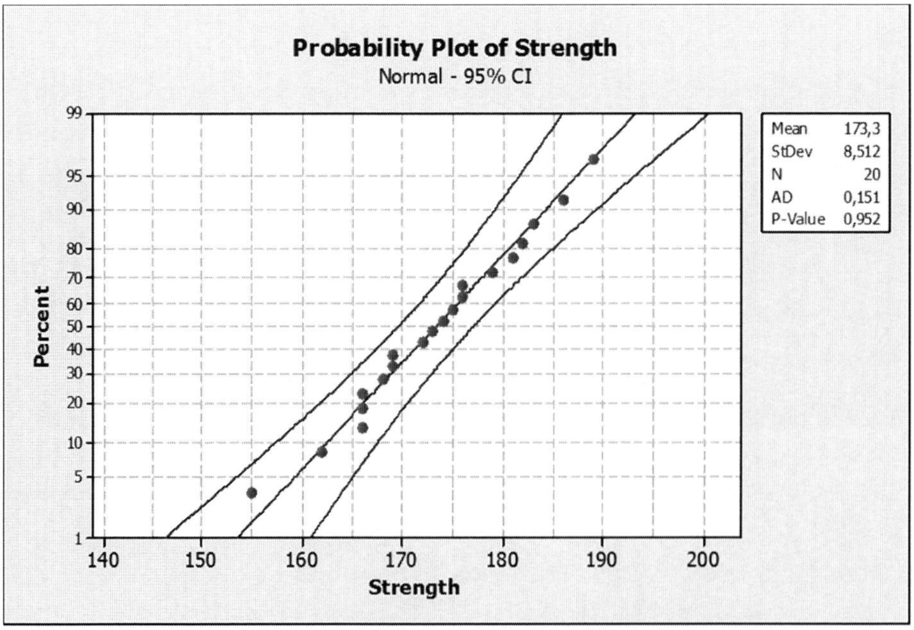

FIGURE 3.7 Probability plot for Example 3.1 data

This probability plot and information on it, briefly tell us the following; the data almost follow the straight line, if we place the thick pen on the line, all data will be nearly covered, this is the first indication of normality.

On the other hand, we can evaluate the *P-value,* which is greater than α=0.05 (level of confidence), then the *null* hypothesis is rejected! Note, Hypotheses are introduced in later chapters of this book. Also, notice that the *Anderson-Darling* statistic is *low*! Then, this tells us that the Example 3.1 data follow a normal distribution sufficiently, (i.e. the Anderson-Darling statistic tells us how good the data follow an assumed distribution, i.e. the smaller this statistic, the better the distribution fits the data). Therefore, the normal distribution appears to fit the sample data pretty well.

- The *mean* strength is *173.3* kN/mm^2, the *standard deviation* is *8.512* kN/mm^2.

3.2 Variation or Variability

During our experiments, although most of the time we carry out so-called "*identical*" successive runs (repetitive runs) of a system of our interest, we usually do not get the same exact results for the same experimental conditions. This is true, even if these experiments are performed by the same person using the same experimental instrumentations. How come? This can be attributed to the concept of *variations* or *variability,* which we also experience in our real-life. Statistics may help us in understanding and take control of variability by incorporating it into our decision-making processes.

For example, consider the machining process in basic cutting(material removal) operation. Assume, we focus on tool life in *hours*, whatever we do in estimating tool life, and we use exactly the same identical cutting tool and conditions, i.e. the same depth of cut, the feed, material removal rate, cutting speed, we may still get fluctuations in tool wear, thus in tool life in *hours*.

This can be attributed to work part material-impurities, ambient temperature, the condition of the CNC machine, cutting apparatus wear, tool holder, friction, temperature levels, coolant type, chattering etc. These factors are explained by potential **sources of variability** in the machining operations (i.e. turning, drilling, milling etc.). *Statistics* can offer us a framework for describing/understanding and reducing this variability.

Since the tool life exhibit variability, we can consider the tool life as a random variable. Assume X is a *random variable*, shown in upper case, which represents the tool life measurements.

Now, we go ahead and state the X, as a sum of (constant mean) μ and (random disturbance) ϵ as in Equation 3.1,

$$X = \mu + \epsilon \qquad 3.1$$

The constant μ remains the same with every measurement, but variations/disturbances may change the value of ϵ. Note, X would always be equal to the constant μ if ϵ was zero i.e. no disturbance case! However, in essence, this can never be the case in practice, so the actual measurements of X exhibit some level of variations.

The source of variability needs to be identified and minimized for a successful step forward in handling real-life challenges in *quality control* of products as would be studied later chapters.

Example 3.2: Gas Mileage Ratings

Automobile manufacturer gas mileage ratings are tested for 4-door mid-size passenger cars for 10 different brands of domestic-use by a well-known automobile magazine. The testing route is 55 miles long with curves, upgrade, downgrade etc. A total of 10 different drivers rides each car according to their driving habits without enforcing any rule on them. These drivers are chosen from age groups of 18-55 years old. The gas mileage readings are recorded and presented for $n=100$ data in Table 3.3.

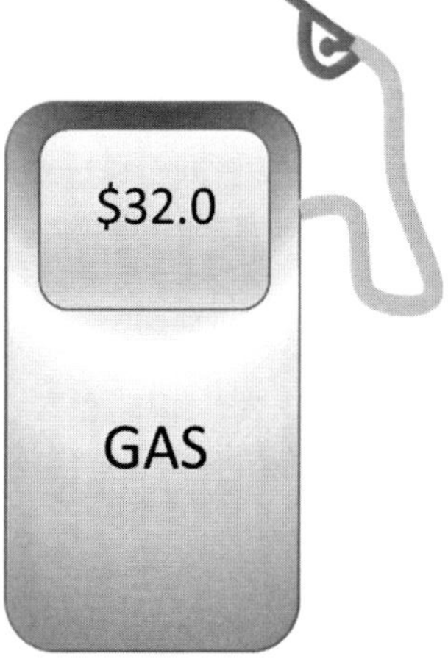

TABLE 3.3 Gas mileage (mpg) data

39.6	28.9	37.5	34.3	29.0	24.9	24.9	34.3	21.7	37.5
34.3	34.3	39.6	28.0	37.5	28.0	32.2	28.0	35.4	24.9
34.3	27.0	35.4	36.4	30.0	41.7	32.2	25.9	25.9	25.9
37.5	32.0	41.7	32.0	35.4	30.5	28.0	31.2	19.6	36.4
33.3	28.0	29.1	33.3	37.5	24.9	40.6	35.4	32.2	31.5
23.8	30.1	36.4	22.8	19.6	30.0	40.6	22.8	28.0	29.1
33.3	27.0	37.5	27.0	18.6	27.0	22.8	24.9	40.6	30.1
21.7	30.1	33.3	35.4	31.2	20.7	34.3	30.1	36.0	29.1
31.2	31.2	31.2	31.2	31.2	31.2	31.2	31.2	31.2	31.2
31.2	18.6	33.3	32.2	31.2	29.0	37.5	28.0	40.6	19.6

Since we have *n=100* good looking data, let's start with the *dot* diagram display as presented in Figure 3.8, where we observe more concentrated dots in the 28-35 mpg interval, which is "good" fuel consumption level, but we also notice some data below 21 mpg and above 35 mpg as well, which highlights the importance of driving habit and age, as well.

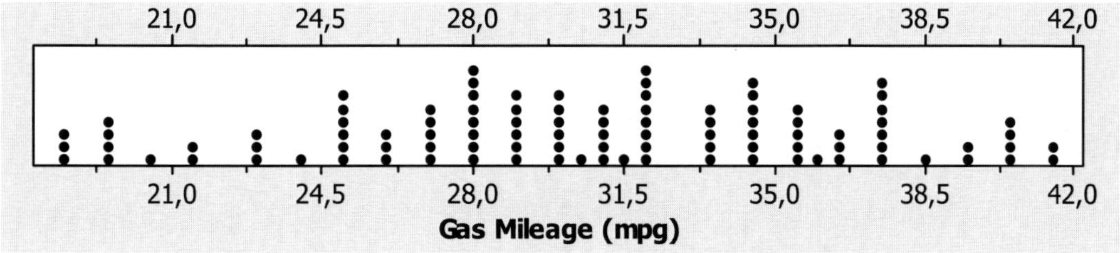

FIGURE 3.8 The *dot* diagram for Example 3.2 data

We can manipulate these data by adjusting gas mileage axis (i.e. X-axis) to be comprised of 13 bins as to make the distribution more visible and see its trend better, as also shown in Figure 3.9.

Remember that we can always control the bin intervals as per the data characteristics. The data distribution is now quite visible! with higher concentration in the *middle peak* and lighter concentration on the *tails*. Here *mean* can be estimated around 32 mpg.

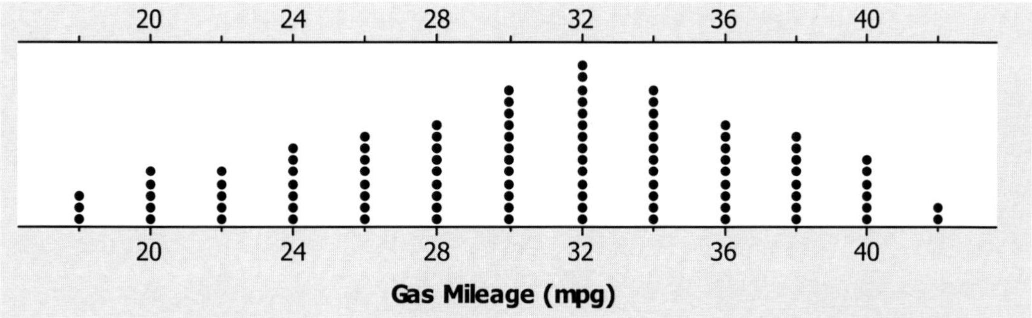

FIGURE 3.9 The *dot* diagram for Example 3.2 data with 13 bins

By examining the scatter plot as presented in Figure 3.10, we also confirm that the mean is around 32 mpg and upper and lower specification limits are also "*almost*" visible. The example data look almost balanced around the *mean*. But lower half is slightly larger in dispersion than upper half when we consider the *mean.*

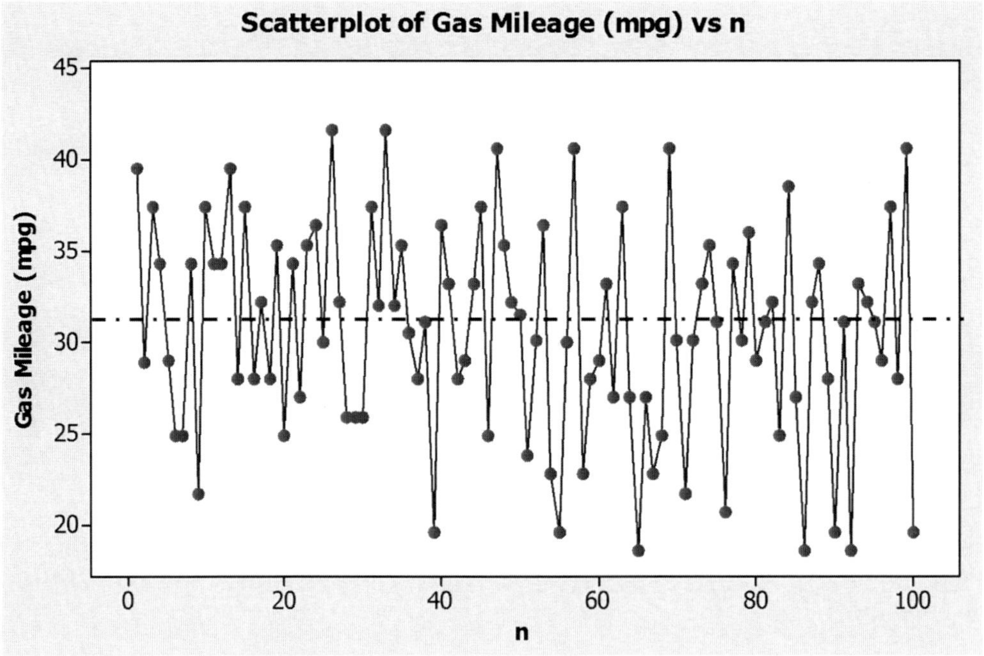

FIGURE 3.10 Scatterplot for Example 3.2 data

Box plot is also very informative about the distribution of data and we easily see the trends and data concentration interval (i.e. IQRange = 8.1375 mpg), as presented in Figure 3.11, note, we can easily locate the *mean* shown as **dot** vs. *median=31.15 mpg* line.

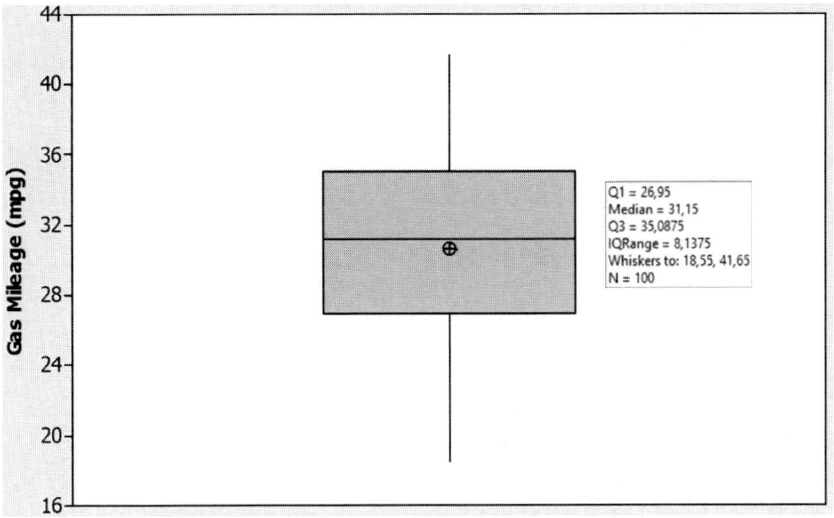

FIGURE 3.11 Scatterplot for Example 3.2 data

In fact, Figures 3.11 and 3.12 are interrelated, i.e. one is the different mapping or explanation of the another. The size of the box (wide-narrow-symmetric) in box plot clearly tells us the central characteristics of the histogram plot. Histogram plots also reveal us very valuable information about the probability density characteristics of the data. Notice that, in Figure 3.12, the data distribution is a little bit inclined to the right but having/displaying pretty good normal distribution plot characteristics.

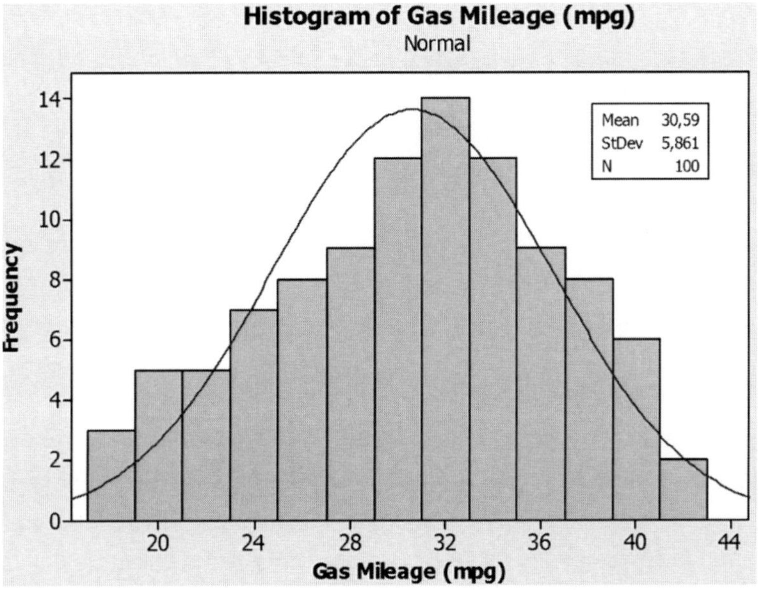

FIGURE 3.12 Histogram plot for Example 3.2 data

In short, the *mean* gives the center location of the data and the *standard deviation* determines the *height* and *width* of the data distribution. For example, a large standard deviation forms a distribution/curve that is short and wide while a small standard deviation forms a tall and narrow looking distribution/curve.

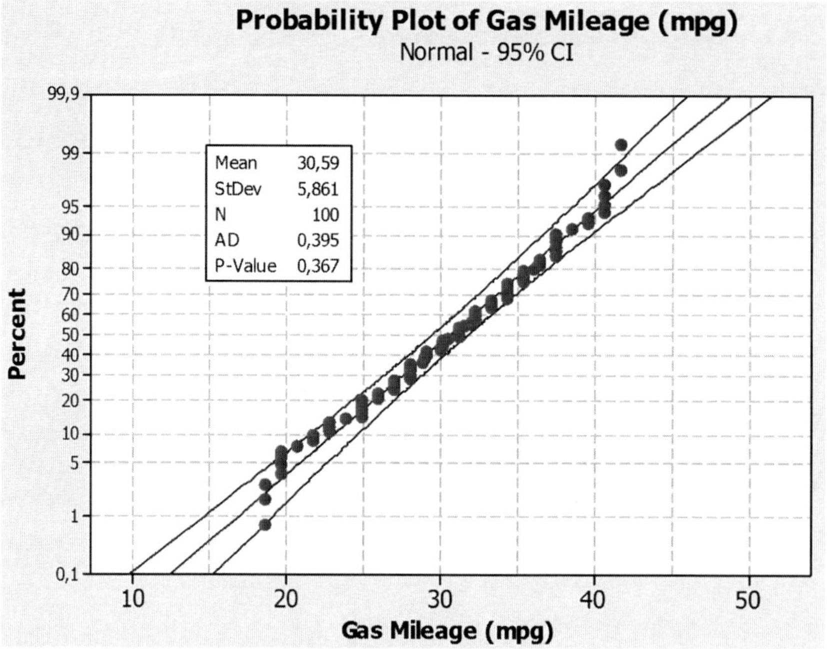

FIGURE 3.13 Probability plot for Example 3.2 data

This point is further proven/verified in the probability plot as shown in Figure 3.13, which demonstrates consistent distribution and confirms that our data is normally distributed, i.e. all dots are following a straight line with very little variation at both ends, which is fine!

Statistics studies the *collection, analysis, interpretation,* and *presentation* of the *"data"*. Data are our very valuable asset, which can carry/contain important characteristics of the "*event*" to which it belongs to. Data is a "*key*" and statistics is a discipline of study which uses this *key* to open the doors of exploration. Therefore, data and statistics are both coupled and co-exist. Where do we use statistics? Well almost in any area, ranging from; art to history, business to industry, social studies to medical science, education to sports, weather reports to species etc. In this context, an introduction to data collection is introduced in the following section with examples.

3.3 Data Collection - Population vs. Sample

When we talk about the *statistics,* first, the concept of "*data collection*" quickly comes to our mind. Why? Because we cannot differentiate statistics and the data from each other. Nevertheless, the data collection involve "*sound*" and "*well-planned*" collection and/or experimental data generation strategies. Once all requirements are met during the data collection or measurement, then we can go ahead and do the followings,

- We employ applicable graphical display methods using our data first, as explained earlier in this chapter.
- Then, we go ahead and evaluate the collected/measured data, with *descriptive statistics* (and *inferential statistics* if required!)
- In case we use our *designed/generated* input data in the analysis of our existing system or problem, Then, input these data into your system or problem and then obtain the output/response of it. Next, perform data analysis by employing the well-founded statistical methods and their metrics to identify the relationship/ dependency between inputs and output.
- Next comes, both *understanding* and *interpreting* the related system or problem itself, using all statistical data. Because this part could be quite elusive i.e. information that we are looking for might be quite hidden.
- Later comes the *presentation* and *organization* of the result data for the sake of transmitting knowledge to the other platforms. The person who is capable of doing such data design/analysis and judgments(inference) etc. is called statistician, this person does not have to be a pure statistician, but can be anyone(engineer, scientist) who has the skills to do these inferences.

In definition, statistics mainly deals with all aspects of data collection and the analysis of data and aims at obtaining a statistical model for representing the underlying system/process. Statistics enables us to further study the data so that we can derive any useful information out of this data to study scientific, industrial/engineering, or social phenomena. Keep in mind that all of this voyage is rather *non-deterministic* in nature.

Data collection in a certain domain is key to valid statistical inference, note that we might have to deal with the huge number of data called *"population"*, whereas we are limited in terms of obtaining/handling/interpreting these huge data; further we have no means whatsoever to evaluate/obtain "*population*" data. Therefore, we need to use some useful portion i.e. a subset of the *"population"* which is called *"sample"* data.

For example, in a urine lab. analysis study, while the population is the whole urine in the bladder of an individual, the *sample* is the amount of urine collected and given to the lab. for the analysis. In Figure 3.14, this concept is depicted as follows,

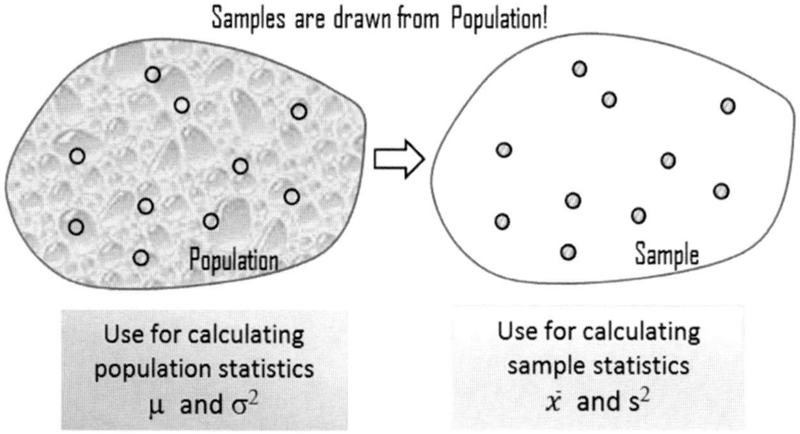

FIGURE 3.14 Population and sample concepts in statistics

For another example, let's assume a research company would like to forecast the national up-coming election results using polls. All citizens, who are going to vote in the elections are considered as *"population"*. Whereas there is no way that this company can find out *"population"* voting statistics since this information is the result of the national elections, therefore it remains elusive.

The remedy is to use "*sampling*" or "*sample*" of voters, i.e. sub-division of "*population*", now a sample of citizens' voting polls are evaluated. Based on this information, this research company does its best to forecast the national voting results by using available statistical techniques. Obviously, this is not an easy task! Choosing a sample domain as being the most representative part of the *"population"* requires good insight of people (i.e. education, culture, trends, geography etc. of voting people) and statistics. The more realistic sampling gives us, the better the forecast. As far as the *parameter* vs. *statistic* are concerned, we define them as

- a characteristic of a *population* is called "*parameter*"
- a characteristic of a *sample* is called "*statistic*"

Considering the applicational/practical usage, we have no choice but to use a *sample*, which is considered and assumed to be a good/valid/reasonable representation of the *population.* In the literature, mainly four basic methods of collecting data are mentioned as follows,

- *A historical data,* is archived data over a long period of time about events and circumstances from past to present. Keep in mind that it may cover either all or available part of the historical data. For example, national raindrop per m^2 or seasonal temperature readings per ^oF, in the last 200 years in a certain region.
- *An observational data,* the researcher observes the process and records the certain input/output measures of his/her interest for a shorter time period without disturbing the system or process. Using this observational data, in the end, one can try to improve any objective of his/her using all the data obtained during this time interval of observation. Further, the impact of the input factors on output might also be investigated, but this data still do not offer a complete solution to all our needs.
- *A survey data,* conducted by a group of researchers on a sample of people, and their views are recorded on a certain issue, i.e. side effects of cholesterol drugs can be surveyed.
- *A designed experiments data,* using statistically valid/well-established techniques, one could smartly/purposefully design the least number of experiments, required to efficiently and accurately assess all relationship and effects of input parameters on output/response. Thus, better performance under variation and lower experiment/simulation costs are both acquired in analyzing/interpreting the system or process of our interest.

This book is designed to introduce the use of all types of data utilizing collection methods introduced earlier. In the following section, very important measures which are frequently used in the preliminary statistical analysis are introduced to lay upon the statistics fundamentals for use in quality engineering.

3.4 Introduction to Statistical Data Analysis

Experimental (or process) *sample* data may tell us so many things about its *population*, yet we are to get/comprehend as much as our container can hold/understand. This process is obviously deemed as extraction of any meaningful information or *inference*. In other words, statistics can help translate the meaning/essence of data into our way of comprehension. Yet, the extent of uncertainty/*confidence* involved in these inferences is also to be investigated.

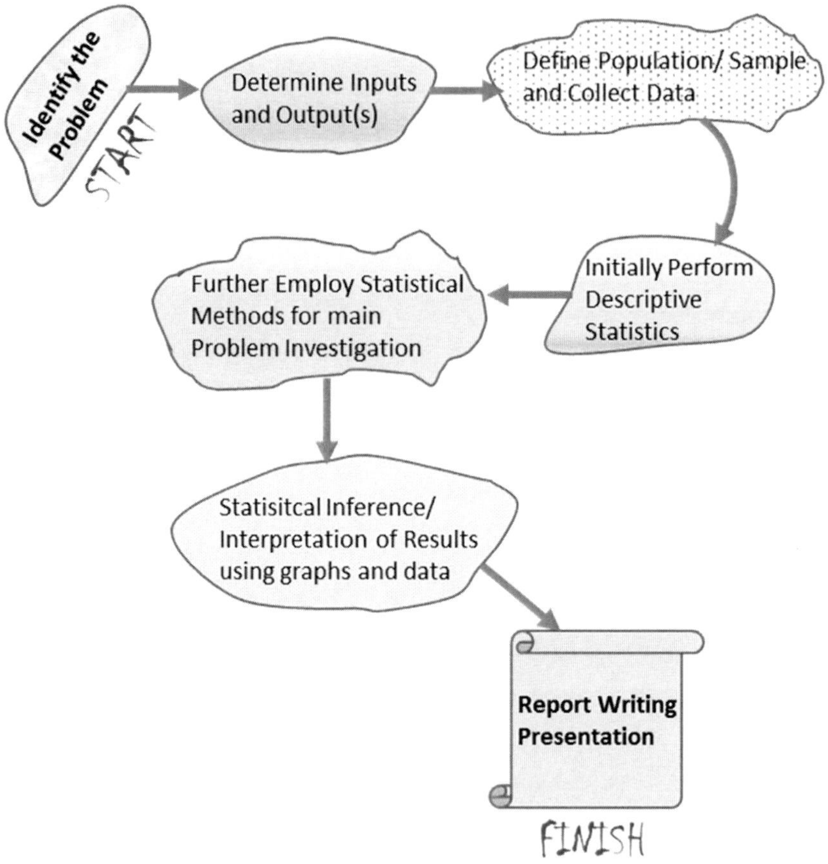

FIGURE 3.15 Flow chart for statistical data analysis

Under this explanation, any successful implementation of the data analysis should follow a self-explanatory flow chart, as depicted in Figure 3.15, i.e. this flow chart constitutes all required steps from problem identification to report writing and presentation. In brief, once we defined our problem, we can go ahead and employ applicable statistical methods through which we can come up with,

- the kind of sample data containing inputs/output data, where inputs are selected to be the most influential factors on the system's output
- the input/output form/pattern of the sample data
- the amount and location of sample data to be collected
- the initial descriptive statistical analysis for exploring the sample data

- object-oriented statistical methods are chosen/employed (i.e. inferential statistics) the inference/interpretation of the above results for getting insight into the population characteristics
- the evaluation of the results under an assumed degree of uncertainty/confidence
- the written report

Some example of contemporary statistical practices,

- ✓ *Educational* challenges in catching up national standards
- ✓ *Technological* problems in meeting current quality standards
- ✓ *Agricultural* problems in crop yield grades
- ✓ *Medical* treatments in exploring the drug side effects for age groups
- ✓ *Political* elections in forecasting election results
- ✓ *Economics*, the effects of currency rate changes on inflation
- ✓ *Philosophy*, the effects of social media on teenager behaviors
- ✓ *Technology,* which communication app is dominating its rivals

3.5 Descriptive Statistics

Descriptive statistics are used to describe and summarize the basic features of any small/medium/large data set so that we can have some idea of what's going on with the data. *Descriptive statistics* provide straightforward, very useful statistical summaries about the sample data, along with simple graphical plots and their evaluations. Keep in mind that, descriptive statistics should be distinguished from *inferential statistics,* which is one step further, where we are trying to reach conclusions beyond descriptive statistics data can offer, i.e. we use *inferential statistics* to infer some useful knowledge from the sample data about the pertinent characteristics of the *population.*

In *descriptive statistics*, a visual examination of sample data is the initial step we should take. For this purpose, we use diagrams, graphs, charts, tables, and the various key calculations to characterize/find out the characteristics/distribution of the data. All these preliminary "*executed*" statistics are called *descriptive statistics.*

There are three major characteristics of a single variable data (univariate analysis) that we focus at;

- the distribution (visual examination), *i.e. introduced* earlier in this chapter
- the central tendency (mean, median, mode)
- the dispersion (standard deviation, variance)

all these statistical markers are presented in detail in the next section and demonstrated on the following example problem.

Example 3.3: Math Final Scores

In a class of *seven* students, at the end of the spring semester, the math final scores are listed as,

$$X=\{21,34,52,60,89,10,71\}$$

where *X* is a *discrete random* design *variable*. We are asked to calculate *descriptive statistics* for this data as presented in the following.

Mean:

The "*statistical mean*" or "*arithmetic mean*" or simply *mean (or average)* is the most well-known measure of *central tendency* of *n: the total number of data* in question. In real-life applications, remember, we mostly use *sample data,* so the *sample mean* is a reasonable measure/estimate of the *population*, because it allows us to estimate the *population* characteristics, where *n* number of data collected/generated and its points are denoted by $x_1, x_2, x_3,,, x_n$. Then, the sample mean $\bar{x}$, is calculated by summing up all the data points (i.e. $x_i, i = 1,2, \ldots, n$) and then dividing the total sum by *n*.

$$\text{mean} = \bar{x} = \frac{x_1 + x_2 + x_3 + \cdots + x_n}{n} = \frac{1}{n}\sum_{i=1}^{n} x_i \quad 3.2$$

$$\bar{x} = \frac{82 + 69 + 60 + 60 + 34 + 21 + 10}{7} = \frac{\text{Sum} = 336}{n = 7} = 48$$

As far as the *population* is concerned, the *population mean* is the average of all the measurements in the population, of course, if it is known! Remember this is not the case for most of the times, because generally it is not known!

For a *finite population* with *N* number of data, the *population mean* is calculated and denoted by μ, as given in Equation 3.3,

$$\mu = \frac{1}{N}\sum_{i=1}^{N} x_i \quad 3.3$$

Notice that, **residual** $= \mathbf{x_i} - \bar{\mathbf{x}}$, is the distance from any x_i point to the *mean*, $\bar{x}$. The data to the left of the *mean* are balanced by the data to the right of the *mean*. The residuals sum to zero and the *mean* minimizes the sum of the squares $(\mathbf{x_i} - \bar{\mathbf{x}})^2$ of the *residuals*.

Remember that, the *mean* is a very useful measure of data distribution but it can be affected by extreme values in the data set and therefore be biased. *This is considered the main disadvantage of using mean.*

Example 3.3 data and its mean value are plotted in *bar and line* as shown in Figure 3.16. Remember that, the *sample mean* is our useful starting statistical metrics, but we need more measures telling us more information/insight about the sample data and its distribution.

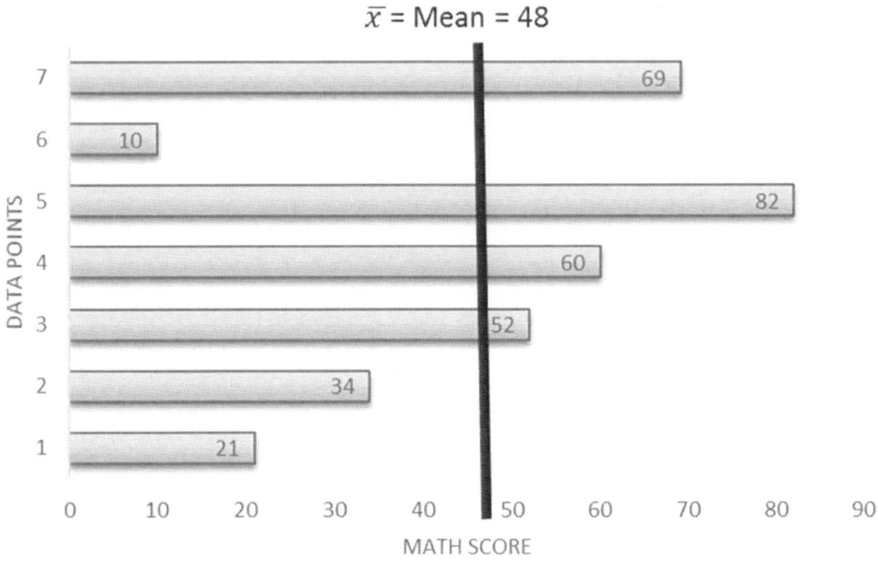

FIGURE 3.16 Math scores and its distribution

Then, the two statistics used to measure the central tendency of data are *median* and *mode*. In a normal distribution, the statistical *mean* is equal to *median* and *mode*. In the next, performing *sample mean* calculation using an *Excel* sheet is demonstrated in detail.

In the following *Excel* sheet, specific *Excel commands* are shown for *n, Sum, Sample Mean* calculations, where note that **B2:B8** cells corresponds to final math score data. Note, it is also shown that the sum of *residuals* in column C (i.e. cell *C9*) *is equal to zero* (i. e. residual sum $= \sum x_i - \bar{x} = 0$).

	A	B	C	D	E	F
1	**i**	**Math Score**	**Residual**			
2	1	21	-27.0			**Excel Commands**
3	2	34	-14.0	n=	7	= COUNT(B2:B8)
4	3	60	12.0	Sum=	336	= SUM(B2:B8)
5	4	60	12.0	Sample Mean=	48.0	= E4/E3
6	5	82	34.0	**Sample Mean=**	48.0	= AVERAGE(B2:B8)
7	6	10	-38.0			
8	7	69	21.0			
9		Residual Sum =	0.00000			

where, all *Excel* commands in **Column F** are shown, which are exactly the same ones used to find specific values in **Column E**. Note, all *Excel* commands are executed after **"="** sign in *Excel* command window as demonstrated on the following example, i.e. in **column E in Cell E3**, *n* (number of data) is counted using an *Excel* command of,

= COUNT(B2:B8)

Excel command format/description is also presented in **cell F3,** which is the same as the one used in the *Excel* command window for cell *E3*. Note: **Residual column** is not shown below!

Notice, in the *Excel* command execution window below, how COUNT command is entered! i.e. = ***Excel* Command Description**

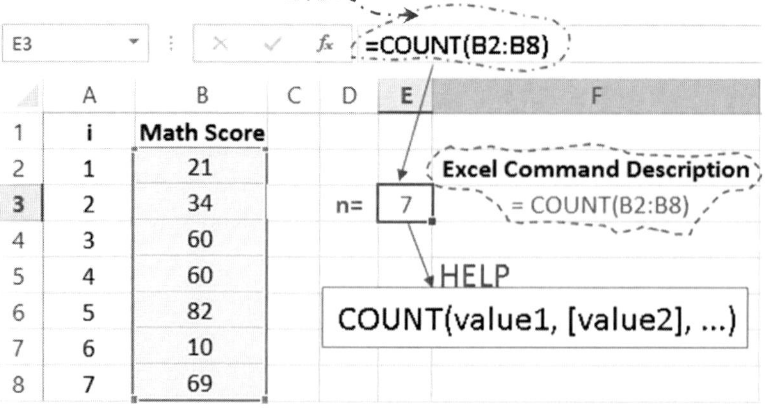

Variance and Standard Deviation:

The *variance* is a measure, which quantifies the spread of data around its *mean* value. It gives us a good sense of the *dispersion* of our data by looking/evaluating at all the data points in the set. Moreover, *zero variance* tells us that there is no *variability*; i.e. all the data points are identical in our data.

A variance of a *population* (mostly unknown and formulated/shown using upper case letters) is denoted by σ^2. It is calculated as the average of the sum squared differences of "each finite population of *n* number of data points from the *mean*", i.e. $x_i, i = 1,2, \ldots, n$ as,

$$\sigma^2 = \frac{1}{N}\sum_{i=1}^{N}(x_i - \mu)^2 \qquad 3.4$$

Sample variance is also a measure of the spread in a data set, measuring how far each number in the set is far (dispersed, scattered) from the *sample mean* and the variance of a *sample* is denoted by s^2. The sample standard deviation, *s*, is the positive square root of the sample variance. By looking at these quantities, *the smaller s^2 means relatively little variability* which we (engineers) mostly prefer, while the higher s^2 means relatively larger variability in the data.

The variance of a *sample* is defined by a slightly different formula and this formula tells us, the variance of the *sample* is an unbiased estimate of the variance of the population as given in Equation 3.5,

$$s^2 = \frac{1}{(n-1)}\sum_{i=1}^{n}(x_i - \bar{x})^2 \qquad 3.5$$

$$= \frac{(21-48)^2 + (34-48)^2 + (60-48)^2 + (60-48)^2 + (82-48)^2 + (10-48)^2 + (69-48)^2}{(7-1)}$$

$$= \frac{4254}{6} = 709$$

Note that in the previous s^2 calculation divisor of *(n-1)* is used for compensation (correction) to the approximations/representations, when using sample data. While as far as the population is concerned divisor of *n* is directly used. We will suffice with this much explanation, but more detailed information, in-depth statistical point of view, can be found in the literature.

The standard deviation is denoted by σ and s for population and sample, respectively. The standard deviation is equal to the square root of the variance or sample variance, $\sigma = \sqrt{\sigma^2}$ or $s = \sqrt{s^2}$, respectively. Also note that the unit of *standard deviation* is the same as the data *unit*, this is desirable in the plotting etc.

While *mean* and *variance* both provide simple and very useful information about the population (and its distribution), these two measures do not solely identify all population characteristics, yet we need further statistics and methods. We can find *s*,

$$s = \sqrt{709} = 26.6 \qquad 3.6$$

In Figure 3.17, the line plot of the math final scores is displayed with a *mean* plus/minus *standard deviation* around *mean*. Notice that when *s* decreases, the dotted line showing the variation, approaches to *mean*, or vice versa.

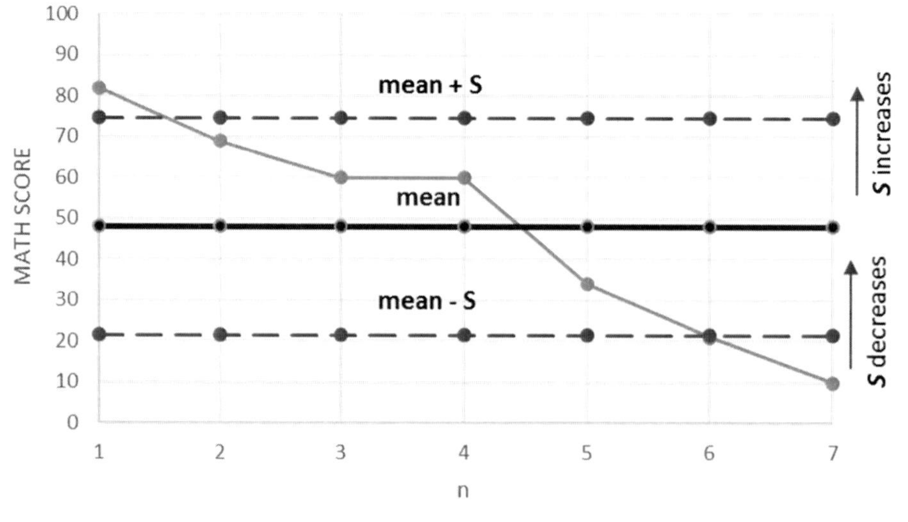

FIGURE 3.17 Mean vs. standard deviation for math final scores

In the *Excel* sheet presented next, specific *Excel commands* are again shown for *n, Sample Mean, Sample Variance, Standard Deviation, and SE Mean* calculations, where note that **B2:B8** (selected cells) corresponds to math final score data. In case you do not know these *Excel* commands, you may add columns C and D with associated calculations and calculate all the related statistics by using formulations given before. as demonstrated on the next page!

	A	B	C	D	E	F	G
1	i	Math Score	x - mean	(x - mean)2			
2	1	21	-27.0	729.0			Excel Commands
3	2	34	-14.0	196.0	n=	7	= COUNT(B2:B8)
4	3	60	12.0	144.0	Sample Mean=	48.0	= AVERAGE(B2:B8)
5	4	60	12.0	144.0	Sample Variance=	709.0	= VAR.S(B2:B8)
6	5	82	34.0	1156.0	Standard Deviation=	26.6	= STDEVA(B2:B8)
7	6	10	-38.0	1444.0	SE Mean=	10.1	= D7/SQRT(F3)
8	7	69	21.0	441.0			
9			Sum(x-mean)^2 =	4254.0			= SUM(D2:D8)
10			Variance =	709.0			= SUM(D2:D8)/(F3-1)
11			Standard Deviation =	26.6			= SQRT(SUM(D2:D8)/(F3-1))

Median:

The sample *median* is a measure of *central tendency*, which divides the sample data into two equal i.e. *half below* and *half above* parts. To obtain median, first, we sort the sample data in order from smallest to largest value (or vice versa) for *n* data points to locate/calculate the *median* as,

$$x_1 \leq x_2 \leq \cdots \leq x_n \tag{3.7}$$

Then, for the *odd* number of data, the median is the middle value. If there is an *even* number of data, the median is the average of the two middle data values.

$$\text{median} = \hat{x} = \begin{cases} x_{\frac{n+1}{2}} & n \text{ odd} \\ (x_{\frac{n}{2}} + x_{\frac{n}{2}+1})/2 & n \text{ even} \end{cases} \tag{3.8}$$

Sometimes *mean* and *median* can be very different, depending on the outliers, i.e. distribution of data points.

In the following *Excel* sheet, specific *Excel commands* are shown for *Median* calculation, using high to low sorted data. Notice that $n = 7$ *(odd number),* we directly get (n+1)/2 = 4th sorted data as median = 60.

	A	B	C	D	E
1	**i**	**Math Score**			
2	1	82			
3	2	69	High to Low sorted data		
4	3	60			**Excel Commands**
5	4	60	**Median=**	60	= MEDIAN(B2:B8)
6	5	34			
7	6	21			
8	7	10			

In the following *Excel* sheet, let's assume 7th data is deleted and we have 6 data, in this case, specific *Excel commands* are shown for *Median* calculation again using high to low sorted data. Now, notice that *n = 6 (even number),* we directly get n/2= 3rd and (n+1)/2 = 4th sorted data to calculate the median=(60+60)/2 = 60. For this task, a specific *Excel* command and visual verification result are both shown next.

	A	B	C	D	E
1	**i**	**Math Score**			
2	1	82	High to Low sorted data		
3	2	69			**Excel Commands**
4	3	60	Median=	60	(B4+B5)/2
5	4	60	**Median=**	60	= MEDIAN(B2:B8)
6	5	34			
7	6	21			
8	~~7~~	~~10~~			

Finally, note that when the data is skewed, we can prefer the use of *median* over the *mean*, because the mean is not anymore good representative for the central location of the data. In this case, we get a better measure of central tendency with the median.

Mode:

If the same data are encountered for more than *m>1,a number of times* in the sample data, then *m* is called *mode*. Note that sample data may have multiple modes for the same number of occurrences of different data. If no number occurs more than once in the data, then there is no *mode* for this data.

As far as this data is involved, we see two math scores are the same, and this makes our *mode* number as 60. For this task, specific *Excel* command and visual verification results are both shown next!

	A	B	C	D	E
1	i	Math Score			
2	1	21			
3	2	34			Excel Commands
4	3	60	Mode=	60	= MODE.SNGL(B2:B8)
5	4	60			
6	5	82			
7	6	10			
8	7	69			

Minimum, Maximum and Range:

The *minimum* is the smallest, the *maximum* is the largest data values. Whereas, the *range* of the data is basically the difference between the largest and smallest values. This information gives us an approximate idea about the boundaries of sample data that we are studying as formulated in the following,

$$\max(x_i) = \{ x_1, x_2, \ldots, x_n\} = 82 \qquad 3.9$$

$$\min(x_i) = \{ x_1, x_2, \ldots, x_n\} = 10$$

$$\text{range} = r = \max(x_i) - \min(x_i) = 72$$

	A	B	C	D	E
1	i	Math Score			
2	1	21			Excel Commands
3	2	34	Maximum=	82	= MAX(B2:B8)
4	3	60	Minimum=	10	= MIN(B2:B8)
5	4	60	Range=	72	= D3-D4
6	5	82			
7	6	10			
8	7	69			

Also, note that for this task-specific *Excel* commands and visual verification results are both shown above.

Note that, generally, when the range increases for sample data, the relative variation for that also increases, or vice versa, but keep in mind that the range offers us very limited bounding information since it does not say anything about the data in between.

Kurtosis:

Our fundamental statistical task is to characterize the location and variability of the data set. Therefore we are going to study *kurtosis* and *skewness.*

Kurtosis tells us how the peak and tails of the distribution of our interest differ from the normal distribution. Initially, we use kurtosis to understand general characteristics about the distribution of our data, while the positive kurtosis indicates a relatively peaked distribution, the negative kurtosis indicates a relatively flat distribution.

For univariate (one variable) data $x_1, x_2, \ldots x_n$ then, *sample excess Kurtosis* formula (consisted with *Excel*) is given in Equation 3.10. As for our math final score example, we can calculate this measure as,

$$\text{Kurtosis} = \frac{n(n+1)}{(n-1)(n-2)(n-3)} \frac{\sum_i^n (x_i - \bar{x})^4}{S^4} - \frac{3(n-1)^2}{(n-2)(n-3)} \qquad 3.10$$

$$\text{Kurtosis} = \frac{7(7+1)}{(7-1)(7-2)(7-3)} \frac{4227282}{502681} - \frac{3(7-1)^2}{(7-2)(7-3)} = -1.476$$

where $\bar{x}$ is the mean, s is the standard deviation, and n is the number of data points. The first part of the equation is 0 when Kurtosis is 0. *Excess Kurtosis* is calculated by subtracting -3 from it. (when n is very large, the second part of the formula becomes almost –3). These formulae can also be used in the calculations using *Excel* Sheet below.

	A	B	C	D	E	F	G
1	i	Math Score	x - mean	(x - mean)^4			
2	1	21	-27.0	531441.0			Excel Commands
3	2	34	-14.0	38416.0	n=	7	= COUNT(B2:B8)
4	3	60	12.0	20736.0	Sample Mean=	48.0	= AVERAGE(B2:B8)
5	4	60	12.0	20736.0	Sample Variance=	709.0	= VAR.S(B2:B8)
6	5	82	34.0	1336336.0	Standard Deviation=	26.6	= STDEVA(B2:B8)
7	6	10	-38.0	2085136.0	Kurtosis=	-1.476	= KURT(B2:B8)
8	7	69	21.0	194481.0			
9			SUM=	4227282.0			= SUM(D2:D8)
10			S^4=	502681.0			= F6^4
11			$\frac{n(n+1)}{(n-1)(n-2)(n-3)}$	0.46666667			= F3*(F3+1)/(F3-1)/(F3-2)/(F3-3)
12			$\frac{3(n-1)^2}{(n-2)(n-3)}$	5.4			= 3*(F3-1)^2/(F3-2)/(F3-3)
13			Kurtosis =	-1.476			= D9*D11/D10-D12

Initially, we can use *Kurtosis* to understand general distribution characteristics of our data. When *Kurtosis* is calculated as 0 or **Excess Kurtosis = -3**, our data follow a normal distribution perfectly. So, normally distributed data establishes the baseline for "*normal kurtosis*" as shown in Figure 3.18. Note, *sample excess Kurtosis* that significantly deviating from -3, maybe the indication of *non-normally* distributed data. This point needs further investigation.

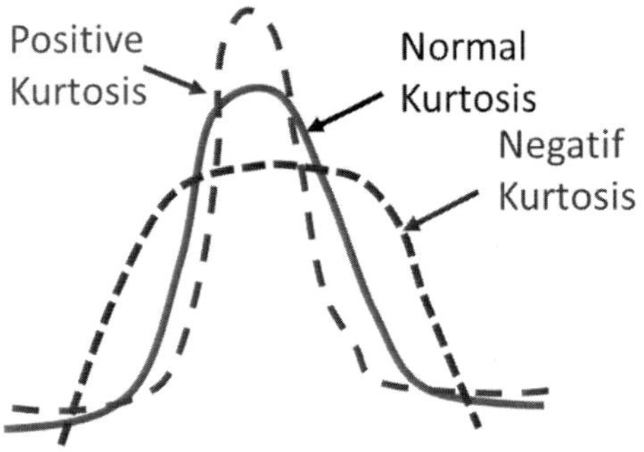

FIGURE 3.18 Kurtosis types

Positive kurtosis:

Excess Kurtosis > **3**, a distribution with a positive kurtosis value indicates that the distribution has a sharper peak(pointy) and heavier tails than the normal distribution

Negative kurtosis:

Excess Kurtosis < **3,** a distribution with a negative kurtosis value indicates that the distribution has a flatter peak and lighter tails than the normal distribution.

Skewness:

Skewness is a measure, indicating the degree of not being symmetrical of data around its mean, i.e. inclined. Skewness may take the value of *zero, positive*, or *negative* and gives us very useful information about the *shape* of the data. Three forms of skewness are listed below,

- Symmetrical or non-skewed distributions
- Positive or right-skewed distributions.
- Negative or left-skewed distributions

For univariate data $x_1, x_2, \ldots x_n$ then the *Skewness* formula (consisted with *Excel*) is given by Equation 3.11 as,

$$\text{Skewness} = \frac{n}{(n-1)(n-2)} \sum \left[\frac{(x_i - \bar{x})}{s} \right]^3 \qquad 3.11$$

where $\bar{x}$ is the mean, s is the standard deviation, and n is the number of data points. The more symmetrical data, the less *skewness* value is, as in the normal distribution in Figure 3.19 and also see Figure 3.20 for other types. Please keep in mind *that low skewness alone doesn't guarantee normality!*

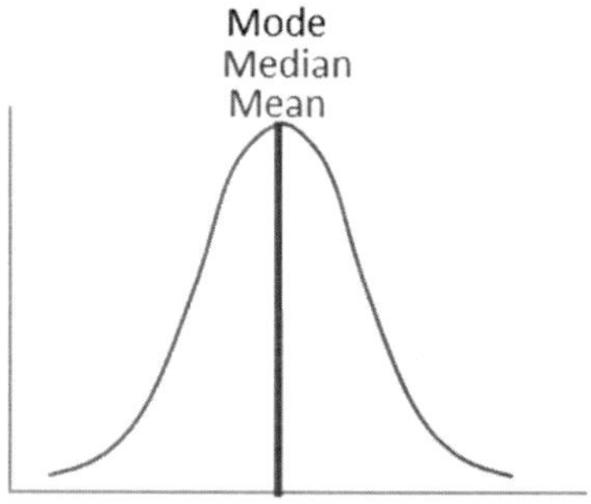

FIGURE 3.19 Symmetrical or non-skewed (normal) distributions

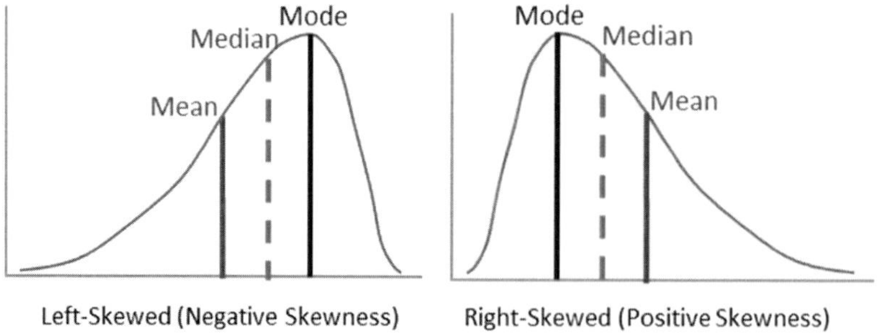

FIGURE 3.20 Positive and negative (normal) distributions

Excel sheet for skewness calculation is also shown on the figure for both using formulations and *Excel* commands. Finally, we are going to introduce how to calculate the confidence intervals for *unknown mean* and *unknown standard deviation* in the next section.

	A	B	C	D	E	F	G
1	i	Math Score	x - mean	$[(x - mean)/S]^3$			
2	1	21	-27.0	-1.0			Excel Commands
3	2	34	-14.0	-0.1	n=	7	= COUNT(B2:B8)
4	3	60	12.0	0.1	Sample Mean=	48.0	= AVERAGE(B2:B8)
5	4	60	12.0	0.1	Sample Variance=	709.0	= VAR.S(B2:B8)
6	5	82	34.0	2.1	Standard Deviation=	26.6	= STDEVA(B2:B8)
7	6	10	-38.0	-2.9	Skewness=	-0.312	= KURT(B2:B8)
8	7	69	21.0	0.5			
9			SUM=	-1.3			= SUM(D2:D8)
10			$\frac{n}{(n-1)(n-2)}$	0.233			= F3/(F3-1)/(F3-2)
11			Skewness=	-0.312			= D10*D9

3.6 Confidence Intervals for Unknown Mean and Unknown Standard Deviation:

Since the *population* standard deviation σ, is not known in most of the statistical problems, σ is replaced by the *sample* standard deviation *s*. In this case, the distribution of the sample mean $\bar{x}$ follows the *t*-distribution with mean μ and standard deviation $s/\sqrt{n}$. (Notice, since we are using *s*, the *sample* mean $\bar{x}$ no longer follows the normal distribution with mean μ and standard deviation $\sigma/\sqrt{n}$).

The *t*-distribution is also defined by its *dof*, i.e. *n* is being the sample data size, then, the *t*-distribution will have n-1 *dof*. As the sample size *n* increases, the *t*-distribution gets closer to the normal distribution as the *s* approaches to the population standard deviation σ, for large sample size, *n*.

Thus, confidence interval for unknown μ and unknown standard deviation σ, using *sample* standard deviation *s*, the confidence interval for *n* random *sample* data is calculated using Equation 3.12 where α=0.05, n=7, *dof=n-1=6* and using *t*-table given in Appendix 3 for Example 3.3, i.e. $t_{\alpha/2,n-1}=t_{0.05/2,7-1}=2.4469$

=T.INV(0.05/2,6) (*Excel* t-distribution formula)

dof\α	0.4	0.25	0.1	0.05	0.025	0.01	0.005	0.0025	0.001	0.0005
5	0.2672	0.7267	1.4759	2.0150	2.5706	3.3649	4.0321	4.7733	5.8934	6.8688
6	0.2648	0.7176	1.4398	1.9432	2.4469	3.1427	3.7074	4.3168	5.2076	5.9588
7	0.2632	0.7111	1.4149	1.8946	2.3646	2.9980	3.4995	4.0293	4.7853	5.4079

$$\bar{x} - \mathbf{t}_{\alpha/2,n-1}\frac{s}{\sqrt{n}} \le \mu \le \bar{x} + \mathbf{t}_{\alpha,n-1}\frac{s}{\sqrt{n}} \quad 3.12$$

$$48 - t_{0.05/2,7-1}\frac{26.63}{\sqrt{7}} \le \mu \le 48 + t_{0.05/2,7-1}\frac{26.63}{\sqrt{7}}$$

$$48 - \mathbf{24.63} \le \mu \le 48 + \mathbf{24.63}$$

$$23.37 \le \mu \le 72.63$$

Confidence Level (95.0%) is calculated as 24.63, which is quite big since the mean is 48. Summary of Example 3.3 *descriptive statistics* calculations and their corresponding *Excel* commands are presented on the next page.

	A	B
1	**i**	**Math Score**
2	1	82
3	2	69
4	3	60
5	4	60
6	5	34
7	6	21
8	7	10

	F	G	H
1	*Math Score*		**Excel Commands**
2			
3	Mean	**48.00**	= AVERAGE(B2:B8)
4	Standard Error	**10.06**	= G7/SQRT(G15)
5	Median	**60.00**	= MEDIAN(B2:B8)
6	Mode	**60.00**	= MODE.SNGL(B2:B8)
7	Standard Deviation	**26.63**	= STDEV.S(B2:B8)
8	Sample Variance	**709.00**	= VAR.S(B2:B8)
9	Kurtosis	**-1.48**	= KURT(B2:B8)
10	Skewness	**-0.31**	= SKEW(B2:B8)
11	Range	**72.00**	= G13-G12
12	Minimum	**10.00**	= MIN(B2:B8)
13	Maximum	**82.00**	= MAX(B2:B8)
14	Sum	**336.00**	= SUM(B2:B8)
15	Count	**7.00**	= COUNT(B2:B8)
16	Confidence Level(95.0%)	**24.63**	= CONFIDENCE.T(0.05,G7,G15)

Where notice that in the last row in the *Excel* sheet, the confidence interval for α=0.05 is also calculated using the *Excel* command given below,

=CONFIDENCE.T(α,Standard_dev, Size)

=CONFIDENCE(0.05,G7,G15)

=CONFIDENCE(0.05,26.6,7)

3.7 Exercises

1) Please express why the visual examination of the data is important? Why do we employ it first?

2) In your class or school collect *n=30* of sample data of your choice. First choose the subject (grade, travel distance to school, school fees etc.) and then you may ask survey questions related to your subject, to your friends! You are expected to do the following;

- Give brief information about your data? i.e. what subject are they belong to? Units of your data etc.
- Briefly, explain what are you going to do with your data? What kind of information are you looking for out of your data?
- Do useful plots of your choice and analyze these plots,(you can use any software of your choice).
- State your conclusions about your subject/data (i.e. distribution characteristics)

3) A research paper is going to be written about blood sugar levels of 8th grade students at a private high school. This paper will be focusing on the effects of current meal habits on blood sugar levels on teenagers. The blood samples are drawn out of *n=50 students* when they were fasting for the test (glucose test). The blood sugar results data in (*mg/dl*) are presented in Table 1.4.

- Use all plot alternatives and investigate the data characteristics
- Draw histogram and probability plot and comment on both!
- What is the percentage of students you think there is a risk for *pre-diabetes* (Unit per measurement is *mg/dl)*, please do some research for this critical level or interval, then state the risk in quantitative fraction or values
- What is the *HbA1c* test? Please do some research and give brief information.

TABLE 3.4 Blood sugar test data (*mg/dl*) for students

103.2	116.8	110.5	117.9	106.3	116.0	127.3	111.6	118.9	103.2
93.7	96.9	113.7	100.0	114.7	93.7	96.9	96.9	116.0	128.4
101.1	117.9	111.0	118.9	116.8	110.5	115.8	109.5	103.2	129.4
106.3	107.4	91.6	114.7	96.9	125.0	102.1	100.0	128.4	112.6
110.5	111.6	112.6	104.2	113.7	89.5	128.4	110.5	102.1	90.6

4) Discuss descriptive statistics and inferential statistics with examples.

5) Discuss the advantages of using sample data? What problems do you see if the improper usage of the sample is done? Please explain.

6) Plot the X data given below and also calculate the *mean*, *median*, *mode*, and *range* for this data:

X={8, 18, 8, 15, 8, 17, 13, 19, 8, 8,12,11, 5}

Note: Although numbers in the set of X are all integer (whole) numbers, as for mean, median you might end up with real (decimal) numbers, which is expected and fine, there is no need to round up these numbers.

7) The following weather temperature data in (°C) are collected in an area for two consecutive weeks in a season,

TABLE 3.6 Weather temperature data

Week Days	1	2	3	4	5	6	7	8	9	10	11	12	13	14
Temp. (°C)	25,5	29,0	33,0	35,0	36,5	40,5	42,0	33,0	33,0	30,0	30,1	20,0	18,0	19,0

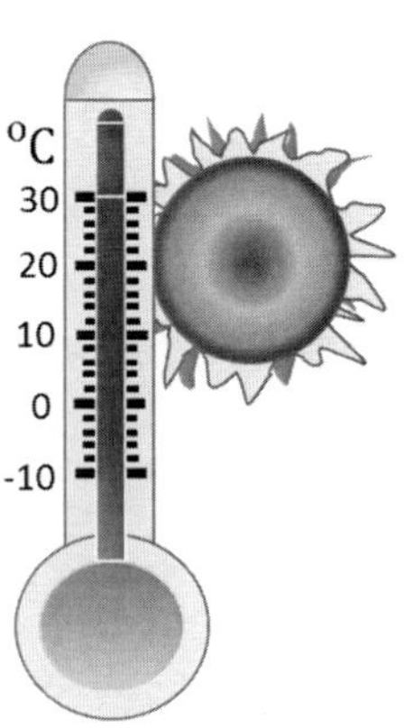

➢ Plot the data, comment a few words on the visual plot.

➢ Calculate the sample *mean, median, mode, range, standard deviation*

➢ Comment on data distribution and interpret your results.

8) Assume you run a small restaurant, where you have 11 staff working, including you and your wife, i.e. owners making big bucks as presented in the following Staff vs. Salary ($) table.

TABLE 3.7 Restaurant salary data

Staff	1	2	3	4	5	6	7	8	9	10	11
Salary ($)	21k	29k	28k	24k	20k	20k	20k	30k	32k	90	99k

- Plot the data, comment a few words on the visual plot.
- Calculate *mean, median, mode, range, standard deviation, Kurtosis* and *skewness values.*
- Comment on data distribution and interpret your results.

Chapter 4

Normal Distribution

Everybody loves math! Is that really so? Although many people would argue and think the other way around, I think the answer is yes! Love of Math is hidden in us. Because, math is an excellent "*way/tool*" to put things into "*nice behaving*" format and order in life, for our understanding and finding common ground among human beings with wisdom. We plan and live with math, we count, weigh, plan with math, therefore, math can talk to us in so many ways, yet we might think we are doing fine without it. Math is good for the one who knows and uses it. It never lets that individual down, always helps us understand, judge and evaluate many things. Events all around us may find some explanation/evaluation with math.

We think/assume, events of our interest, take place with some patterns, in possible/defined/feasible/meaningful ways, which might be referred to as chances or probabilities, all collected in a big container, where everything is mixed and put into some form of equation which represents a math distribution of the event. This distribution is valid only in that container. There is a further container i.e. design space, beyond that container, and this goes on. As a human being, we need to put/explain things with math to proceed to one step further, otherwise, we get stuck and cannot use our wisdom, and feel unfulfilled/locked. Probability distributions are our math meanings/forms/correspondence of representation of events' taking place format/way /formulation", and these events are always happening around us.

In this context, the term *"statistical experiment"* is a random or *non-deterministic* trial/test. Then, in a broad sense, a *probability distribution* is a *look-up table* or *an equation* that believed/assumed to map each outcome of a statistical experiment with its probability of occurrence for a random (input) variable that we are interested in.

Consider a group/set of patients. We are asked to find out their blood types as *A,B,AB* and *O* in groups. Here all possible outcomes of our blood types search experiment makes up a *sample space, S* and can be stated as a set, *S={A,B,AB,O}*.

Here, we are looking for some kind of (numerical) description of the outcome. Assume a random variable *X (which is* determined by the outcome of a random experiment)*,* in this case representing the chance/probability of blood type A, is being drawn, if 3 patients are tested i.e. $x_1=0$, $x_2=1$, $x_3=2$ and $x_4=3$ *(*Note, *X,* becomes *x* when assigned a number*)*. This case can be extended to more patients for other blood types. The next question is to inquire about the *type of random variables* in our usage, which would be explored in the next section.

4.1 Random Variable Types

We use several types of random variables to explain or model events that are taking place around us. These random variables are also defined within some bounds (i.e. upper, lower) of our *design space*, which looks like a *heated-pot* using which, we cook our meals, as depicted in Figure 3.1,

FIGURE 4.1 Random variable spaces

where, *three* types of random variables are encountered in our practical applications, as briefly outlined below,

i. *Text random variable* (description*)*, for example, *'yes, no, undecided','in, out', 'white, black, green, yellow', 'full, empty'* etc.

ii. *Discrete random variable* (countable), for example, *Week days={1,2,3,4,5,6,7}*, *Bolt sizes={5, 8,10, 12 and 15 in mm}*, etc. Note that bolt size 6.5 mm is not possible, you need to select either 6 or 7.

iii. *Continuous random variable* (uncountable), for example, real numbers that we use frequently. i.e. no distinction between numbers. Note that bolt size 6.5 mm is now possible with custom made products for all real numbers. Of course, practical usages are always limiting the bolt size selection.

In the following section, more information about distribution functions is presented inconsistent with the content of this book.

4.2 Distribution function

A (discrete or continuous) probability distribution is a some defined form of table or formulation, returning associated probabilities of any "*event*" when all possible values of a random (discrete or continuous) variable presented to it. In this context, we talk about

P.m.f.: *Probability mass function* for *discrete* variables (i.e. mass is a *point load*) and

P.d.f.: *Probability density function* for *continuous* variables (i.e. density is a *distributed load*).

Note, *pmf* or *pdf* is defined according to the type of random variable used and *load* term is only for our description and understanding. Also note that a histogram (i.e. frequency vs. random variable plot) overlapped to *f(x)* function below in Figure 4.2, is a practical approximation to probability density function of *f(x)* as shown below,

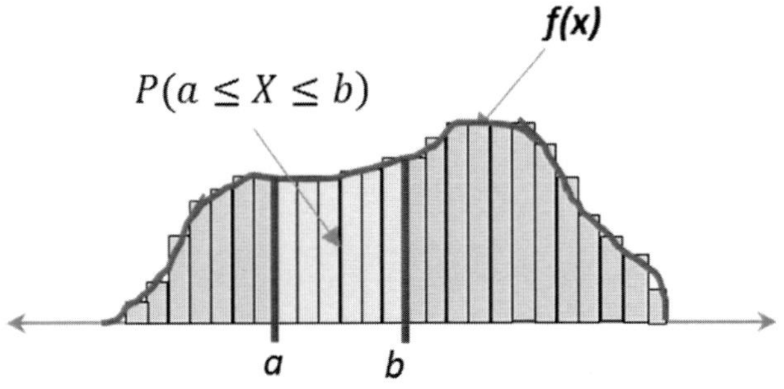

FIGURE 4.2 Sample probability density function, *f(x)*

In Figure 4.2, the $f(x) \geq 0$, the function is shown called *pdf:* probability density function for the continuous random variable *X*. The area under an example *f(x)* curve is 1 (unity) as given in Equation 4.1,

$$\int_{-\infty}^{+\infty} f(x)dx = 1 \qquad 4.1$$

The area, under the curve between $x = a$ and $x = b$ ordinates, gives the probability that *X* lies between *a* and *b* as follows,

$$\int_{a}^{b} f(x)dx = P(a \leq X \leq b) \qquad 4.2$$

In the next section, the widely used *Normal (probability) distribution* is going to be introduced in the context of this book.

4.3 Normal Distribution

Also called Gaussian distribution, is the most widely used mathematical model for the distribution of a random variable, say X. Remember that, we *frequently measure things such as;*

- Weight, height, age, life span, behaviors
- Peoples income, tax, medical records
- Machine parts dimensions, i.e. nominal diameter, counts
- Outcomes of any physical or chemical process
- etc.

Pdf graph of these data often follows a *normal distribution* curve characteristics, i.e. symmetric bell-shaped curve. A random variable *X,* whose probability density distribution has the shape of a **normal curve of *f(x)*,** is called a ***normal random variable* with parameters** μ *and* σ; thus *f(x)* is shown as $X \sim N(\mu, \sigma)$,

$$f(X) = \frac{1}{\sigma\sqrt{2\pi}} e^{\frac{-(x-\mu)^2}{2\sigma^2}} \qquad -\infty < x, \mu < \infty \text{ and } \sigma > 0 \qquad 4.3$$

where **mean,** μ is a balance point (average) of all distributed data, around which these data are (i.e. normally) distributed. σ, the ***standard deviation*** of the probability distribution is a number describing the **spread** of the distribution with respect to or around mean, μ. While, small standard deviation σ, means the small spread, large standard deviation means the large spread.

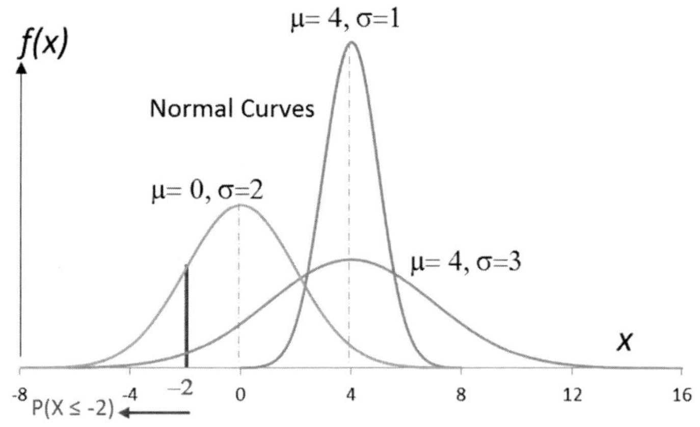

FIGURE4.3 Different normal probability (density function) curves

In Figure 4.3, *three* different normal distributions are shown (i.e. each *f(x)* distribution data per *X*) where, we have two different **means** (μ = 0 and 4), but three standard deviations as σ = 1,2 and 3, i.e. when σ becomes larger, the spread of the specific distribution goes wider, i.e. it looks like smashed on the top.

By examining Figure 4.3, we can talk about the following properties of the normal distribution;

Normal Distribution (Curve) Properties,

- The mean, μ is at the middle and divides the area into halves,
- The curve is symmetrical about its mean, μ;
- μ, mean and standard deviation σ, characterizes the curve,
- The area under the curve is equal to 1;

Note, we *always* use *Cumulative Probability Distribution,* which is considered as a combination of probabilities where a random variable value falls within a specified range, i.e. in Figure 4.3, a probability of $P(X\leq 2)$ corresponds to the area on the *always left* side of the normal curve! And these values are given in the Appendix of the statistics textbooks in ***Z-tables*** (i.e. Appendix 2) of *Cumulative standard normal distribution tables* for easy probability readings as would be introduced in the next sections.

Example 4.1:

In the manufacturing of automobile shafts, based on our observations, we experienced that the machined shaft diameter measurements are following a normal distribution with a mean of *20 mm* and standard deviation of *0.5 mm* (or variance of $0.25\ mm^2$). As far as the quality requirements impose, we are going to only accept, if $d=20\pm0.05$ *mm,* as also shown in Figure 4.4. What is the probability that machined shaft diameters will be between *19.95-20.5 mm?*

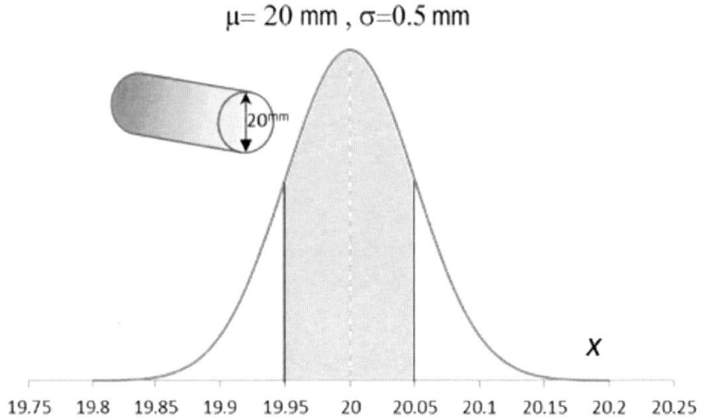

FIGURE 4.4 Machined shaft diameters and their distributions

When we machine the shaft parts, we always get variations in dimensions, here in *diameter*! although we use exactly the same turning machine, the same technician, tooling, and procedure. Why? Because we always observe fluctuations due to uncontrollable effects! (i.e. noise effects). For example; vibrations, temperature differences, tool/machine wears, work part irregularities etc. These variations can be assumed to follow normal distribution curve with more probabilities or having similar part diameters concentrated in the middle (or Target diameter *d=20 mm*) of the curve.

Let *X* denote the part shaft diameter in *mm*. The requested probability can be shown as *P(19.95<X<20.05)* and we use Cumulative Probability Distribution to find out the *area or requested probability*, using standardized *Z-tables,* as also given in Appendix 2. This probability is shown as the shaded area under the normal curve in Figure 4.4.

4.4 Area Divisions under the Normal Curve

Very noticeable and useful information about a *normal distribution* area percentages (or probabilities) is given in Equation 4.4. These areas depend on sample variance increments about the mean. Interestingly, these areas can be stated and depicted for any *random variable, i.e.* the area between two ordinates (of x) for populations in terms of σ,

$$\textbf{Area 1}: P(\mu - \sigma < X < \mu + \sigma) = 0.68269 \qquad 4.4$$

$$\textbf{\textit{Area}}\ \mathbf{2}: P(\mu - 2\sigma < X < \mu + 2\sigma) = 0.9545$$

$$\textbf{\textit{Area}}\ \mathbf{3}: P(\mu - 3\sigma < X < \mu + 3\sigma) = 0.9973$$

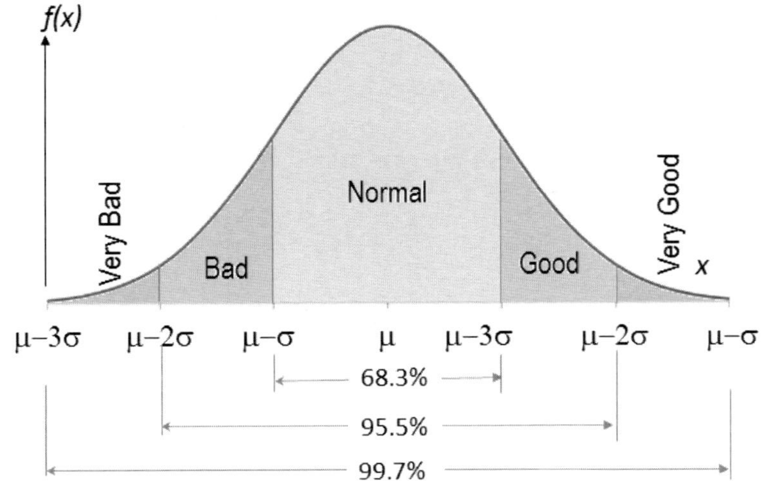

FIGURE 4.5 Area divisions under normal distribution per σ

- **Area 1:** Approximately 68% of the measurements (corresponding to the area under the curve) falls within *one-*σ ($\mu \pm \sigma$) of the mean for population or one-*s* ($\bar{x} \pm s$) for samples.
- **Area 2:** Approximately 95% of the measurements (corresponding to the area under the curve) falls within *two-*σ ($\mu \pm 2\sigma$) of the mean for population or one-*s* ($\bar{x} \pm 2s$) for samples.
- **Area 3:** Approximately 99.7% of the measurements (corresponding to the area under the curve) falls within *three-*σ ($\mu \pm 3\sigma$) of the mean for population or one-*s* ($\bar{x} \pm 3s$) for samples.

For example, we can interpret this normal distribution curve given in Figure 4.5 for % of students in a class,

☺ Area "*Very Good*" for % of students with **very good** grades
☺ Area "*Good*" for % of students with **good** grades
😐 Area "*Normal*" for % of students with **normal** grades
☹ Area "*Bad*" for % of students with **bad** grades
☹ Area "*Very Bad*" for % of students with **very bad** grades

The area under a normal distribution curve beyond 3σ, can be thought as negligible small, so a total 6σ length of the area is commonly referred as the width of the normal distribution curve, remember the area under the curve is *unity*.

Chebyshev's Rule:

This rule also gives how many measurements fall within 1,2 and 3 *standard deviations* of the mean and applies to *any* set of data.

- **Area *1-fraction*:** For $k = 1$, no estimation, *for the fraction of measurements*, is provided!
- **Area *k-fraction*:** For any $k > 1$, ***at least***

$$\textbf{Area } \textit{k-fraction} = 1 - \frac{1}{k^2} \qquad 4.5$$

of the measurements will fall within $k\sigma$ of the μ i.e. $(\mu \pm k\sigma)$ for population or ks $(\bar{x} \pm ks)$ for samples, i.e. consider for $k=2$ and 3, *Area k fraction=3/4* and *8/9* or 75% and 89%.

Example 4.2: Helicopter Design

In a paper helicopter design competition, a class of 30 students, each student is expected to design and build a paper helicopter with different upper and lower bounds of their choice, on selected design variables. Maximum of A4 size paper is to be permitted in the design. Each helicopter is released from free fall of 3-meter high, and their flight times are recorded. Maximum recorded flight times for each student's helicopter are given in the following.

3.8	2.0	4.1	3.5	2.1	4.5	4.2	3.4	3.2	3.7
1.8	2.8	1.8	5.1	4.6	1.8	2.2	2.3	3.2	2.9
2.5	2.7	2.9	4.4	2.6	2.2	1.7	1.6	1.8	3.3

- Determine the fraction of the 30 measurements in the intervals of $\bar{x} \pm s, \bar{x} \pm 2s$, and $\bar{x} \pm 3s$.
- Compare these results with normal distribution characteristics.

By using the corresponding formulations and *Excel* sheet for calculation of descriptive statistics. The mean and standard deviations, i.e.

- $\bar{x} = 2.9567$ secs and $s = 1$ sec are obtained.

n=	30
$\bar{x}$ = mean=	2.9567
Standard Deviation	1.00
Sample Variance=	1.00
Median=	2.85
Minimum=	1.60
Maximum=	5.10
Range=	3.50

- We can calculate the intervals for $\bar{x} \pm s$, $\bar{x} \pm 2s$ and $\bar{x} \pm 3s$ as follows.

- $(\bar{x} - s, \ \bar{x} + s) = (2.9567 - 1, 2.9567 + 1) = (1.96, 3.96)$, and there are 18 data points in this interval, 18/30 = 0.6 or 60% of the samples fall in this interval.
- $(\bar{x} - 2s, \ \bar{x} + 2s) = (2.9567 - 2 * 1, 2.9567 + 2 * 1) = (0.96, 4.96)$, and there are 29 data points in this interval, 29/30 = 0.97 or 97% of the samples fall in this interval.
- $(\bar{x} - 3s, \ \bar{x} + 3s) = (2.9567 - 3 * 1, 2.9567 + 3 * 1) = (-0.05, 5.96)$, and there are 30 data points in this interval, 30/30 = 1.0 or 100% of the samples fall in this interval.

- As it was summarized in the below, data distributions %, i.e. 60, 96.7, 100% are consistent with 68.3, 95.5, 99.7% normal distribution characteristics as also calculated and compared in the following *Excel* table.

Intervals	Max. F. Time	# of data	Data Dist.	Normal Dist.
$\bar{x}$ - s =	1.96	18	**60.0**	68.3
$\bar{x}$ + s =	3.96			
$\bar{x}$ - 2s =	0.96	29	**96.7**	95.5
$\bar{x}$ + 2s =	4.96			
$\bar{x}$ - 3s =	-0.05	30	**100.0**	99.7
$\bar{x}$ + 3s =	5.96			

These results are also consistent with Chebyshev's Rule since it guarantees <u>*at least*</u> 75%(≤96.7%) and 89%(≤ 100%) of values are within 2 and 3 *standard deviations* of the mean.

For this data, the *sample mean* and *standard deviation*, histogram plot and normal curve are both shown as overlapped in Figure 4.6 which gives a quite clear idea about the distribution characteristics.

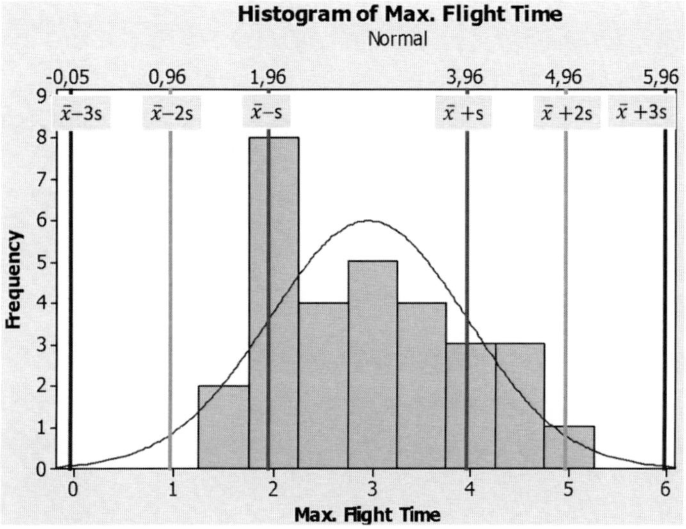

FIGURE 4.6 Histogram for maximum flight time

4.5 The Standard Normal Distribution

The probability calculation through the use of normal distribution requires a lot of effort in math since no closed-form of expression for the solution of Equation 4.3 for the normal random variable function is available! Therefore numerical solutions have to be given in tables. But, it is impossible to add an appendix for all possible combinations of normal probability distribution tables.

Therefore, we can make it easier for students and professionals by standardizing (i.e. normalizing) the normal curve, abbreviated as $X \sim N(\mu, \sigma)$, with mean μ =0 and a standard deviation of σ =1, abbreviated as $Z \sim N(0,1)$. Then, using the basic *Z to X* transformation given in Equation 4.6, we come up with *f(Z)* as given in Equation 4.7,

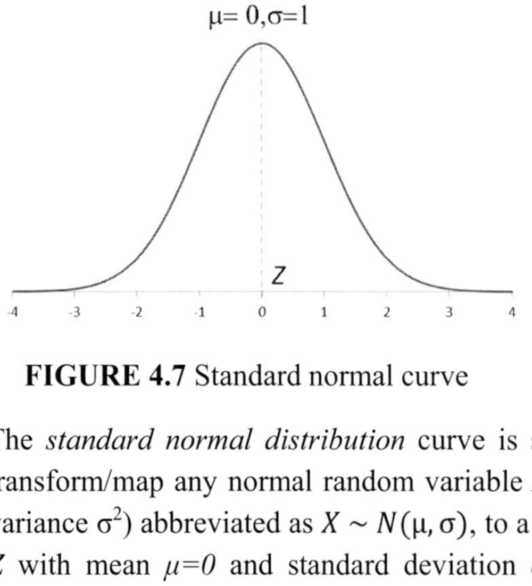

$$Z = \frac{X - \mu}{\sigma} \qquad 4.6$$

$$f(Z) = \frac{1}{\sqrt{2\pi}} e^{\frac{-z^2}{2}} \qquad 4.7$$

FIGURE 4.7 Standard normal curve

The *standard normal distribution* curve is shown in Figure 4.7, where we are able to transform/map any normal random variable *X* with mean μ and standard deviation σ (or variance σ^2) abbreviated as $X \sim N(\mu, \sigma)$, to a new set of another standard random variable Z with mean $\mu=0$ and standard deviation $\sigma=1$, abbreviated as $Z \sim N(0,1)$, using the transformation given in Equation 4.6.

The *cumulative standard normal distribution function* is also denoted by Φ in the literature defined as follows;

$$\Phi(z) = P(Z \leq z) = \int_{-\infty}^{z} \frac{1}{\sqrt{2\pi}} e^{-\frac{1}{2}y^2} dy \text{ (i.e. } y \text{ is any variable)} \qquad 4.8$$

Notice that, by employing standardization, cumulative standard normal distribution scores for all *Z*, can be given in two pages in any statistical textbook's appendix as ***cumulative standard normal distribution tables*** (i.e. Appendix 2) for easy direct probability readings without worrying about the mathematical solution.

Example 4.3:

Transform $\mu=2$ and $\sigma=0.4$ normal distribution curve into a standardized normal distribution curve using both Equations 4.6 and 4.7. Also remember that these two curves have different μ and σ, but they have the same *area* under the curves!

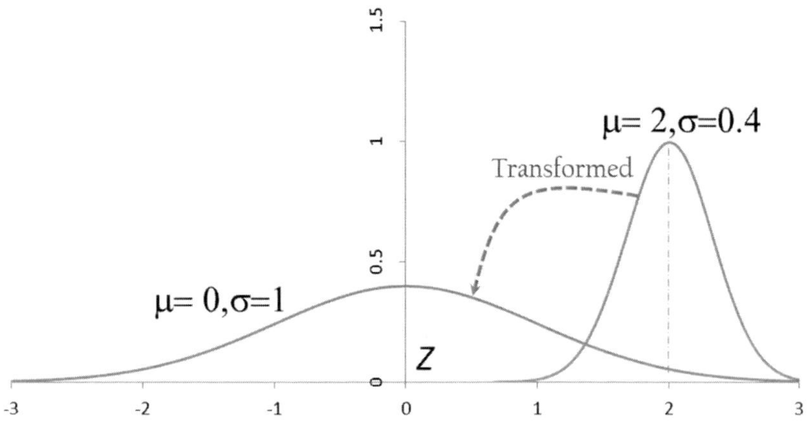

FIGURE 4.8 Before and after standardization of normal curve

We are obtaining *Z*-score first for $\mu=2$ and $\sigma=0.4$ using the following transformation and also an example transformation ordinate, assume $x=2$, then

$$Z = \frac{X - \mu}{\sigma} = \frac{X - 2}{0.4} = \frac{2 - 2}{0.4} = 0$$

As demonstrated in Figure 4.8, after the transformation, $\mu=x=2$ score has moved to $\mu=z=0$ score. Once we do similar operations for all *X* scores, then, we would have all corresponding *Z* scores. Using related equations, we would be transforming the curves as well, for example, notice to the maximum *f* value at the specified ordinate below,

$$f_{max}(X = 2) = \frac{1}{\sigma\sqrt{2\pi}} e^{\frac{-(x-\mu)^2}{2\sigma^2}} = \frac{1}{0.4\sqrt{2\pi}} e^{\frac{-(2-2)^2}{2.0.4^2}} = 0.9974$$

Equivalent *Excel* command to calculate *f(x)* values is presented below, **note that " = " sign accompanies all the equations because it executes all formulas typed in after itself, " = " sign is also shown in Excel Formula column for allowing associated formula visualization.**

= NORMDIST(x,mean,standard_dev,cumulative)

Where x is the Z=2, mean=2, standard_dev=0.4 and cumulative is a logical operator. If the cumulative term is TRUE, NORMDIST returns the cumulative distribution function, i.e. *cdf*; if FALSE, then, it returns the probability density function i.e. *pdf*.

Using the *Excel* command below, we get the same results as $f_{max}(X = 2) = 0.9974$. Note " $ " sign locks the *Excel* cell to the specific value, i.e. B$1 is fixed at 0.4!

	A	B	C	D	E
1	**Mean=**	**2**	***X***	***f(X)***	**Excel Formula**
2	**St. dev=**	**0.4**	2	0.9974	= NORMDIST(C2,B$1,B$2,FALSE)

As for the second ordinate, we obtain the same result by using Equation 4.7 and *Excel* command,

$$f_{max}(Z = 0) = \frac{1}{\sqrt{2\pi}} e^{\frac{-z^2}{2}} = \frac{1}{\sqrt{2\pi}} e^{\frac{-0^2}{2}} = \frac{1}{\sqrt{2\pi}} e^{0} = 0.3989$$

	A	B	C	D	E
1	**Mean=**	**0**	***Z***	***f(Z)***	**Excel Formula**
2	**St. dev=**	**1**	0	0.3989	= NORMDIST(C2,B$1,B$2,FALSE)

Very important to note that during this transformation, an area under any two ordinates when transformed from X_a, X_b to Z_a, Z_b, remains *the same*, as the probabilities of occurrence remain *the same* as well, i.e. $P(x_a \leq X \leq x_b)$= $P(z_a \leq Z \leq z_b)$.

Example 4.4:

Find the area under the standard normal curve i.e. $z \sim N(0,1)$ between the following two ordinates using the *Z-table* and *Excel* commands.

1. $z_{low} = 0$ and $z_{high} = 0.89$
2. $z = -0.76$ and $z = 0$
3. $z = -1.03$ and $z = 1.69$
4. $z = 0.33$ and $z = 1.86$
5. $z = -1.33$ and $z = \infty$

1. We are going to use both *Z-table* and *Excel* spread sheet, which is more accurate and easy to use for this task! First use *Z-table* for reading the *cdf* results, as shown in the following, remember, $z_{low} = 0$ and $z_{high} = 0.89$

z	0	0.01	0.02	0.03	0.04	0.05	0.06	0.07	0.08	0.09
0	0.5	0.503989	0.507978	0.511966	0.515953	0.519939	0.523922	0.527903	0.531881	0.535856
0.1	0.539828	0.543795	0.547758	0.551717	0.55567	0.559618	0.563559	0.567495	0.571424	0.575345
0.2	0.57926	0.583166	0.587064	0.590954	0.594835	0.598706	0.602568	0.60642	0.610261	0.614092
0.3	0.617911	0.62172	0.625516	0.6293	0.633072	0.636831	0.640576	0.644309	0.648027	0.651732
0.4	0.655422	0.659097	0.662757	0.666402	0.670031	0.673645	0.677242	0.680822	0.684386	0.687933
0.5	0.691462	0.694974	0.698468	0.701944	0.705401	0.70884	0.71226	0.715661	0.719043	0.722405
0.6	0.725747	0.729069	0.732371	0.735653	0.738914	0.742154	0.745373	0.748571	0.751748	0.754903
0.7	0.758036	0.761148	0.764238	0.767305	0.77035	0.773373	0.776373	0.77935	0.782305	0.785236
0.8	~~0.788145~~	~~0.79103~~	~~0.793892~~	~~0.796731~~	~~0.799546~~	~~0.802337~~	~~0.805105~~	~~0.80785~~	~~0.81057~~	0.813267
0.9	0.81594	0.818589	0.821214	0.823814	0.826391	0.828944	0.831472	0.833977	0.836457	0.838913

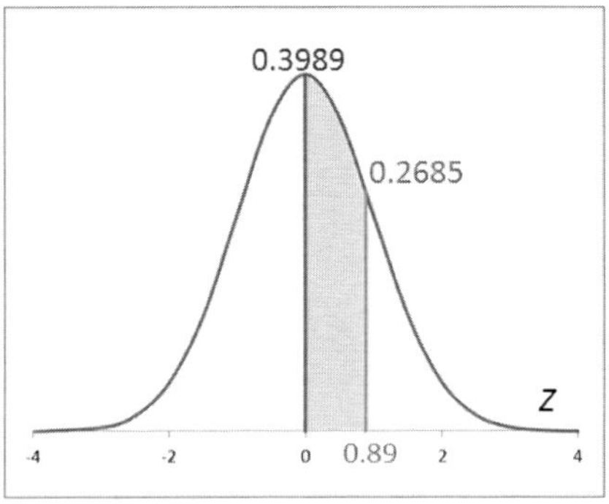

Area = 0.3133, between 0 < z < 0.89 in the normal curve below!

$$f_{max}(Z=0) = \frac{1}{\sqrt{2\pi}} e^{\frac{-z^2}{2}} = \frac{1}{\sqrt{2\pi}} e^{\frac{-0^2}{2}} = \frac{1}{\sqrt{2\pi}} e^{0} = 0.3989$$

$$f(Z=0.89) = \frac{1}{\sqrt{2\pi}} e^{\frac{-z^2}{2}} = \frac{1}{\sqrt{2\pi}} e^{\frac{-0.89^2}{2}} = 0.2685$$

First, refer to Z-table and read $P(Z_{low} < 0) = 0.5$, i.e. half of the curve and $P(Z_{high} < 0.89) = 0.813267$,

$$P(0 < z < 0.89) = \text{Shaded Area}$$

$$P(0 < z < 0.89) = P\left(Z_{high} < 0.89\right) - P(Z_{low} < 0)$$

$$= 0.813267 - 0.5 = 0.3133$$

We are going to use the following *Excel* command for the same task below, i.e. *cdf* calculation,

NORMDIST(x,mean,standard_dev,cumulative)

Where x is the Z=0 and 0.89, mean=0, standard_dev=1 and the cumulative term is a logical operator. If cumulative is TRUE, NORMDIST returns the cumulative distribution function, i.e. *cdf*; if FALSE, then, it returns the probability density function i.e *pdf.* i.e. area calculation, exactly the same as *Z-table* reading.

z_{low} = 0 (Enter C2 cell) and z_{high} = 0.89 (Enter C3 cell),

	A	B	C	D	E	F	G
1	**Mean=**	**0**	***Z***	***f(Z)***	**Excel Formula**	**Area**	**Excel Formula**
2	**St. dev=**	**1**	0	0.3989	=NORMDIST(C2,B$1,B$2,FALSE)	0.5000	=NORMDIST(C2,B1,B2,TRUE)
3			0.89	0.2685	=NORMDIST(C3,B$1,B$2,FALSE)	0.8133	=NORMDIST(C3,B1,B2,TRUE)
4						**0.3133**	=F3 - F2 **(Area in between)**

As you would notice *Excel* calculations are so easy and requires no *Z-table* reading need.

2. $z = -0.76$ and $z = 0$

z	-0.09	-0.08	-0.07	-0.06	-0.05	-0.04	-0.03	-0.02	-0.01	0
-0.7	0.214764	0.217695	0.22065	0.223627	0.226627	0.22965	0.232695	0.235762	0.238852	0.241964
-0.6	0.245097	0.248252	0.251429	0.254627	0.257846	0.261086	0.264347	0.267629	0.270931	0.274253
-0.1	0.424655	0.428576	0.432505	0.436441	0.440382	0.44433	0.448283	0.452242	0.456205	0.460172
0	0.464144	0.468119	0.472097	0.476078	0.480061	0.484047	0.488034	0.492022	0.496011	0.5

$$P(-0.76 < z < 0) = P(Z_{high} < 0) - P(Z_{low} < -0.76)$$

$$= 0.5 - 0.223627 = 0.2764$$

	A	B	C	D	E	F	G
1	**Mean=**	**0**	***Z***	***f(Z)***	**Excel Formula**	**Area**	**Excel Formula**
2	**St. dev=**	**1**	-0.76	0.2989	=NORMDIST(C2,B$1,B$2,FALSE)	0.2236	=NORMDIST(C2,B1,B2,TRUE)
3			0	0.3989	=NORMDIST(C3,B$1,B$2,FALSE)	0.5000	=NORMDIST(C3,B1,B2,TRUE)
4						**0.2764**	=F3 - F2 **(Area in between)**

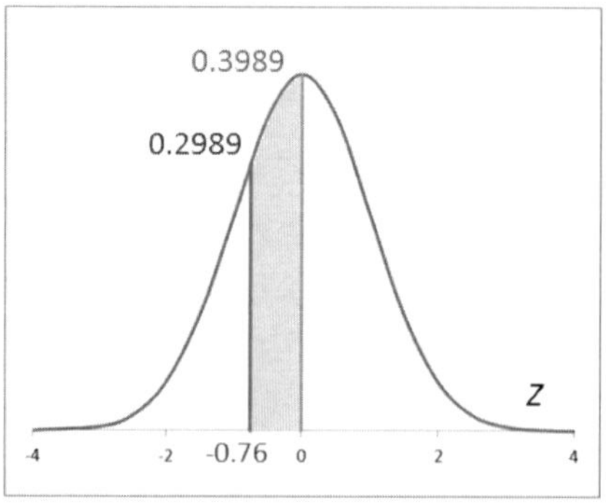

$$\text{Area} = 0.2764, \text{between} - 0.76 < Z < 0 \text{ in the normal curve}$$

3. $z = -1.03$ and $z = 1.69$

z	-0.09	-0.08	-0.07	-0.06	-0.05	-0.04	-0.03	-0.02	-0.01	0
-1	0.137857	0.140071	0.14231	0.144572	0.146859	0.14917	0.151505	0.153864	0.156248	0.158655

z	0	0.01	0.02	0.03	0.04	0.05	0.06	0.07	0.08	0.09
1.6	0.945201	0.946301	0.947384	0.948449	0.949497	0.950529	0.951543	0.95254	0.953521	0.954486

$$P(-1.03 < z < 1.69) = P(Z_{high} < 1.69) - P(Z_{low} < -1.03)$$

$$= 0.9545 - 0.1515 = 0.8030$$

	A	B	C	D	E	F	G
1	Mean=	0	Z	f(Z)	Excel Formula	Area	Excel Formula
2	St. dev=	1	-1.03	0.2347	= NORMDIST(C2,B$1,B$2,FALSE)	0.1515	= NORMDIST(C2,B1,B2,TRUE)
3			1.69	0.0957	= NORMDIST(C3,B$1,B$2,FALSE)	0.9545	= NORMDIST(C3,B1,B2,TRUE)
4						**0.8030**	= F3 - F2 (Area in between)

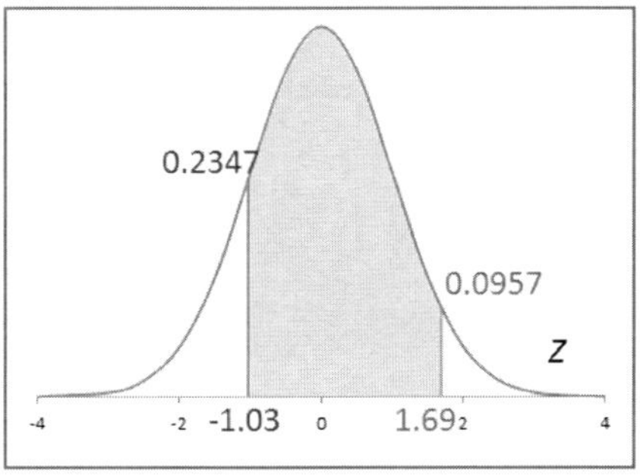

$$\text{Area} = 0.8030, \text{between} - 1.03 < Z < 1.69 \text{ in the normal curve}$$

4. $z = 0.33$ and $z = 1.86$

z	0	0.01	0.02	0.03	0.04	0.05	0.06	0.07	0.08	0.09
0	0.5	0.503989	0.507978	0.511966	0.515953	0.519939	0.523922	0.527903	0.531881	0.535856
0.3	0.617911	0.62172	0.625516	0.6293	0.633072	0.636831	0.640576	0.644309	0.648027	0.651732
1.8	0.96407	0.964852	0.96562	0.966375	0.967116	0.967843	0.968557	0.969258	0.969946	0.970621

$$P(0.33 < z < 1.86) = P(Z_{high} < 1.86) - P(Z_{low} < 0.33)$$

$$= 0.9686 - 0.6293 = 0.3393$$

	A	B	C	D	E	F	G
1	Mean=	0	Z	f(Z)	Excel Formula	Area	Excel Formula
2	St. dev=	1	0.33	0.3778	= NORMDIST(C2,B$1,B$2,FALSE)	0.6293	= NORMDIST(C2,B1,B2,TRUE)
3			1.86	0.0707	= NORMDIST(C3,B$1,B$2,FALSE)	0.9686	= NORMDIST(C3,B1,B2,TRUE)
4						**0.3393**	= F3 - F2 (Area in between)

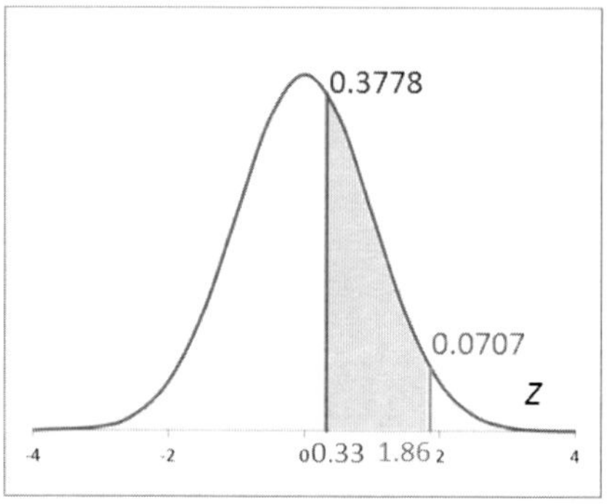

Area = 0.3393, between 0.33 < Z < 1.86 in the normal curve

5. $z = -1.33$ and ∞ (i.e. enter 99 as big number for ∞)

z	-0.09	-0.08	-0.07	-0.06	-0.05	-0.04	-0.03	-0.02	-0.01	0
-1.3	0.082264	0.083793	0.085343	0.086915	0.088508	0.090123	0.091759	0.093418	0.095098	0.0968
-1.2	0.098525	0.100273	0.102042	0.103835	0.10565	0.107488	0.109349	0.111232	0.113139	0.11507
0.0	0.464144	0.468119	0.472097	0.476078	0.480061	0.484047	0.488034	0.492022	0.496011	0.5
∞										1

$$P(-1.33 < z < \infty) = P(Z_{high} < \infty) - P(Z_{low} < -1.33)$$

$$= 1 - 0.0918 = 0.9082$$

	A	B	C	D	E	F	G
1	Mean=	0	Z	f(Z)	Excel Formula	Area	Excel Formula
2	St. dev=	1	-1.33	0.1647	= NORMDIST(C2,B$1,B$2,FALSE)	0.0918	= NORMDIST(C2,B1,B2,TRUE)
3			99	0	= NORMDIST(C3,B$1,B$2,FALSE)	1.0000	= NORMDIST(C3,B1,B2,TRUE)
4						**0.9082**	= F3 - F2 **(Area in between)**

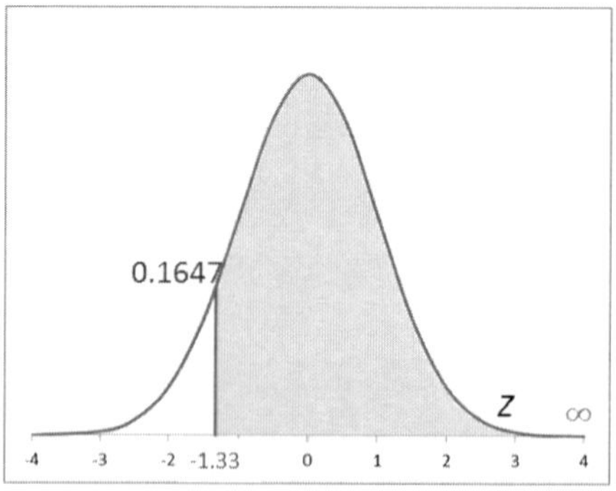

Area = 0.9082, between 1.33 < Z < ∞ in the normal curve

The advantage of using *Excel* sheet and commands, we can directly calculate the shaded area shown, whereas, in the classic *Z-table* readings, we read always the value covering left or the un-shaded portion of the normal curve.

Example 4.5:

Now, find the following probabilities using the standard normal curve i.e. $N \sim (0,1)$. You can use Z-table or *Excel* sheet!

a. P(Z > 1.13)

b. P(Z < -1)

c. P(0.53 < Z < 2.5)

a. P(Z > 1.13), this expression is analogous to ask for the area to the right of $Z=1.13$ under the standard normal curve. So, we can use the same *Excel* sheet as presented before. We are going to calculate the area between z = ∞ to 1.13 (i.e. z=1.13 is the left side of the normal curve which we read from Z-table and quite large value, i.e. 99 is entered in *Excel* for ∞),

z	0	0.01	0.02	0.03	0.04	0.05	0.06	0.07	0.08	0.09
0	0.5	0.503989	0.507978	0.511966	0.515953	0.519939	0.523922	0.527903	0.531881	0.535856
1.1	0.864334	0.8665	0.868643	0.870762	0.872857	0.874928	0.876976	0.879	0.881	0.882977

$$P(Z > 1.13) = P(Z < \infty) - P(Z < 1.13) = 1 - 0.870762 = 0.1292$$

	A	B	C	D	E	F	G
1	Mean=	0	Z	f(Z)	Excel Formula	Area	Excel Formula
2	St. dev=	1	1.13	0.2107	= NORMDIST(C2,B$1,B$2,FALSE)	0.8708	= NORMDIST(C2,B1,B2,TRUE)
3			99	0	= NORMDIST(C3,B$1,B$2,FALSE)	1.0000	= NORMDIST(C3,B1,B2,TRUE)
4						**0.1292**	= F3 - F2 **(Area in between)**

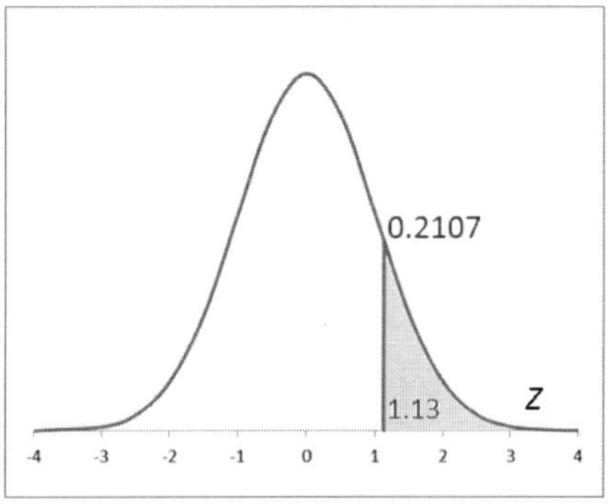

Area $= 0.1292$, between $1.13 < Z < \infty$ in the normal curve

b. P(Z < -1), this expression is analogous to ask for the area to the left of Z = -1 under the standard normal curve. We are going to read this area/probability i.e. z < -1, from Z-table, (-99 is entered in *Excel* for -∞),

z	-0.09	-0.08	-0.07	-0.06	-0.05	-0.04	-0.03	-0.02	-0.01	0
-1	0.137857	0.140071	0.14231	0.144572	0.146859	0.14917	0.151505	0.153864	0.156248	0.158655

$$P(Z < -1) = 0.1587$$

	A	B	C	D	E	F	G
1	Mean=	0	Z	f(Z)	Excel Formula	Area	Excel Formula
2	St. dev=	1	-99	0	= NORMDIST(C2,B$1,B$2,FALSE)	0.0000	= NORMDIST(C2,B1,B2,TRUE)
3			-1	0.242	= NORMDIST(C3,B$1,B$2,FALSE)	0.1587	= NORMDIST(C3,B1,B2,TRUE)
4						**0.1587**	= F3 - F2 **(Area in between)**

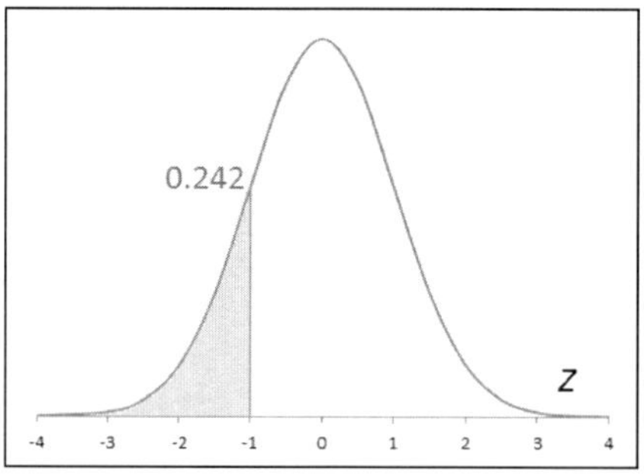

$$\text{Area} = 0.1587, \text{between} - \infty < Z < -1 \text{ in the normal curve}$$

c. P(0.53 < Z < 2.5), this expression is analogous to ask for the area between z = 0.53 and z = 2.5 under the standard normal curve. We are going to calculate the area between P(Z< 2.5) and P(Z < 0.53) i.e. because, area to the left side of the normal curve, which we can read from Z-table,

z	0	0.01	0.02	0.03	0.04	0.05	0.06	0.07	0.08	0.09
0.5	0.691462	0.694974	0.698468	0.701944	0.705401	0.70884	0.71226	0.715661	0.719043	0.722405
2.5	0.99379	0.993963	0.994132	0.994297	0.994457	0.994614	0.994766	0.994915	0.99506	0.995201

$$P(0.53 < Z < 2.5) = P(Z < 2.5) - P(Z < 0.53) = 0.9938 - 0.7019 = 0.2918$$

	A	B	C	D	E	F	G
1	**Mean=**	**0**	***Z***	***f(Z)***	**Excel Formula**	**Area**	**Excel Formula**
2	**St. dev=**	**1**	0.53	0.3467	=NORMDIST(C2,B$1,B$2,FALSE)	0.7019	=NORMDIST(C2,B1,B2,TRUE)
3			2.5	0.0175	=NORMDIST(C3,B$1,B$2,FALSE)	0.9938	=NORMDIST(C3,B1,B2,TRUE)
4						**0.2918**	=F3 - F2 **(Area in between)**

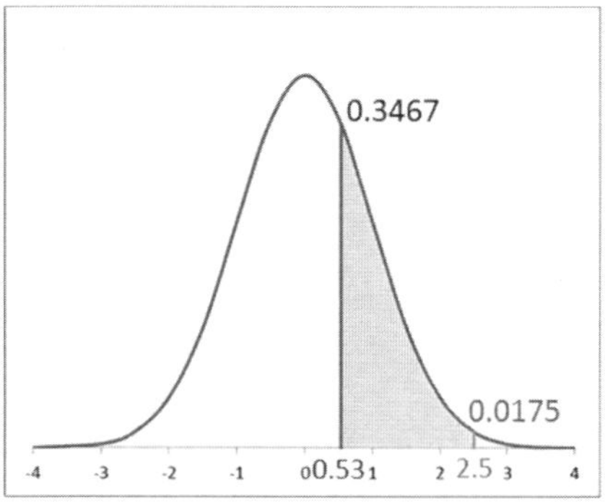

$\text{Area} = 0.2918, \text{between } 0.53 < Z < 2.5 \text{ in the normal curve}$

Example 4.6:

A tire company pays an hourly wage of $10 with a standard deviation of $2 since experience levels are different per employee. Assume the wage follows an almost normal distribution. Then, you are to calculate,

a) the count of the workers getting paid between $7.5 and $11.5 an hour,

b) the minimum wage of the highest 3%

Let's directly proceed with using an *Excel* spreadsheet,

a) We can use the same *Excel* Sheet that we used in the previous example. Make sure you enter Mean=10 (B1) and St. dev=2 (B2). Also X=7.5 (C2) and 11.5 (C3). We find the Area, i.e. Probability.

	A	B	C	D	E	F	G
1	**Mean=**	**10**	***X***	***f(X)***	**Excel Formula**	**Area**	**Excel Formula**
2	**St. dev=**	**2**	7.5	0.0913	=NORMDIST(C2,B$1,B$2,FALSE)	0.1056	=NORMDIST(C2,B1,B2,TRUE)
3			11.5	0.1506	=NORMDIST(C3,B$1,B$2,FALSE)	0.7734	=NORMDIST(C3,B1,B2,TRUE)
4						**0.6677**	=F3 - F2 **(Area in between)**

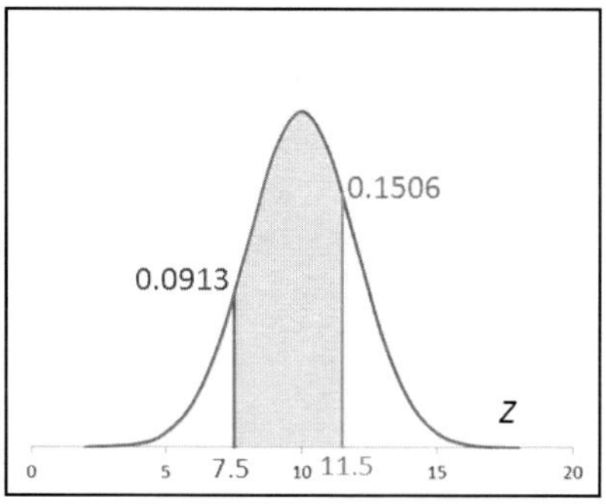

Area = 0.6677, between 7.5 < z < 11.5 in the normal curve

So about 66.8% of the workers get their wages between \$7.5 and \$11.5 per hour. You can see this portion illustrated as the shaded area under the normal curve shown above.

b) Now, we need to find the top 3% employee's wages, we re-state the problem in finding the area or probability between 0.97 and up and X corresponding to this area. This is an *inverse problem* for that we need to use a new *Excel* formula as follows,

NORM.INV(probability,mean,standard_dev)

Where, *Probability = 0.97* (and up), mean and standard deviation is inputted to this *Excel* command. i.e.

x=NORM.INV(0.97,10,2)=13.762

We may further verify the following,

f(x)=NORMDIST(13.762,10,2,FALSE)=0.034

Shaded Area=Probability=NORMDIST(13.762,10,2,TRUE)=0.97

	A	B	C	D	E	F
1	**Mean=**	**10**	***Probability***	***X***	*f(X)*	**Excel Formula**
2	**St. dev=**	**2**	0.97	13.762	0.034	= NORM.INV(C2,B$1,B$2)

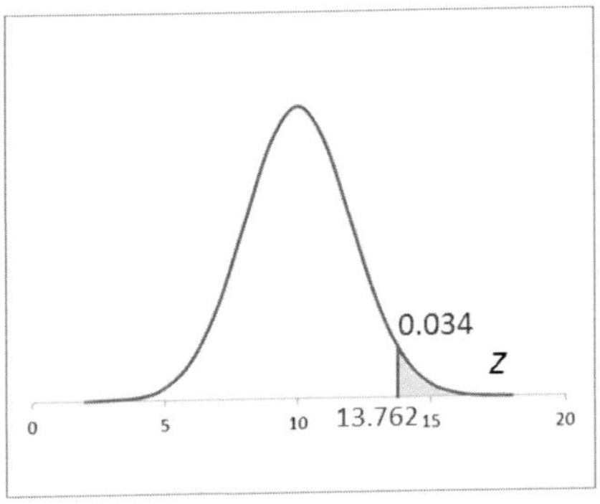

The shaded portion on the far right hand of the normal curve represents the top *3%* wage-paid people in the company. Minimum wage is calculated as \$13.76 per hour for the top *3%* of the people.

Example 4.7:

A refrigerator compressor is designed to operate an average life of 10 years without failure after startup and this is written in the manual. But, due to compressor suppliers and other variations, the compressor life may exhibit fluctuations with a standard deviation of 3.65 years. Assume the manufacturer is ok with replacing 5% of the compressors if they fail. Under this condition, how long a warranty should this manufacturer give to meet this replacement amount? Assuming compressor life follows a normal distribution.

Now, this problem is analogous to finding *x:* warranty period for the *X:* compressor life. We need to calculate the *period in years* that would correspond to left/bottom %5 of the normal distribution curve.

Note, $P(X<x) = 0.05$ area covers the compressors that manufacturer is ok to replace under guarantee.

- Use *Z-table* and find the area that corresponds to $P(X<x) = 0.05$, (Note, we are doing an inverse reading!)

Using *Z-table*, an *Area=0.05* corresponds to $z \cong -1.64$

z	-0,09	-0,08	-0,07	-0,06	-0,05	-0,04	-0,03	-0,02	-0,01	0
-1,7	0,036727	0,037538	0,038364	0,039204	0,040059	0,04093	0,041815	0,042716	0,043633	0,044565
-1,6	0,045514	0,046479	0,04746	0,048457	0,049471	0,050503	0,051551	0,052616	0,053699	0,054799
-1,5	0,055917	0,057053	0,058208	0,05938	0,060571	0,06178	0,063008	0,064255	0,065522	0,066807

Using this *Z-score* in the following transformation equation, we find un-standardized (i.e. original) *x* value for normal distribution curve,

$$Z = \frac{X - \mu}{\sigma} = \frac{X - 10}{3.65} = -1.64 \rightarrow x = 4.014$$

Finally, we can conclude that the guarantee period should be around 4 years. We can graph these outcomes on the normal curve with N ~ (μ=*10*, σ=3,65).

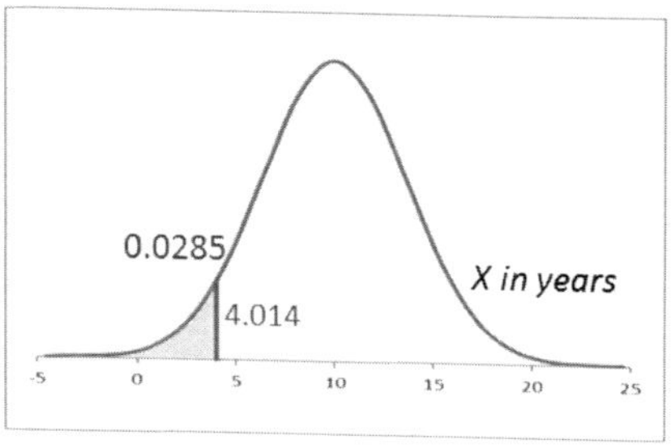

The left/bottom shaded area represents the *5%* compressors that are expected to fail (ok to be replaced by the manufacturer!) within the 4 years of operation.

Now we can do the same thing in our ready to use *Excel* sheet as follows, please see that **the use of *Excel* formulas are so easy and we do not need to standardize anything and use *Z-table*. We do all calculations directly, which is so easy to use!**

	A	B	C	D	E	F
1	**Mean=**	**10**	***Probability***	***X***	*f(X)*	**Excel Formula**
2	**St. dev=**	**3.65**	0.05	4.005	0.0283	=NORM.INV(C2,B$1,B$2)

where we enter for *Probability=Area = 0.05* and our *Excel* formula directly gives us *X=4.005* value which is more accurate and we do not need to deal with going any backward standardization and Z-table reading and calculating *X* or interpolation etc. The use of *Excel* is so neat and straight forward.

4.6 Exercises

1) Determine the following probabilities for the standard normal random variable Z, using *Z-table* given in Appendix 2.

a) P(Z < 1.43)   b) P(Z > 1)   c) P(Z< -1)   d) P(-2< Z < 2)   e) P($-1 < Z < \infty$)

2) Determine the *z* values that correspond to the following cumulative density probability values (i.e. area, probability). Assume a standard normal distribution. Use *Z-table* given in Appendix 2.

a) P(Z < z) = 0.75   b) P(Z < z) = 0.55   c) P(Z > z) = 0.3   d) P(Z > z) = 0.75

3) Assume X has a mean of 20 and a standard deviation of 3 and assume a standard normal distribution. Determine the following probabilities. Use *Z-table* given in Appendix 2.

a) $P(X < 19)$ b) $P(X > 22)$ c) $P(6 < X < 18)$ d) $P(11 < X < 24)$

4) The machined shaft diameter of automobile parts is following a normal distribution with a mean of 20 mm and a standard deviation of 0.5 mm (or variance of 0.25 mm^2). As far as quality requirements imposed on us we are going to only accept if $d=20\pm0.05$ mm, as also shown in Figure 4.9. What is the probability that machined shaft diameters will be between 19.95-20.5 mm? Use *Z-table* given in Appendix 2.

Hint: X denotes the shaft diameter in *mm* and determine $P(19.75 < X < 20.25)$?

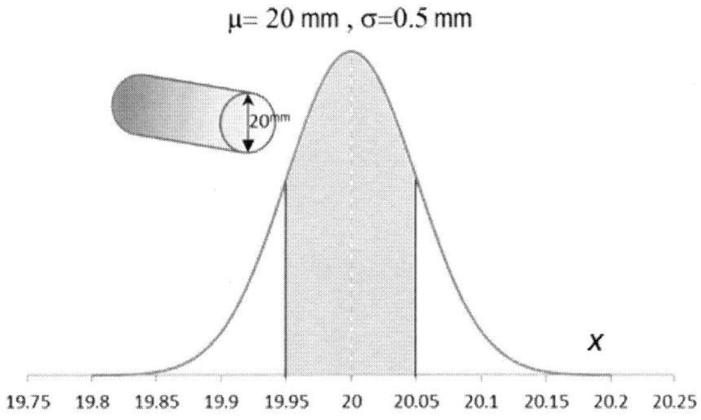

FIGURE 4.9 Machined shaft diameter and its distribution

5) The *SAT exam* is composed of three sections and SAT scores are key to get a scholarship and get acceptance from high-ranking state or private universities. For three sections the SAT scores range from 600 to 2,400. Assume we can use a normal probability curve to estimate probabilities with mean 1500 and a standard deviation of 334.

- Determine the percentage of students who got below 1800.
- What percentage of the students got above 2100 score?

Chapter 5

Statistical Inference - Basics

As for the *inferential statistics*, the comparison of two different *"things"* (i.e. two independent *populations*) is widely studied in statistics. For example, we may want to compare the life of two batteries, the effectiveness of two methods, the success of two teaching methodologies, the strength of two different plastic/cement formulations, the production outputs with the "old" and "new" machining approaches, etc. Here, we usually compare the *mean* of two sample distributions i.e. *populations*. Our primary objective is to test if the "*things*" with new data is better than the "things" with the old data or vice versa or we prove no difference! Hence, we would like to test *the null* (no difference) hypothesis $H_0: \mu_1 = \mu_2$, against *the alternative* hypothesis $H_1: \mu_1 \neq \mu_2$. With this regard, simple comparative experiments are performed to test if *two experiments (*experiments may also be referred as *conditions/treatments/formulations/samples/populations/tests/ measurements/designs*) differ or give equivalent results in *mean* values! Note, these hypotheses encompasses statements about the *population*, under study, not about the *sample*! but we use the *sample* for the *true* judgment of the *population*. Hypothesis-testing procedures check the information in a random *sample* drawn from the *population*. If this information is backed up with the hypothesis, it is proven that the sample, *with a big chance(confidence level, CI)*, represents the *population* and we can make more close to reality judgment by using this *sample.*

In this chapter, we will look at both hypothesis testing and population mean comparison within this context. Clearly, this comparison has a great deal of importance in comparing two different treatments to obtain their statistically sound comparative evaluations for many fields of study. In the following section, the difference in *means* of two distributions is compared when the variances of both distributions σ_1^2 and σ_2^2 are unknown, which is the case for most of the times. Also, assume that a small number of *samples* i.e. around 10-15 are used and both *populations* are assumed to be *normally distributed.* We judge our hypothesis tests and confidence intervals in this case based on the *t-distribution* as introduced in Section 5.1 and 5.2, where this methodology has also been studied and demonstrated in detail on example problems.

5.1 The Two-Sample T-Test

As per the comparison of two experiments and their distributions, we would like to further perform the *inferential statistics.* Because we are still not sure that we have different samples or not! *even* after the *visual examination* and study/analysis of *descriptive statistics* of both samples!

The two-Sample t-test is one step further action to test for *equality/not equality* of the *population* means using completely randomized data *sample* sets (i.e. two measurements/ treatments/designs/experiments etc.), so that we can ***reject*** or ***accept*** the hypothesis testing of our interest. In general, we encounter *commonly* two different assumptions and cases to employ as follows,

a. The variances of the two normal distributions are *unknown* but equal!
b. The variances of the two normal distributions are *unknown* and not-equal!

5.1.1 Variances are Unknown but Equal

Consider our tests of hypotheses on differences in means i.e. $\mu_1 - \mu_2$ of two normally distributed samples, where assume both variances are *unknown* but *equal*! i.e. $\sigma_1^2 = \sigma_2^2 = \sigma^2$. As stated earlier, in this case, *t*-statistics can be used for hypothesis testing, (i.e., when Δ_o=0).

$$H_0: \mu_1 - \mu_2 = \Delta_0 = 0 \quad 5.1$$

$$H_1: \mu_1 - \mu_2 \neq \Delta_0 \neq 0$$

where the null hypothesis H_0 states no difference in means, whereas an alternative hypothesis H_1 rejects H_0 and asserts both means are different based on a level of confidence α. On the following Example 5.1, we are going to go over all related formulations in detail.

Example 5.1: Neighboring Coal Mines

Consider an example problem, where the heating/*calorific* value of coal is defined as the amount of total *heating energy* released, under complete combustion at standard conditions. In this context, a technician is studying the heating value *kcal/kg*, i.e. the bigger is better, more heat *(kcal)* is desired per *(kg)*) of coal which is produced in two adjacent regions (each region is around *5-6 sq-mi*). Call them as *main* and *neighbor* regions as shown in Figure 5.1. The owner of both coal mines believes that the *neighbor* region's coal is better in terms of heating value than that of the *main* region. So, he claims these regions are really different in terms of coal quality. The owner asks his coal technician for support to his claim. The coal technician thinks about the problem and decides to run simple comparative experiments for this task.

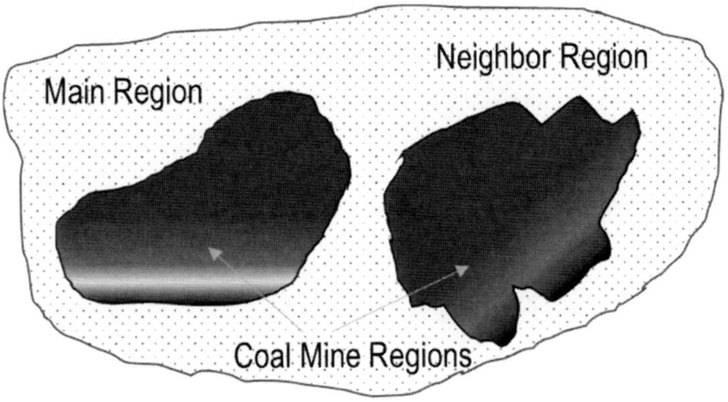

FIGURE 5.1 Coal mine regions for Example 4.1

He goes to both mines and picks 11 samples (each around 25 *lbs*) from different places of both regions, in a random manner and calculate the average coal heating values for both regions by using these samples. These data are presented in Table 5.1,

TABLE 5.1 The data for coal mine regions

	Base (kcal/kg)	Neighbor (kcal/kg)
j	y_1	y_2
1	5505	5630
2	5490	5735
3	5720	5495
4	5810	5890
5	5570	5850
6	5910	5990
7	5690	5690
8	5790	5925
9	5470	5590
10	5390	5780
11	5450	5965

Now, we can go ahead and start exploring the data by following the steps outlined below,

- First plot the data using the most basic display, *the dot diagram* which is a good choice for plotting a small number of data. These data are plotted in Figure 5.2. Visual examination of the data distribution gives the first impression that the coal heating value obtained from the *neighbor* region is greater than the heating value of the *main* region.

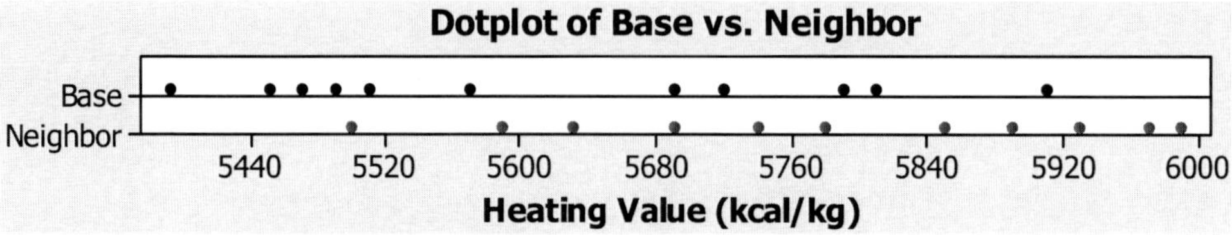

FIGURE 5.2 *The dot plot* for coal mines

- Next, let's draw a *box plot* diagram of two sample data for better 2D visualization of *descriptive statistics*. Visual examination of the *box plot* gives the impression that the *mean* coal heating value obtained from the neighbor region is greater than the heating value of the main region as shown in Figure 5.3. Note the *mean* (*dot* marker) and *median* of neighbor region's coal heating values coincide, those of base region is somewhat off.

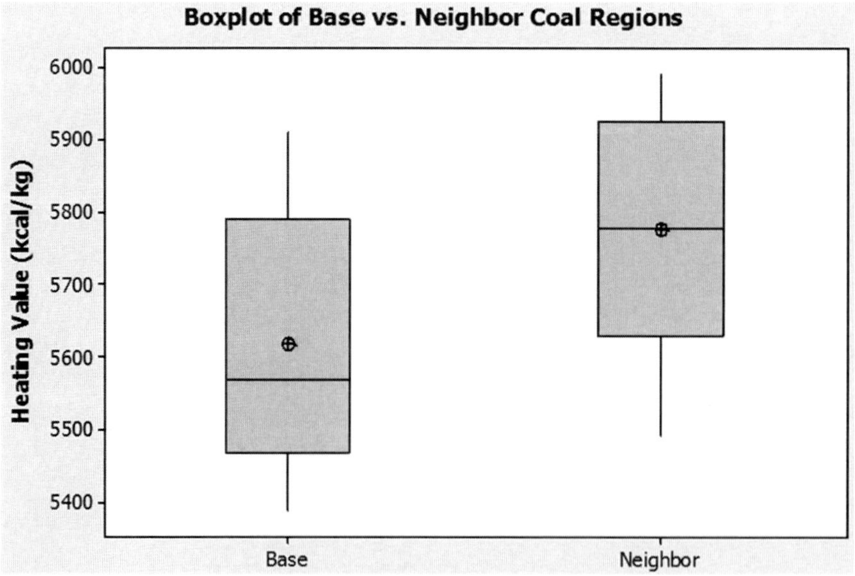

FIGURE 5.3 The *box plot* for both coal mines

- This statement can be supported by comparing the *average* heating value of the main region, $\bar{y}_1 = 5617.73$ kcal/kg and neighbor region, $\bar{y}_2 = 5776.36$ kcal/kg. The average heating value of these two sample regions differs by a *slight* amount. But we are not still sure about our argument. Because,
    a. this difference is not large enough! two samples/populations may not still be different.
    b. one might claim that the observed difference in average heating value is the result of sampling fluctuation and the two formulations are really identical!
    c. Perhaps, another two random data samples might be giving opposite results.
    d. We need no go beyond the *descriptive statistics* i.e. towards to *inferential statistics*

Then, we need to employ further study for statistical inference that we call *hypothesis testing* to compare these two coal region's heating level. Hypothesis testing allows us to compare two types of coals on the basis of their *mean* heating value. But, this is not in a deterministic manner i.e. with the risks associated with reaching the wrong conclusion will always be there (i.e. level of confidence must always be stated when expressing any conclusion!)

If both sample sizes i.e. n_1 and n_2 are fairly small and the underlying distributions are assumed normal, further suppose the variances of coal heating value for both regions were identical, then we use *t-test: Two-Sample Assuming Equal Variances*. Using equal sample sizes is preferred for both populations, why? because, when $n_1=n_2=n$, then *t-test* is more suitable to the assumption of *equal variances*.

Remember, this *t-test* is valid under the assumption that the two data sets come from *distributions* with the same *variances*. Here, we can employ this test to determine whether these two samples are likely to have come from distributions with equal population means. The test statistics, t_0 is given as,

$$t_0 = \frac{\bar{y}_1 - \bar{y}_2 - (\mu_1 - \mu_2) = 0}{S_p\sqrt{\frac{1}{n_1} + \frac{1}{n_2}}} \qquad 5.2$$

where $\bar{y}_1, \bar{y}_2$ are the sample means, n_1 and n_2 are the sample data sizes, S_p^2 is called as pooled variance, i.e. an estimated composite variance, remember again the equal variances assumption, *i.e.* $\sigma_1^2 = \sigma_2^2 = \sigma^2$,

$$S_p^2 = \frac{(n_1 - 1)S_1^2 + (n_2 - 1)S_2^2}{n_1 + n_2 - 2} \qquad 5.3$$

where S_1^2 and S_2^2 are obtained from both sample data variances. We can test the H_0: Null hypothesis against H_1: Alternative hypothesis with three possible cases, (this test is also called as the *pooled t-test*),

- $H_0: \mu_1 = \mu_2$ against *two-sided alternative hypothesis(two tailed),* $H_1: \mu_1 \neq \mu_2$ where we would compare test statistics, t_0 to the t distribution with $n_1 + n_2 - 2$ *dof. If* $|t_0| > t_{\alpha/2,n_1+n_2-2}$, where $t_{\alpha/2,n_1+n_2-2}$ is the upper $\boldsymbol{\alpha = 0.05/2}$ percentage value of the t distribution with $n_1 + n_2 - 2$ *dof.* We would *reject H_o* and conclude that the heating values of the two coal regions differ (i.e. remember α is the level of significance).
- $H_0: \mu_1 = \mu_2$ against $H_1: \mu_1 > \mu_2$ where we want to reject H_o, only if the first mean is larger than the other. Thus, we would specify a *one-sided alternative hypothesis(upper tailed),* $H_1: \mu_1 > \mu_2$ and would reject H_0 only if $t_0 > t_{\alpha,n_1+n_2-2}$
- $H_0: \mu_1 = \mu_2$ against $H_1: \mu_1 < \mu_2$ we want to reject H_o, only if μ_1 is less than μ_2, then the *one-sided alternative hypothesis(lower tailed),* $H_1: \mu_1 < \mu_2$ and one would reject H_o if $t_0 < -t_{\alpha,n_1+n_2-2}$

These conditions can be summarized in Table 5.2, assuming Null hypothesis, $H_0: \mu_1 = \mu_2$ or $\mu_1 - \mu_2 = \Delta_0 = 0$

TABLE 5.2 Summary of statistics and associated criterions

Hypothesis	**Test Statistics**	**Rejection Criterion**
$H_0: \mu_1 = \mu_2$ $H_1: \mu_1 \neq \mu_2$	$t_0 = \dfrac{\bar{y}_1 - \bar{y}_2 - (\mu_1 - \mu_2) = 0}{S_p\sqrt{\dfrac{1}{n_1} + \dfrac{1}{n_2}}}$	$\lvert t_0\rvert > t_{\alpha/2,n_1+n_2-2}$
$H_0: \mu_1 = \mu_2$ $H_1: \mu_1 > \mu_2$		$t_0 > t_{\alpha,n_1+n_2-2}$
$H_0: \mu_1 = \mu_2$ $H_1: \mu_1 < \mu_2$		$t_0 < -t_{\alpha,n_1+n_2-2}$

Now, let's proceed with calculating the required *descriptive statistics* that would be required in this *inferential statistics* evaluations. Using the formulations given in the next page, we can go ahead and calculate $\bar{y}_1$=5617.73, $\bar{y}_2$=5776.36, S_1=173.456, S_2=163.142 kcal/kg and using the Table 5.1 y_i values, where $n_1 = n_2 = n = 11$.

$$\bar{y} = \frac{1}{n}\sum_{i=1}^{n} y_i \text{ and } S^2 = \frac{1}{(n-1)}\sum_{i=1}^{n}(y_i - \bar{y})^2$$

$$S_p^2 = \frac{(n_1 - 1)S_1^2 - (n_2 - 1)S_2^2}{n_1 + n_2 - 2} \text{ and } t_0 = \frac{\bar{y}_1 - \bar{y}_2 - 0}{S_p\sqrt{\frac{1}{n_1} + \frac{1}{n_2}}} \text{ and SE Mean} = \frac{S}{\sqrt{n}}$$

where, we notice that *S*, both standard deviations are close enough, therefore, we can employ the case of *equal variances assumption* in *the two-sample test* in our hypothesis testing which can be put in a nice way as,

$H_0: \mu_1 = \mu_2$

$H_1: \mu_1 \neq \mu_2$ *(two tailed)*

$H_1: \mu_1 > \mu_2$ *(upper tailed)*

$H_1: \mu_1 < \mu_2$ *(lower tailed)*

$$S_p^2 = \frac{(11-1)173.456^2 + (11-1)163.142^2}{11 + 11 - 2} = 28351.1$$

$$t_0 = \frac{5617.73 - 5776.36 - 0}{168.378\sqrt{\frac{1}{11} + \frac{1}{11}}} = -2.2$$

Furthermore, $\text{dof} = n_1 + n_2 - 2 = 20$, *confidence interval for* $\alpha = \mathbf{0.05}$ */ 2=0.025, thus* $\mathbf{t_{0.025,20}} = \mathbf{2.086}$ *is read from t-table in Appendix 3 as follows,*

dof\α	0,4	0,25	0,1	0,05	0,025	0,01	0,005	0,0025	0,001	0,0005
1	0,3249	1,0000	3,0777	6,3138	12,71	31,82	63,66	127,32	318,31	636,62
2	0,2887	0,8165	1,8856	2,9200	4,3027	6,9646	9,9248	14,089	22,327	31,599
19	0,2569	0,6876	1,3277	1,7291	2,0930	2,5395	2,8609	3,1737	3,5794	3,8834
20	0,2567	0,6870	1,3253	1,7247	2,0860	2,5280	2,8453	3,1534	3,5518	3,8495
21	0,2566	0,6864	1,3232	1,7207	2,0796	2,5176	2,8314	3,1352	3,5272	3,8193

Now, we can calculate the same statistics and find the comparison data using *Excel* as well, as shown in Table 5.3,

TABLE 5.3 T-test data for Example 5.1

Coal Mine Regions		
j	Base	Neighbor
Mean =	5617.7	5776.4
S^2 =	30086.8	26615.5
S =	173.5	163.1
n =	11	11
S_p^2 =	28351.1	
S_p =	168.378	
T_o =	-2.2	
SE Mean =	52.3	49.2
CI Lower	$\leq \mu_1 - \mu_2 \leq$	CI Upper
-308.4	$\leq \mu_1 - \mu_2 \leq$	-8.9
$t_{0.025,20}$ =	2.0860	
p-value =	0.039	

Remember, $\alpha/2 = 0.05/2$ and *dof* = $n_1 + n_2 - 2 = 11 + 11 - 2 = 20$. Now, we can evaluate the test statistics using the following rules.

***Decision**: Reject* H_0 *if any of the following is true!*

$$t_0 > t_{0.025,20} \quad or \quad t_0 < -t_{0.025,20}$$

Then, since $t_0 = -2.2 < -t_{0.025,20} = -2.086$, we go ahead and reject H_o and conclude that the mean heating values of the two treatments/regions are *different*! Obviously, this rejection also depends on α i.e. *level of significance*. Keep in mind that, we still have no idea about whether the computed value of the *test-statistics* closeness level to the rejection region. We need to deal with this argument by using ***p-value*** for this judgment.

The *p-value* is widely used in practice. Formally, we define the *p-value* as the *smallest level of significance* i.e. $\boldsymbol{\alpha_{limit}} \leq \alpha$, that would lead to rejection of the *null hypothesis* H_o. The *p-value* bounds smallest level α at which the data are significant. Once the *p-value* is known, we can determine (the border) how significant the data are without formally imposing a preselected α: level of significance.

Since *p-value* can be calculated using *Excel, Minitab* etc. programs, we are going to spend no more effort on the calculation of this value. We also need to make sure our normal probability distribution assumptions are valid for this data sets by evaluating probability plots which prove that our initial normal probability assumption is reasonable as shown in Figure 5.4. Also, notice that both probability lines lies almost parallel to each other. This point also supports our preliminary assumption of *equal variances* for both distributions.

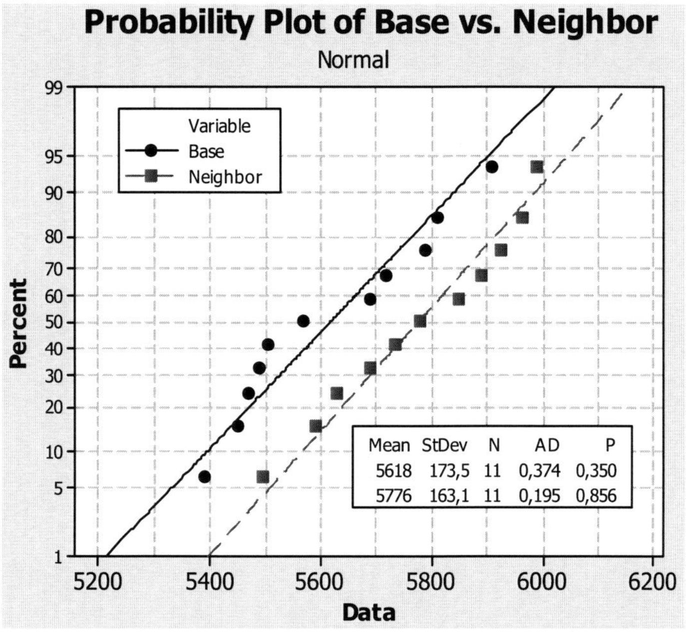

FIGURE 5.4 The probability plots for Example 5.1

Further, based on 100(1-α)% confidence, the interval on the difference in means $\mu_1 - \mu_2$, *can be calculated* from two independent normal populations *for unknown but equal variances case and t-distribution with* $n_1 + n_2 - 2$ *dof in the following interval,*

$$\bar{y}_1 - \bar{y}_2 - t_{\alpha/2,n_1+n_2-2} S_p \sqrt{\frac{1}{n_1} + \frac{1}{n_2}} \leq \mu_1 - \mu_2 \qquad 5.4$$

$$\leq \bar{y}_1 - \bar{y}_2 + t_{\alpha/2,n_1+n_2-2} S_p \sqrt{\frac{1}{n_1} + \frac{1}{n_2}}$$

$$5617.7 - 5776.4 - 2.086 * 168.378 \sqrt{\frac{1}{11} + \frac{1}{11}} \leq \mu_1 - \mu_2$$

$$\leq 5617.7 - 5776.4 + 2.086.168.378 \sqrt{\frac{1}{11} + \frac{1}{11}}$$

$$-308.4 \leq \mu_1 - \mu_2 \leq -8.9$$

Also notice that the interval on the difference in means, $\mu_1 - \mu_2$ *is with α=0.05 level of confidence, never goes above 0, i.e. always between -308.4 and -8.9 which also supports our result.*

In the following, the *Minitab* output for this two-sample *t-test* with 95% CI ($\alpha = 0.05$) difference in means of two mine regions is presented below.

Two-Sample T-Test and CI: Base; Neighbor

```
Two-sample T for Base vs Neighbor

           N   Mean  StDev  SE Mean
Base       11  5618    173     52
Neighbor   11  5776    163     49

Difference = mu (Base) - mu (Neighbor)
Estimate for difference:  -158.6
95% CI for difference:  (-308.4; -8.9)

T-Test of differ. = 0(vs not =):T-Value= -2.21 P-Value= 0.039 DF=20
Both use Pooled StDev = 168.3780
```

Since *p-value* for this problem is $\alpha_{limit} = 0.039 \ll \alpha = 0.05$, we can also conclude that our above judgment is further verified, both population means are *different*!

5.1.2 Variances are Unknown and not-Equal

Consider tests of hypotheses on differences in means i.e. $\mu_1 - \mu_2$ of two normally distributed samples, in some cases, it is not reasonable to assume that unknown variances σ_1^2 and σ_2^2 are equal. In essence $\sigma_1^2 \neq \sigma_2^2$ but still assuming Null hypothesis, $\mathrm{H_0}: \mu_1 = \mu_2$ or $\mu_1 - \mu_2 = \Delta_0 = 0$. As stated, in this case, *t-statistics* can be used for hypothesis testing with approximate *dof* correction/modification as shown in the following where note that t_0^* is also given,

$$t_0^* = \frac{\bar{y}_1 - \bar{y}_2 - (\mu_1 - \mu_2) = 0}{\sqrt{\frac{S_1^2}{n_1} + \frac{S_2^2}{n_2}}} \qquad 5.5$$

with a correction to *t-distribution* of *dof*, given by *v* integer (rounded to the closest integer) as given in Equation 5.6,

$$v = \frac{\left(\frac{S_1^2}{n_1} + \frac{S_2^2}{n_2}\right)^2}{\frac{1}{n_1 - 1}\left(\frac{S_1^2}{n_1}\right)^2 + \frac{1}{n_2 - 1}\left(\frac{S_2^2}{n_2}\right)^2} \quad 5.6$$

where note that $n_1 + n_2 - 2$ *dof is replaced by v, to find out t-distribution as* $t_{\alpha/2,v}$ *using t-table given in Appendix 3 of this book.* In this context Example 5.2 is used for demonstration as presented in the next section.

Example 5.2: Recycled Plastics

In industrial research focusing on the use of parts made with recycled plastics, an engineer is interested in obtaining two pairs of plastic parts each made with different treatments of additives.

Briefly, in this process, recycled plastic parts are collected, processed and crushed into small particles of certain sizes then blended with some additives and a certain amount of original plastics. After some heat treatment and cool down to a certain form, the recycled compound is ready to feed into the injection molding machine. The molten mixture of plastics is molded into tensile strength bars to use in the tensile testing machine.

Finally, the strength of these bars is tested on a tensile testing machine and their strengths are also recorded for each sample. Two different treatments of additives are used in the manufacture of two different tensile strength bars using $n_1=10$ and $n_2=10$ samples. Both treatment's, y_1 and y_2, strength values in *MPa* are given in Table 5.4.

TABLE 5.4 Two sets of recycled plastics data

j	y_1	y_2
1	22.5	24.5
2	28.0	26.0
3	25.0	25.0
4	29.5	27.5
5	25.5	27.3
6	30.5	21.5
7	23.5	20.0
8	20.0	18.7
9	16.0	19.5
10	15.5	21.0

Here, we are interested in if there is any difference in the strength of both parts each made with different treatments of additives. We will also assume that both data samples are normally distributed and find a %95 confidence interval on the differences in means $\mu_1 - \mu_2$, for the two types of treatments of recycled parts.

Examining the normal probability plot given in Figure 5.5, for two samples of recycled plastics treatments reveal that the assumption of normal distribution looks reasonable good and the population variances are unlikely same since the *slopes of these lines* are also quite different. This is good news which supports our initial assumptions i.e. variances are unknown and not-equal!

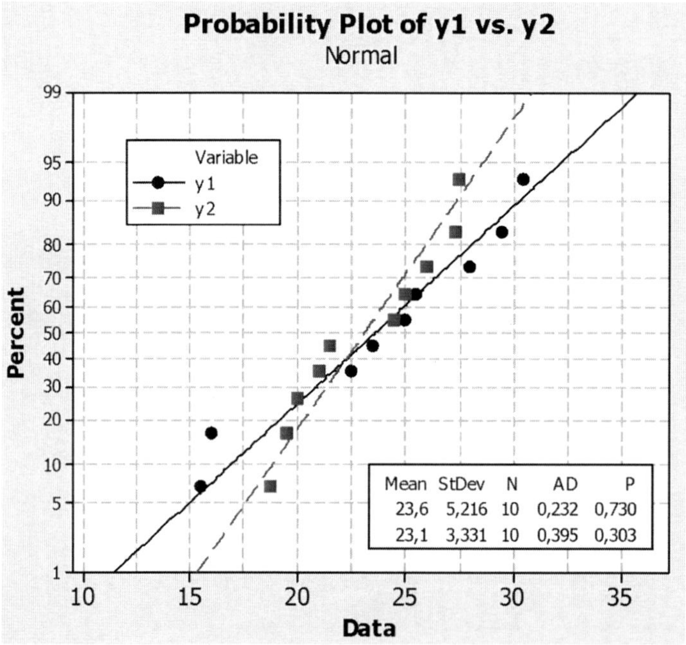

FIGURE 5.5 Normal Probability plot of two sets of recycled plastics

We can go ahead and calculate $\bar{y}_1$=23.6, $\bar{y}_2$=23.1 MPa, S_1^2=27.21, S_2^2=11.10 MPa^2 and using the Table 5.4 y_i values, where $n_1 = n_2 = n = 10$.

Similarly, let's go ahead and calculate t_0^*, test statistics as follows,

$$t_0^* = \frac{23.6 - 23.1 - 0}{\sqrt{\frac{27.21}{10} + \frac{11.1}{10}}} = 0.26$$

Corrected *dof* for *t-test* are calculated using Equation 5.6 as,

$$v = \frac{\left(\frac{27.21}{10} + \frac{11.1}{10}\right)^2}{\frac{1}{10-1}\left(\frac{27.21}{10}\right)^2 + \frac{1}{10-1}\left(\frac{11.1}{10}\right)^2} = 15.29 \approx 15 \text{ dof}$$

dof\α	0.4	0.25	0.1	0.05	0.025	0.01	0.005	0.0025	0.001	0.0005
1	0.3249	1.0000	3.0777	6.3138	12.71	31.82	63.66	127.32	318.31	636.62
14	0.2582	0.6924	1.3450	1.7613	2.1448	2.6245	2.9768	3.3257	3.7874	4.1405
15	0.2579	0.6912	1.3406	1.7531	2.1314	2.6025	2.9467	3.2860	3.7328	4.0728
16	0.2576	0.6901	1.3368	1.7459	2.1199	2.5835	2.9208	3.2520	3.6862	4.0150

Using *t-table*, where $\boldsymbol{\alpha} = \mathbf{0.05}$ and *v=15*, we would reject $H_0 = \mu_1 - \mu_2$ if $t_0^* > t_{0.025,15} = 2.131$ or if $t_0^* < -t_{0.025,15} = -2.131$. Since either of these two conditions are not met, we would not reject H_0 at the confidence level of $\alpha = 0.05$, this means we do not have credible evidence to accept that both *means* are different!

Further, based on 100(1-α)% confidence interval, for the difference in means $\mu_1 - \mu_2$, can be calculated from two independent normal populations for unknown and not equal variances case. Use *t-distribution* with modified v *dof* in the following interval,

$$\bar{y}_1 - \bar{y}_2 - t_{\alpha/2,v}\sqrt{\frac{S_1^2}{n_1} + \frac{S_2^2}{n_2}} \le \mu_1 - \mu_2 \le \bar{y}_1 - \bar{y}_2 + t_{\alpha/2,v}\sqrt{\frac{S_1^2}{n_1} + \frac{S_2^2}{n_2}} \quad 5.7$$

$$23.60 - 23.1 - t_{0.05/2,15}\sqrt{\frac{27.21}{10} + \frac{11.1}{10}} \le \mu_1 - \mu_2$$

$$\le 23.60 - 23.1 + t_{0.05/2,15}\sqrt{\frac{27.21}{10} + \frac{11.1}{10}}$$

$$-3.67 \le \mu_1 - \mu_2 \le 4.67$$

Now, we can calculate the same statistics and find the comparison data using *Excel* as shown in Table 5.5,

TABLE 5.5 T-test data for Example 5.2

Re-cycled Plastics		
j	y_1	y_2
Mean =	23,6	23,1
S^2 =	27,21	11,10
S =	5,22	3,33
n =	10	10
S_p^2 =	19,154	
S_p =	4,377	
T_o =	0,26	
SE Mean =	1,650	1,053
CI Lower	$\leq \mu_1 - \mu_2 \leq$	CI Upper
-3,67	$\leq \mu_1 - \mu_2 \leq$	4,67
v=	15,29	15,0
$t_{0.025,15}$ =	2,131	
p-value =	0,801	

In the following, the *Minitab* output for this two-sample *t-test* with 95% CI ($\alpha = 0.05$) in difference means of two mine regions is also presented.

Two-Sample T-Test and CI: y1; y2

```
Two-sample T for y1 vs y2

     N   Mean  StDev  SE Mean
y1  10  23.60   5.22      1.6
y2  10  23.10   3.33      1.1

Difference = mu (y1) - mu (y2)
Estimate for difference:  0.50
95% CI for difference:  (-3.67; 4.67)

T-Test of difference=0 (vs not =):T-Value =0.26 P-Value=0.802 DF=15
```

Since *p-value* for this problem is $0.8 \gg 0.05$, we can also conclude that our above judgment is further verified, both population means are *not different*!

5.2 Paired T-Test

There is one more special case of the two-sample *t*-test given in Section 5.1, where the observations are collected in pairs i.e. both pairs are taken under some conditions changing from one pair to another. Then, a *paired t-test* is performed to compare the observations on the two population *means* represented by two-pair-samples. These examples might occur in two ways,

- *Before-and-after observations on the same subjects*, for example, the students' final test results before and after a particular training course. Suppose that we are investigating the effect of the quality training course on the class of fifteen engineers, in terms of both understanding and application of this philosophy. We are somehow trying to quantify the effects of this course in numbers and we are observing the changes that this course caused on the engineers. Let the quality index per engineer before the lesson is $\mathbf{Y_{BEFORE}}$ and after the lesson is $\mathbf{Y_{AFTER}}$. Verify if there was an improvement before and after the course.

- *A comparison of two different methods/way of measurement,* where both measurements are applied to the same subjects (e.g. blood sugar level measurements with different brand test equipment).

Now, we can compare these two pairs by defining the *"distance"* measure as $D_j = X_{1j} - X_{2j}$, $j=1,2,3,...,n$ where the D_j's are assumed to be normally distributed with mean μ_D and variance σ_D^2,

$$\mu_D = \mu_1 - \mu_2$$

Now, we go ahead and do hypothesis testing on the difference between $\mu_1 - \mu_2$ by using a one-sample *t-test* on μ_D. And we are testing $H_0: \mu_1 - \mu_2 = \Delta_0 = 0$ *against* $H_1: \mu_1 - \mu_2 \neq \Delta_0 \neq 0$ *as follows,*

$$H_0: \mu_D = \Delta_0 = 0$$
$$H_1: \mu_D \neq \Delta_0 \neq 0$$

The test statistic is given by,

$$T_0 = \frac{\bar{D} - (\boldsymbol{\Delta_0 = 0})}{\frac{S_D}{\sqrt{n}}} \qquad 5.8$$

where $\Delta_0 = \mu_1 - \mu_2$, **mostly checked** for being $\Delta_0 = 0$ and $\overline{D}$ *is the sample average of the differences for data.* S_D *is the sample standard deviation of these differences. We evaluate* T_0 *and* we reject $\mathbf{H_0}$: $\boldsymbol{\mu_D} = \Delta_0 = \mathbf{0}$, when

- H_1: $\mu_D \neq \Delta_0 \neq 0 \rightarrow t_0 > t_{\alpha/2,n-1}$ *or* $t_0 < -t_{\alpha/2,n-1}$
- H_1: $\mu_D > \Delta_0 > 0 \rightarrow t_0 > t_{\alpha,n-1}$
- H_1: $\mu_D < \Delta_0 < 0 \rightarrow t_0 < -t_{\alpha,n-1}$

As far as the confidence interval for $\mu_D = \mu_1 - \mu_2$ is concerned,

$$T_0 = \frac{\overline{D} - (\boldsymbol{\mu_D})}{\frac{S_D}{\sqrt{n}}}$$

where T_0 test statistics follows a *t-distribution* with n-1 dof with CI% or 100(1-α)% confidence interval on $\mu_D = \mu_1 - \mu_2$ is given as,

$$\bar{D} - t_{\alpha/2,n-1} S_D/\sqrt{n} \leq \mu_D \leq \bar{D} + t_{\alpha/2,n-1} S_D/\sqrt{n} \qquad 5.9$$

where again, $t_{\alpha/2,n-1}$ *is the upper* $\alpha/2$ *% value of the t-distribution curve with n-1 dof and can be read from t-table as given in Appendix 3. Again, under both assumptions that the populations of differences in means is normally distributed and the sample differences are randomly selected from the population of differences.*

Example 5.3: Math Scores of Sixth Grade

Math scores for elementary middle school sixth-grade students fall below the nation's average for two consecutive years. School principal asks for professional aid courses in weekends for two consecutive months, ending just before the national exams.

Thus, a sample class of *n=15* students is given a diagnostic test before taking weekend courses and after taking weekend courses. We would like to see if professional-aid courses in the weekends lead to any improvements in students' math test scores.

We are going to use our sample classes' test scores to draw conclusions about the impact of this course, in general. The before and after test scores for n=15 students are presented in Table 5.6,

TABLE 5.6 Test scores example before/after training

j	y_1	y_2	Diff., D_j
1	52	56	-4
2	57	57	0
3	61	65	-4
4	56	57	-1
5	66	67	-1
6	77	74	3
7	72	81	-9
8	41	59	-18
9	69	68	1
10	79	78	1
11	80	85	-5
12	23	38	-15
13	31	45	-14
14	33	44	-11
15	44	50	-6

Let y_1 = test score *before* the test, y_2 = test score *after* the test and assume $\alpha = 0.05$. To test the null hypothesis that the true mean difference is *zero*, the procedure that we are going to follow is as follows:

Note, $\mu_D = \mu_1 - \mu_2 = 0$ and

$$H_0: \mu_D = 0$$

$$H_1: \mu_D \neq 0$$

We reject H_0 if $t_0 > t_{0.05/2,15-1} = 2.145$, *or if* $t_0 < -t_{0.05/2,15-1} = -2.145$. We can calculate the standard deviation, $S_D = 6.51$, and the mean, $\overline{D}$=-5.53 where n=15, using the same equations used before. Eventually, the test statistics can be calculated as,

$$T_0 = \frac{\overline{D} - (\Delta_0 = 0)}{\frac{S_D}{\sqrt{n}}} = \frac{-5.53 - 0}{\frac{6.51}{\sqrt{15}}} = -3.29$$

Note that we make the reading of $t_{0.05/2,15-1} = 2.145$, using *t-table* as shown below.

dof\α	0.4	0.25	0.1	0.05	0.025	0.01	0.005	0.0025	0.001	0.0005
1	0.3249	1.0000	3.0777	6.3138	12.71	31.82	63.66	127.32	318.31	636.62
13	0.2586	0.6938	1.3502	1.7709	2.1604	2.6503	3.0123	3.3725	3.8520	4.2208
14	0.2582	0.6924	1.3450	1.7613	2.1448	2.6245	2.9768	3.3257	3.7874	4.1405
15	0.2579	0.6912	1.3406	1.7531	2.1314	2.6025	2.9467	3.2860	3.7328	4.0728

- H_1: $\mu_D \neq 0 \rightarrow t_0 < -t_{\alpha/2,n-1}$ *or* $-3.29 < -t_{0.05/2,15-1} = -2.145$ *and p-value = 0.005 < 0.05, we conclude that there is a visible* ***success after taking*** *quality training course in the class, so H_0: Null hypothesis is rejected!*

TABLE 5.7 T-test data for Example 5.3

	Test Scores		
j	Before y_1	After y_2	Diff., D_j
Mean =	56,07	61,60	**-5,53**
S^2 =	337,50	198,97	**42,41**
S =	18,37	14,11	**6,51**
n =	15	15	**15**
T_o =			-3,29
SE Mean =	4,743	3,642	**1,681**
	CI Lower	$\leq \mu_D \leq$	CI Upper
	-9,14	$\leq \mu_D \leq$	-1,93
		$t_{0.025,14}$ =	2,145
		p-value =	0,005

In the following the Minitab output for this paired *t-test* with 95% CI ($\alpha = 0.05$) difference in *means* as *score differences* are presented below. If you also examine CI intervals on μ_d with α=0.05, which is between -9.14 and -1.93, never go above 0. This indicates that our argument/judgment about increasing success after taking the training course is *reasonable*!

Paired T-Test and CI: Before Course; After Course

```
Paired T for Before Course - After Course

                 N   Mean   StDev  SE Mean
Before Course   15  56.07   18.37     4.74
After Course    15  61.60   14.11     3.64
Difference      15  -5.53    6.51     1.68

95% CI for mean difference: (-9.14; -1.93)

T-Test of mean difference=0(vs not=0):T-Value= -3.29 P-Value=0.005
```

5.3 Single Mean Comparison to a Specified Value

In the previous sections, we have considered the comparison of two experiments/ measurements in *population* means. What if we have a *single population* (i.e. single data set) mean μ and we would like to compare it to μ_0, note that μ_0 is determined based on past experiences on data, critical values, customer specified limits etc. Then, the hypotheses can be stated as,

$$H_0: \mu = \mu_0 \tag{5.10}$$

$$H_1: \mu \neq \mu_0$$

The hypothesis may be tested using a direct application of the normal distribution, using the test statistic Z_0 with following assumptions as given,

$$Z_0 = \frac{\bar{y} - \mu_0}{\frac{\sigma}{\sqrt{n}}} \tag{5.11}$$

when,

- if the data population is *normal* and variance is *known!*
- or, the data population is not normal but the data sample size is large!

✓ Note that if H_o is true! then Z_o follows $N(0,1)$, i.e. normal distribution so that we make our rejection for H_o when $|Z_0| > Z_{\alpha/2}$

✓ CI: Confidence interval of 100(1-α)% on the population mean can be written as in Equation 5.12 (two-tailed)

$$\bar{y} - Z_{\alpha/2}\frac{\sigma}{\sqrt{n}} \leq \mu \leq \bar{y} + Z_{\alpha/2}\frac{\sigma}{\sqrt{n}} \tag{5.12}$$

What if the variance of the population is *unknown*! which is true for most of the cases where n is large! (say, n≥ 30), then we must make a "*brave*" assumption that the population is normally distributed. Sample variance, S would replace population variance, μ for H_0: $\mu = \mu_0$ testing, then our test statistics is

$$t_0 = \frac{\bar{y} - \mu_0}{\frac{S}{\sqrt{n}}} \tag{5.13}$$

Then H_0: $\mu = \mu_0$ would be rejected when $|t_0| > t_{\alpha/2,n-1}$ where $t_{\alpha/2,n-1}$ denotes the upper $\alpha/2$ percentage value of the *t-distribution* with n-1 *dof.* The 100(1- α) percent confidence interval is given in Equation 5.14, next!

$$\bar{y} - t_{\alpha/2,n-1}\frac{S}{\sqrt{n}} \leq \mu \leq \bar{y} + t_{\alpha/2,n-1}\frac{S}{\sqrt{n}} \qquad 5.14$$

TABLE 5.8 Summary of statistics and criterions

Hypothesis	Test Statistics	Rejection Criterion
$H_0: \mu = \mu_0$ $H_1: \mu \neq \mu_0$	$Z_0 = \frac{\bar{y} - \mu_0}{\frac{\sigma}{\sqrt{n}}}$	$\lvert Z_0 \rvert > Z_{\alpha/2,n-1}$
$H_0: \mu = \mu_0$ $H_1: \mu < \mu_0$		$Z_0 < -Z_\alpha$
$H_0: \mu = \mu_0$ $H_1: \mu > \mu_0$		$Z_0 > Z_\alpha$
$H_0: \mu = \mu_0$ $H_1: \mu \neq \mu_0$	$t_0 = \frac{\bar{y} - \mu_0}{\frac{S}{\sqrt{n}}}$	$\lvert t_0 \rvert > t_{\alpha/2,n-1}$
$H_0: \mu = \mu_0$ $H_1: \mu < \mu_0$		$t_0 < -t_{\alpha,n-1}$
$H_0: \mu = \mu_0$ $H_1: \mu > \mu_0$		$t_0 > t_{\alpha,n-1}$

Example 5.4: High-Pressure Valve

A supplier manufacturers the high-pressure valves to the industry. These valves are designed to operate at no lower than (greater than) 150 psi, therefore each valve is tested for minimum operating pressure and we only accept the valves which can hold test pressure over $\mu_0 = 150$ psi. Note: As far as the accepted industry standards, the *variance* to be imposed on valve pressure is σ^2=81 (psi)^2 and n=10 specimens are selected randomly, the average valve pressure is found to be 158 psi.

In this example, we test the following hypotheses,

$$H_0: \mu = 150\ psi$$

$$H_1: \mu > 150\ psi$$

where we would accept the valves *only if,* the *null* hypothesis $H_0: \mu = 150\ psi$ is to be rejected! in favor of H_1.

$$Z_0 = \frac{\bar{y} - \mu_0}{\frac{\sigma}{\sqrt{n}}} = \frac{158 - 150}{\frac{9}{\sqrt{10}}} = 2.81$$

Using α=0.05, we read/find Z_α=$Z_{0.05}$ = 1.645 as the average of two scores from the *z-table* as shown below.

z	-0.09	-0.08	-0.07	-0.06	-0.05	-0.04	-0.03	-0.02	-0.01	0
-1.7	0.036727	0.037538	0.038364	0.039204	0.040059	0.04093	0.041815	0.042716	0.043633	0.044565
-1.6	0.045514	0.046479	0.04746	0.048457	0.049471	0.050503	0.051551	0.052616	0.053699	0.054799
-1.5	0.055917	0.057053	0.058208	0.05938	0.060571	0.06178	0.063008	0.064255	0.065522	0.066807

- Here, our objective is to test that if the mean coefficient of valve pressure (assumed to be a normally distributed) exceeds 150 psi,
- The *one-sided* alternative hypothesis is appropriate!
- Looking at test statistics, since $Z_0 = 2.81 > 1.645$, we reject $H_0: \mu = 150$ psi at the α=0.05 level of significance. The mean valve pressure exceeds 150 psi, using calculations, based on a sample of 10 measurements.

Also note that 95% lower bound (153.32 psi) on μ, exceeds 150 psi, so we would reject the null hypothesis, H_0 : μ=150 psi and conclude that the $H_1: \mu > 150$ psi is ***true*** for given confidence interval!

CI: Confidence interval of 100(1-α)% on the population mean, i.e. rewriting Equation 5.12 for *upper-tailed* hypothesis can be written as, (use Z_α instead of $Z_{\alpha/2}$).

$$\bar{y} - Z_\alpha \frac{\sigma}{\sqrt{n}} \leq \mu \leq \bar{y} + Z_\alpha \frac{\sigma}{\sqrt{n}}$$

$$158 - 1.645 \frac{9}{\sqrt{10}} \leq \mu \leq 158 + 1.645 \frac{9}{\sqrt{10}}$$

$$153.32 \leq \mu \leq 162.68$$

*where note that **95% Lower Bound** $\mu \geq \mathbf{153.32}$ is critical and evaluated.*

In the following the *Minitab* output for this paired t-test with 95% CI ($\alpha = 0.05$) in mean valve pressure μ, is presented below.

One-Sample Z

```
Test of mu = 150 vs > 150
The assumed standard deviation = 9


                          95% Lower
N     Mean   SE Mean        Bound       Z       P
10  158,00      2,85       153,32    2,81   0,002
```

Also notice that *p-value=0.002* is much less than $\alpha = 0.05$, which further supports our hypothesis judgment!

5.4 Supplementary Material for *Excel* Commands

Example 5.1: Neighboring Coal Mines:

Note, in the following *Excel* sheet, the first three commands are ready to use *Excel* commands and the remaining ones are coded in *Excel,* using formulations those are given in Column D. Note *all* formulations that are shown below are given for **Column-B only**!

	A	B	C	D
1	j	y_1	y_2	
2	1	5505	5630	
3	2	5490	5735	
4	3	5720	5495	
5	4	5810	5890	
6	5	5570	5850	
7	6	5910	5990	
8	7	5690	5690	
9	8	5790	5925	
10	9	5470	5590	
11	10	5390	5780	
12	11	5450	5965	**EXCEL FORMULA**
13	**Mean =**	**5617.7**	**5776.4**	= AVERAGE(B2:B12)
14	**S^2 =**	**30086.8**	**26615.5**	= VAR.S(B2:B12)
15	**S =**	**173.5**	**163.1**	= STDEV.S(B2:B12)
16	**n =**	**11**	**11**	= COUNT(B2:B12)
17	**S_p^2 =**	**28351.1**		= ((B16-1)*B14+(C16-1)*C14)/(B16+C16-2)
18	**S_p =**	**168.378**		= B17^0.5
19	T_o =	-2.2		= (B13-C13)/(B18*((1/B16+1/C16)^0.5))
20	**SE Mean =**	**52.3**	**49.2**	= B15/B16^0.5
21	**CI Lower**	$\leq \mu_1 - \mu_2 \leq$	**CI Upper**	
22	-308.4	$\leq \mu_1 - \mu_2 \leq$	-8.9	= B13-C13±TINV(0.05,2*A12-2)*B18*SQRT(1/A12+1/A12)
23	**$t_{0.025,20}$ =**	2.0860		= TINV(0.05,B16+C16-2)
24	**p-value =**	0.039		= TDIST(ABS(B19),B16+C16,2)

Example 5.2: Re-cycled Plastics

Note that all formulations are given *only* for **Column B**, and corresponding values are used in formulations.

	A	B	C	D
1	j	y_1	y_2	
2	*1*	22.5	24.5	
3	*2*	28.0	26.0	
4	*3*	25.0	25.0	
5	*4*	29.5	27.5	
6	*5*	25.5	27.3	
7	*6*	30.5	21.5	
8	*7*	23.5	20.0	
9	*8*	20.0	18.7	
10	*9*	16.0	19.5	
11	*10*	15.5	21.0	**EXCEL FORMULA**
12	**Mean =**	**23.6**	**23.1**	= AVERAGE(B2:B11)
13	**S^2 =**	**27.21**	**11.10**	= VAR.S(B2:B11)
14	**S =**	**5.22**	**3.33**	= STDEV.S(B2:B11)
15	**n =**	**10**	**10**	= COUNT(B2:B11)
16	**S_p^2 =**	**19.154**		= ((B15-1)*B13+(C15-1)*C13)/(B15+C15-2)
17	**S_p =**	**4.377**		= B16^0.5
18	T_o =	0.26		= (B12-C12)/(((B13/B15+C13/C15)^0.5))
19	**SE Mean =**	**1.650**	**1.053**	= B14/B15^0.5
20	**CI Lower**	$\leq \mu_1 - \mu_2 \leq$	**CI Upper**	
21	-3.67	$\leq \mu_1 - \mu_2 \leq$	4.67	= B12-C12±TINV(0.05,C22)*SQRT(B13/A11+C13/A11)
22	v=	15.29	15.0	= (B13/A11+C13/A11)^2/((B13/A11)^2/(A11-1)+(C13/A11)^2/(A11-1))
23	**$t_{0.025,15}$ =**	2.131		=TINV(0.05,C22)
24	**p-value =**	0.801		=TDIST(ABS(B18),B15+C15-2,2)

Example 5.3: Math Scores of Sixth grade

Note that all formulations are given *only* for **Column D**, and corresponding values are used in formulations.

	A	B	C	D	E
1	j	Before y_1	After y_2	Diff., D_j	
2	1	52	56	-4	
3	2	57	57	0	
4	3	61	65	-4	
5	4	56	57	-1	
6	5	66	67	-1	
7	6	77	74	3	
8	7	72	81	-9	
9	8	41	59	-18	
10	9	69	68	1	
11	10	79	78	1	
12	11	80	85	-5	
13	12	23	38	-15	
14	13	31	45	-14	
15	14	33	44	-11	
16	15	44	50	-6	**EXCEL FORMULA**
17	**Mean =**	56.07	61.60	**-5.53**	= AVERAGE(D2:D16)
18	**S^2 =**	337.50	198.97	**42.41**	= VAR.S(D2:D16)
19	**S =**	18.37	14.11	**6.51**	= STDEV.S(D2:D16)
20	n =	15	15	**15**	= COUNT(D2:D16)
23	T_0 =			-3.29	= D17/(D19/SQRT(A16))
24	**SE Mean =**	4.743	3.642	**1.681**	= D19/D20^0.5
25		**CI Lower**	$\leq \mu_D \leq$	**CI Upper**	
26		-9.14	$\leq \mu_D \leq$	-1.93	= D17±D27*D19/SQRT(A16)
27			$t_{0.025,14}$ =	2.145	= TINV(0.05,D20-1)
28			p-value =	0.005	= TDIST(ABS(D23),D20-1,2)

Example 5.4: High Pressure Valve

Note that all formulations are given *only* for **Column B**, and corresponding values are used in formulations.

	A	B	C	D
1	**High Pressure Valve**			
2	**Mean =**	**158**		
3	σ =	**9**		
4	$\mu_0 =$	**150**		
5	**n =**	**10**		
6	**SE Mean =**	**2.846**		= B3/SQRT(B5)
7	z_o=	**2.81**		= (B2-B4)/B3*SQRT(B5)
8	**CI Lower**	$\leq \mu_D \leq$	CI Upper	
9	153.32	$\leq \mu_D \leq$	162.68	= B2-B11*B3/SQRT(B5)
10	$\alpha =$	0.05		
11	z_α=	1.645		= NORM.INV(1-B10,0,1)
12	**p-value=**	0.002		= 1-NORMDIST(B7,0,1,TRUE)
13				
14	p-value(Low.Tail)=	0.998		= NORMDIST(B7,0,1,TRUE)
15	p-value(Upp.Tail)=	0.002		=1-NORMDIST(B7,0,1,TRUE)
16	p-value(Two Tail)=	0.005		= 2*MIN(B14:B15)

5.5 Supplementary Material for Minitab Results

Example 5.1: Neighboring Coal Mines:

As for to demonstrate how to do **2-Sample t-test** in *Minitab*, we are going to show all the steps below, the other examples can be solved similarly.

- ✓ Start-up *Minitab* and enter both data in C1 as *Base* and C2 as *Neighbor* columns. Then using **Stat > Basic Statistics > 2t 2-Sample t...** option, click here as shown in the following,

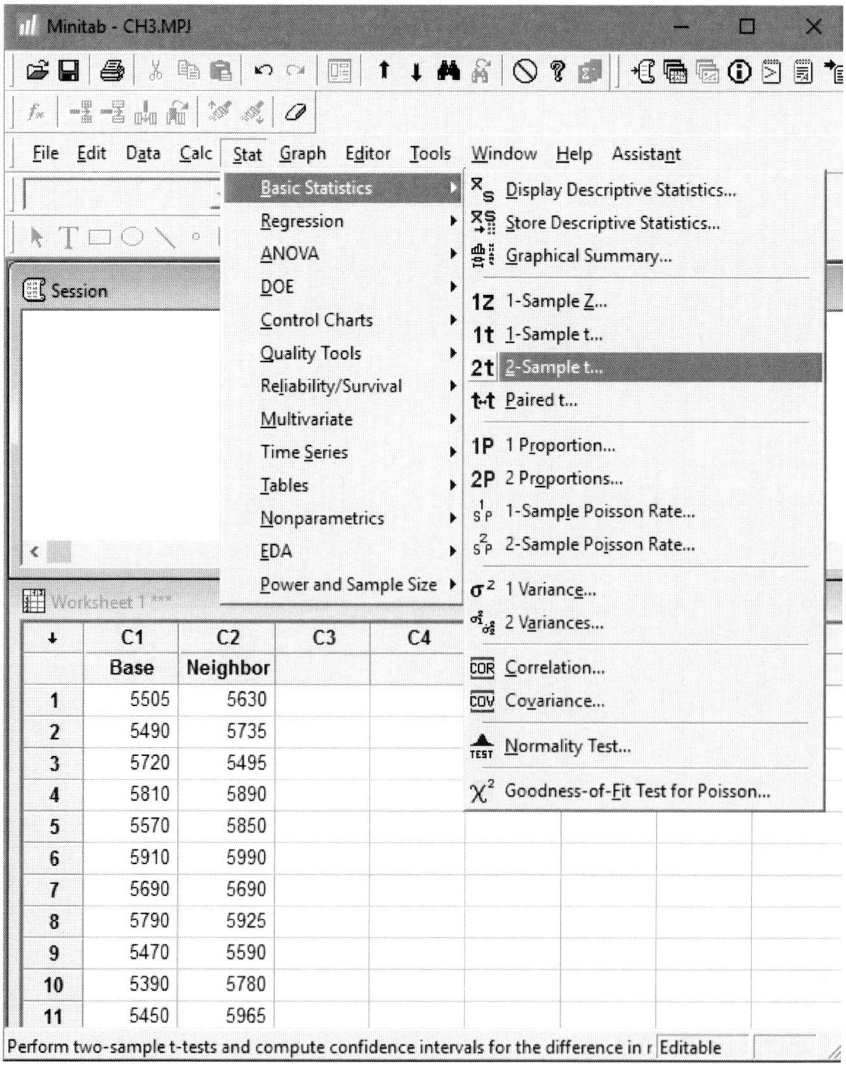

- ✓ You will get, **2-sample (Test and Confidence Interval)** window, where click on *Samples in different column* option. *(Note:* **Hidden C1 Base** *and* **C2 Neighbor** *data will* **show up,** *once you click in the small box in front of* **First:**, *now you can go ahead select* **C1 Base** *and* **C2 Neighbor** *data to* **First:** *and* **Second:** *boxes, respectively).* Also click on *Assume equal variances* option, since we assume equal variances in this problem. Click on OK to close the window!

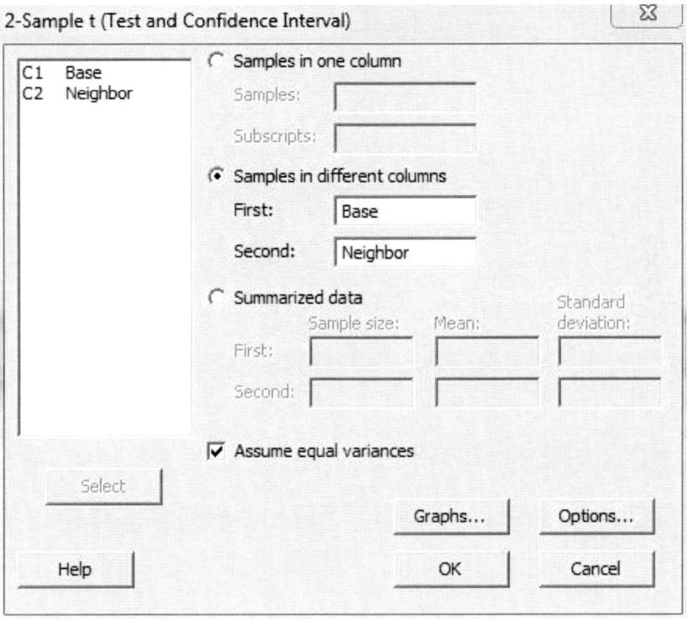

- ✓ Finally, you will get the following output results.

Two-Sample T-Test and CI: Base; Neighbor

```
Two-sample T for Base vs Neighbor
                              SE
           N  Mean  StDev  Mean
Base      11  5618    173    52
Neighbor  11  5776    163    49

Difference = mu (Base) - mu (Neighbor)
Estimate for difference:  -158.6
95% CI for difference:  (-308.4; -8.9)

T-Test of diff. =0 vs not =): T-Value= -2.21 P-Value=0.039 DF = 20
Both use Pooled StDev = 168.3780
```

- ✓ Now, let's go ahead and do Probability Plot, using **Stat > Graph >Probability Plot…** option and click here,

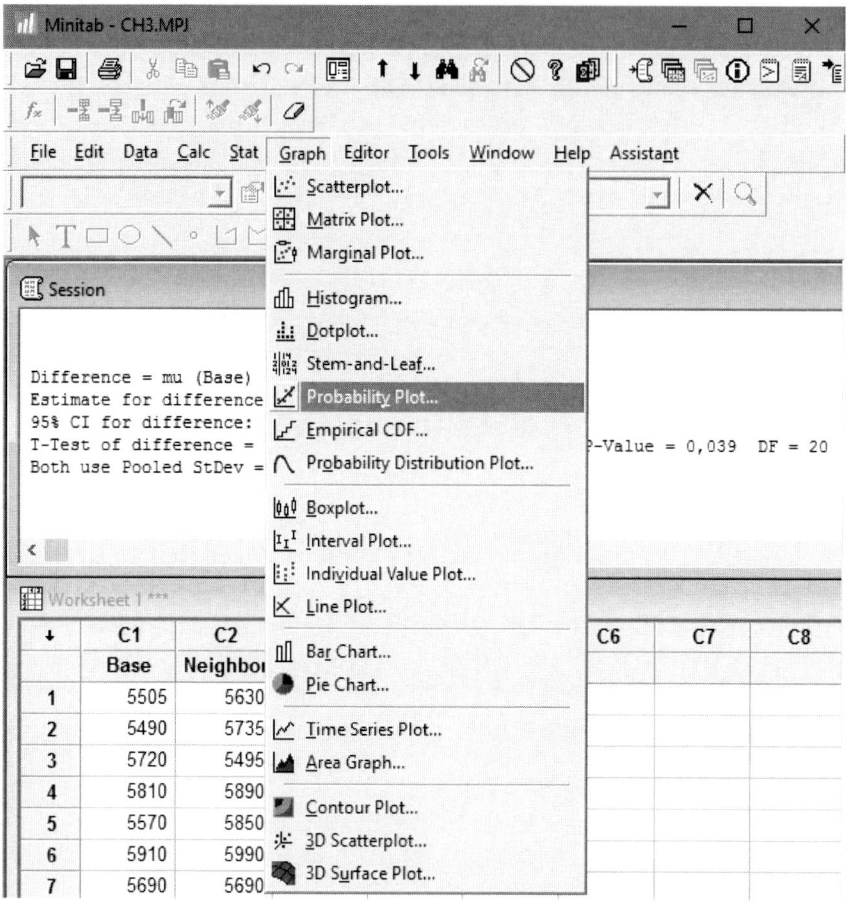

✓ You will get **Probability Plots** window and choose *Multiple* option, click on OK!

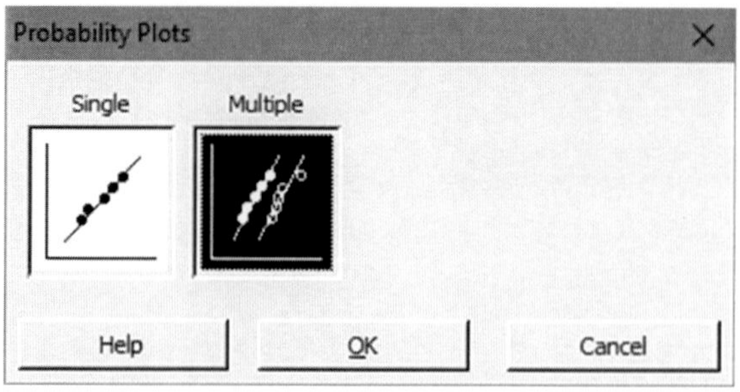

✓ You will get **Probability Plot – Distribution** window, where click on *Data Display* tab as shown,

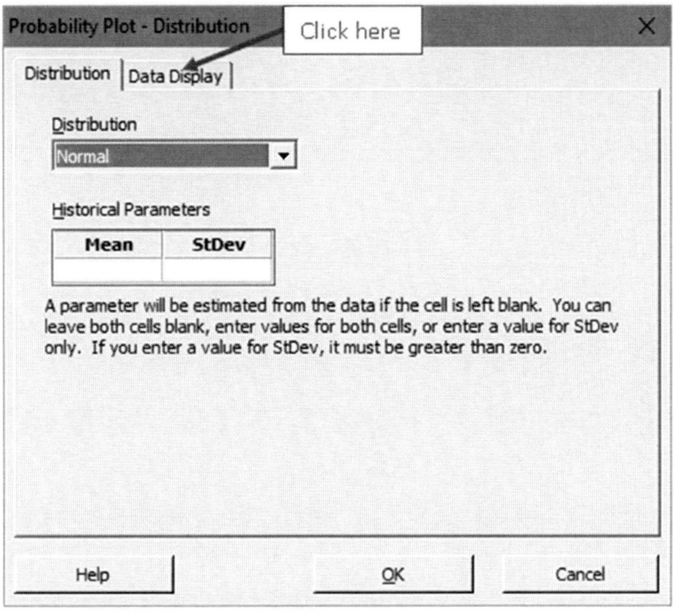

✓ You will get to Data Display window options where un-check *Show confidence interval* option then click on OK! As shown.

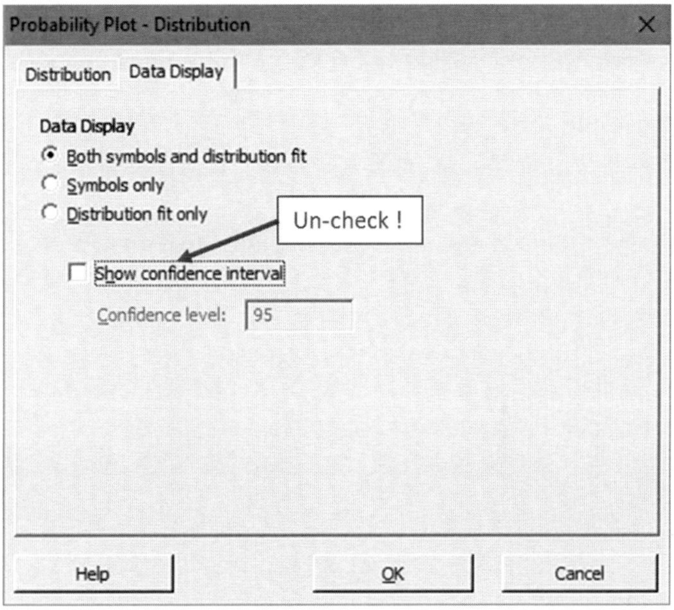

✓ Now, You will get probability plot as follows, some minor editing is performed in the following plot.

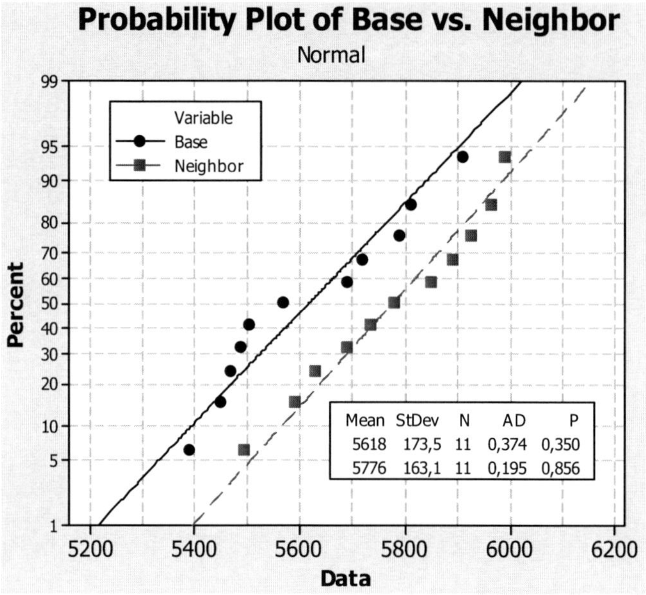

Example 5.2: Recycled Plastics:

In this example the same steps shown above are followed with only,

✓ In **2-sample (Test and Confidence Interval)** window, do not click on *Assume equal variances* option, since we do not assume equal variances in this problem. Click on OK to close the window!

✓ Remaining options are the same including the normal probability graph.

Example 5.3: Math Scores of Sixth grade

In this example the same steps shown above are followed with only,

✓ Using **Stat > Basic Statistics > t-t Paired t...** option, click tis option and follow the same steps,

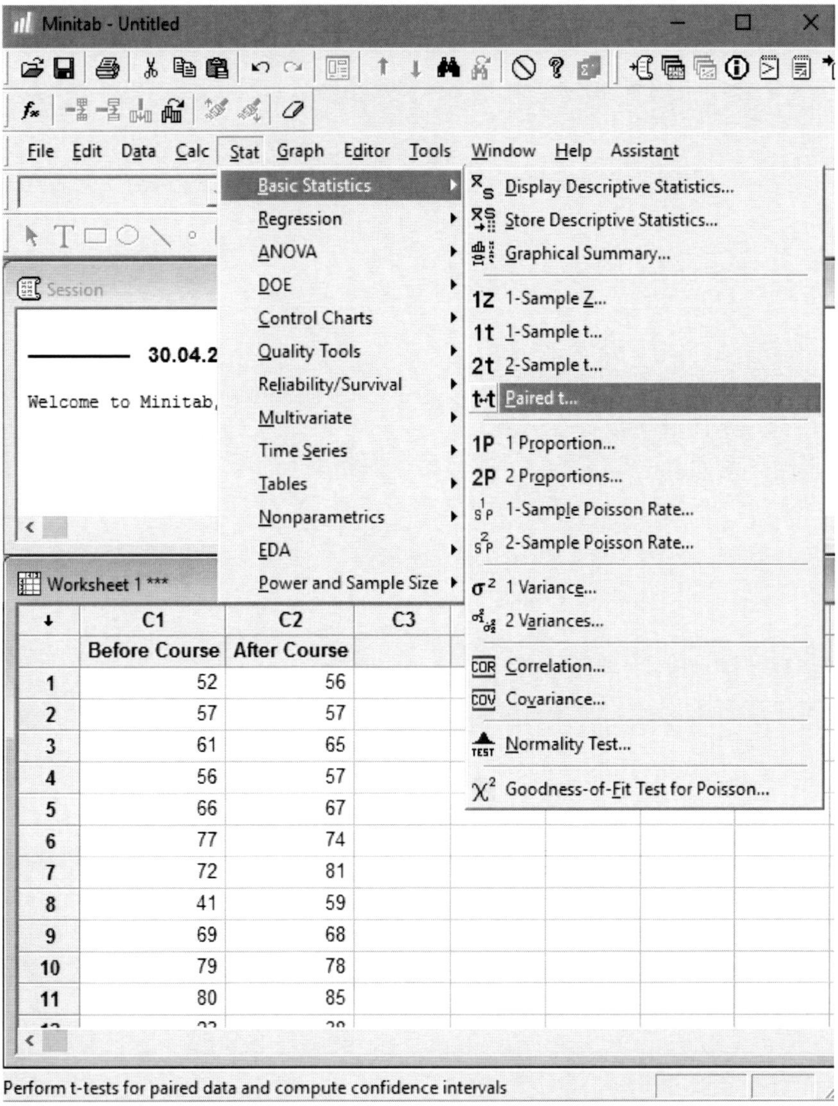

✓ You will get, **Paired t (Test and Confidence Interval)** window, where click on *Samples in different column*, and choose the appropriate columns. Click OK to close the window!

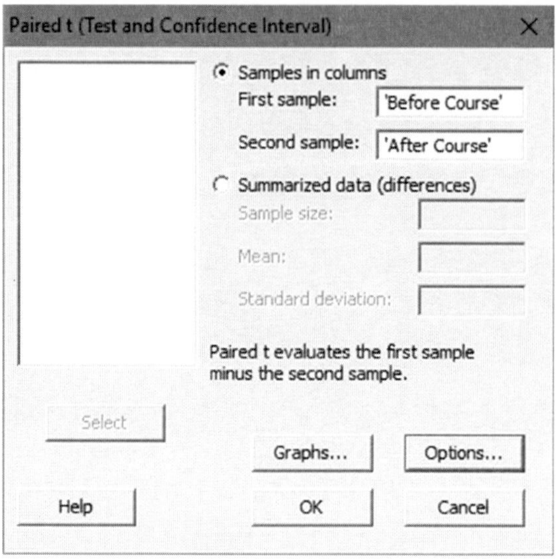

- ✓ You will get the following output results.

Paired T-Test and CI: Before Course; After Course

```
Paired T for Before Course - After Course

                  N    Mean   StDev  SE Mean
Before Course    15   56.07   18.37     4.74
After Course     15   61.60   14.11     3.64
Difference       15   -5.53    6.51     1.68

95% CI for mean difference: (-9.14; -1.93)
T-Test of mean diff. =0 (vs not = 0):T-Value= -3.29 P-Value= 0.005
```

- ✓ The same steps are followed to do normal probability plot as explained before.

Example 5.4: High Pressure Valve

As for to demonstrate how to do **1-Sample Z-test** in *Minitab*, we are going to show all the steps below, any other examples can be solved similarly.

- ✓ Start-up *Minitab* and then using **Stat > Basic Statistics > 1Z 1-Sample ...** option, click here as shown in the following,

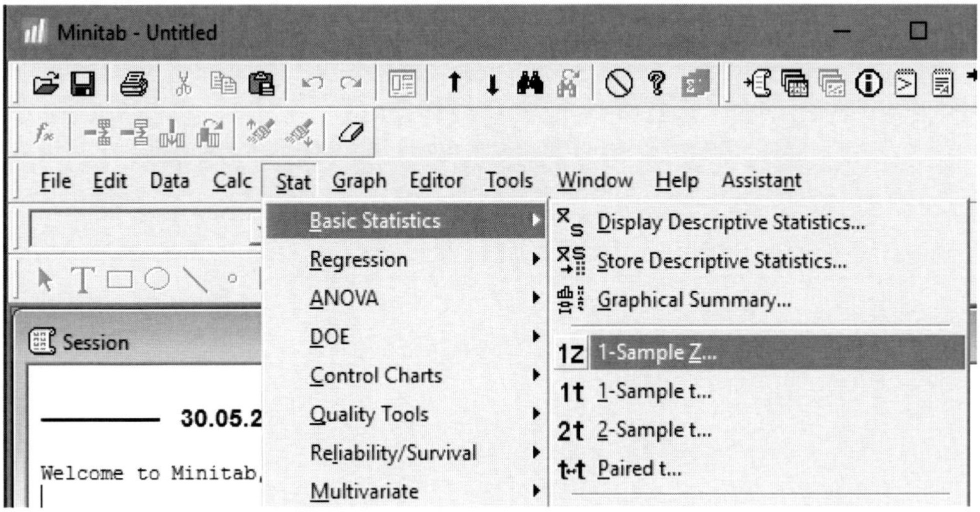

- ✓ You will get, **1-Sample Z (Test and Confidence Interval)** window, where click on *Summarized data* option and enter all the data as shown,

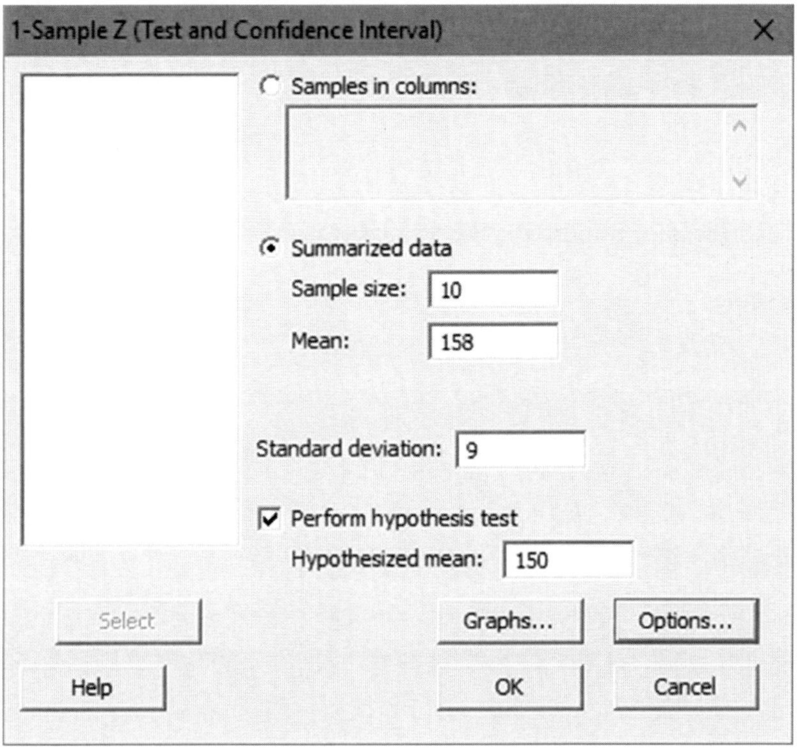

✓ Next click on **Options...** tab, **1-Sample Z – Options** window will open where choose the "greater than" option with Confidence level of 95%. Click on OK to close the window!

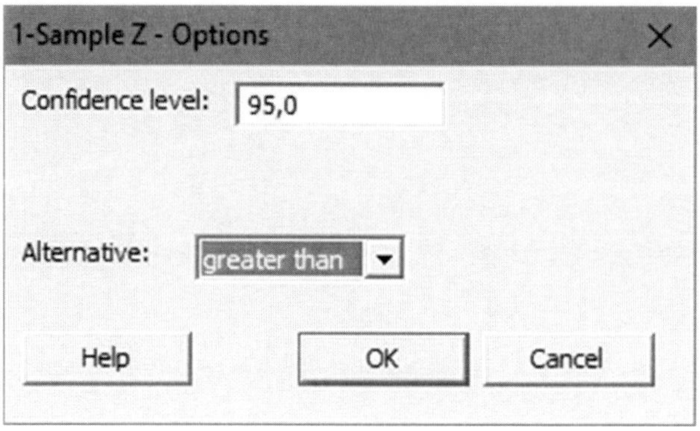

✓ Finally, you will get the following output results.

One-Sample Z

```
Test of mu = 150 vs > 150
The assumed standard deviation = 9


                        95% Lower
 N     Mean  SE Mean       Bound     Z       P
10   158.00     2.85      153.32  2.81   0.002
```

✓ The same steps are followed to do normal probability plot as explained before.

5.6 Exercises

1) Assume we developed a new method for foreign language teaching to international students. Using a "new method" we trained 11 foreign students and using "traditional method" we also trained 11 different students by the same teachers for 4 months in a fall semester. In the end, all foreign language level scores for 22 students are tested/recorded and the results are presented in Table 5.9, where out coming scores vary between 0-100.

✓ Do comparison plots for both methods using their specific data and tell your preliminary thoughts about these methods using descriptive statistics.

✓ Use the data and estimate the mean difference between both methods. Take a confidence interval of 95%.

- ✓ Interpret your results and prove that you used reasonable assumptions in the calculations.

TABLE 5.9 Test data for Exercise 4.6.1

j	New Method	Traditional M.
1	86	84
2	85	71
3	83	82
4	77	74
5	69	70
6	88	84
7	57	58
8	69	69
9	65	69
10	75	57
11	71	56

2) Let's consider the following hypothesis test:

$$H_0: \mu = 16$$

$$H_1: \mu \neq 16$$

n=40 samples collected with a sample mean of 15 and a sample standard deviation of 3.5 from the production line.

- ✓ Find the test statistics,
- ✓ What is p-value?
- ✓ Given $\alpha = 0.05$, evaluate and comment on the hypothesis.

3) Starting salaries of two competitive university graduate are compared for this purpose pairs are formed for University-A and University-B in computer engineering major and n=12 pairs are formed (with similar GPA) and starting salaries are noted as shown in Table 5.10,

- ✓ You are asked to compare the mean starting salary for University-A to the mean starting salary for University-B graduates, assume $\alpha = 0.05$.
- ✓ Comment on the results.

TABLE 5.10 Starting salaries for both universities

j	University-A	University-B	Diff. D_j
1	65500	66000	-500
2	71000	69800	1200
3	69000	70000	-1000
4	66000	65000	1000
5	64000	64500	-500
6	57000	57600	-600
7	61800	61000	800
8	79000	77500	1500
9	57000	57500	-500
10	62000	61500	500
11	63000	62500	500
12	70000	70400	-400

Chapter 6

Simple Linear Regression

Exploration of dependency/relationship between inputs(factors) and output(response), interaction on inputs and the level of importance of each input on Output(Response), are main challenges for scientists and engineers in any experimental or simulation evaluations using statistics. Consider a cook who is looking for the relationship between the baking temperature and the taste of bread; or the teacher who wants to study the key relationship between the homework and students overall success; or the doctor who wants to relate the effect of a certain dosage of medicine on the faster and better treatment of a disease.

For some instances, the relationships between system inputs/output are well known since inputs and output are functionally related. This kind of relationships rather *deterministic* in nature, since repeated experiments for certain level of inputs will follow the same output/response with very well-known formula i.e. traditional laws of physics, i.e. $F=ma$. On the contrary, in real-life, i.e. in most of the cases, these relationships are not known and might be complicated with many aspects, i.e. experimental errors, outer conditions etc. In such situations, we tend to use *approximation*s to represent the relationship between inputs and output, and we try to develop the best *statistical/empirical* models. If the output/response is called Y (uppercase means random variable) which varies around a certain *mean* value and depends on the input variables. Obviously, the same experiment inputs would not yield the same output/response value every time, even if the same experiment is repeated by the same person and under the same conditions. In such situations, Regression analysis is performed to obtain above-mentioned *approximations* by utilizing data analysis, i.e. available tools of statistics are used to derive causal, logical and mathematical relationship by the formation of mapping (or forming transfer function) between input variables and output(s). Once this mapping is successfully established, now we are dealing with an explicit and more controllable/tractable system of equation(s) from which, we can extract some meaningful information that we can use in solving and understanding the behavior of the problem of our interest. Also, note that we need to further verify the validity of the regression model by using some marker tools of statistics which is entirely another important subject of the statistical data analysis.

In short, regression analysis is a very common and versatile statistical technique for this kind of problem relationship formulation/identification between inputs and output, where one can define *dependent* output/response variable, as a function of *independent* input variables of the problem. In other words, regression can be viewed as fitting a line/surface or user-defined order of curvature/hyperplane to data. As it would be very clear, this operation can be viewed as the *mapping* of input variables on the output/response by using some form of mathematical formulations, using statistical data analysis as depicted in Figure 6.1.

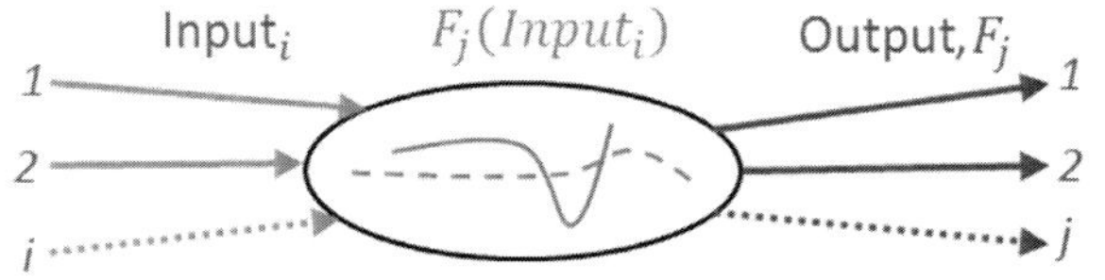

FIGURE 6.1 Representative model for input/output mapping

Here, we are trying to come up with the most useful/representative form of mathematical equations, say, polynomial, exponential, logarithmic etc. formulations for transfer functions F_j, i.e. which generates statically valid output(s). This can be done in many different ways. In this chapter, we will focus on the application of widely used *Simple Linear Regression Analysis* (*SLR*). Clearly, Regression models are good for constructing *empirical models* to explore the aforementioned characteristics of the problem.

As far as the final regression model is involved, we do our best to successfully represent the set of input(s)/output pairs obtained by scientist or engineers during out-coming experiments or simulations (i.e. of course, this representation is established to the certain degree of approximation accuracy with a certain level of confidence). Regression analysis is widely used in many fields, because,

- Regression models are easy to build up since they are some generic form of widely used equations i.e. polynomial, power functions etc.
- They are simple to use, one can obtain any function evaluation, derivative information in a closed form of equations.

- We obtain explicit formulations, they are easy to use for inputs/output evaluations or to identify interactions.
- They are statistically well-established and their use is very common.

Example 6.1

Let's focus on a problem, in which according to one of the university's historical data, it is reported that a point success measure of GPA (Grade Point Average) of the graduating class of students is studied by relating students' attendance (i.e. taking notes in class and listening to the lecturer) to their GPA's *for most of the classes*. Here, university management's objective is to see if attendance to the classes vs. GPA is directly linked? If so how? Obviously, regression analysis can be used for exploring this fact/relation for a given historical data. As an illustration, let's consider such data taken in the year of 2017 as given in Table 6.1. In this table, GPA is the *Grade Point Average* for the graduating class of the year 2017 and y are the *percentage of attendance* of the graduating class to the courses.

The first thing to do in any statistical analysis is to draw/plot associated diagram(s) of the data in as shown in Figure 6.1, which presents a *scatter diagram* of the data presented in Table 6.1.

In Figure 6.2, each n number of (x_i,y_i) pairs are plotted as a point in an x-y coordinate system. Also, note that this scatter diagram is plotted in *Excel.* After inspection of the scatter diagram, it is could be easily seen that no simple curve will pass through all the data points (in essence, this is really not we want! we are more interested in trend rather than having a curve passing through all the (x_i,y_i) pairs) and data points lay scattered along a virtual straight line like poison ivy.

TABLE 6.1 Example 6.1 - %Attendance vs. GPA data

n	% Attendance	GPA (y)	11	79	86
1	91	99	12	52	52
2	89	93	13	25	13
3	90	95	14	55	46
4	85	76	15	44	40
5	88	78	16	80	81
6	81	75	17	29	12
7	59	50	18	68	72
8	60	70	19	75	80
9	48	37	20	49	44
10	65	73	21	33	20

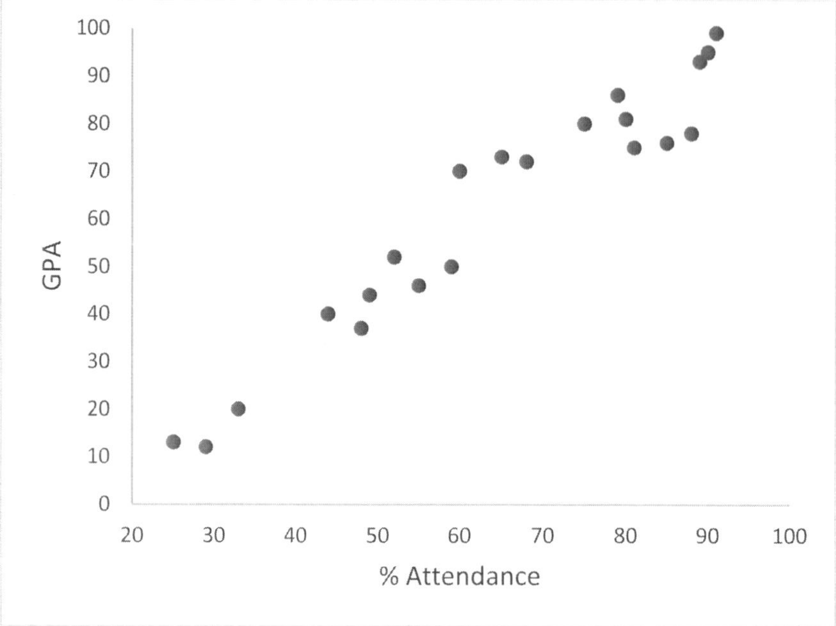

FIGURE 6.2 Scatter diagram of Example 6.1 - %Attendance vs. GPA data

Since the overall behavior of the scatter plot looks like a simple line, we can go ahead and fit a line to this *n* pairs of *(x,y)* input and output data as given in the following (mapping) empirical model of Equation 6.1,

$$Y(x) = \beta_0 + \beta_1 x + \epsilon \qquad 6.1$$

where $Y_i(x)$ is the output/response value, and β_0, β_1 units are *regression* coefficients, i.e. β_0 is the intercept and, β_1 the slope of the line, i.e. a change of one unit in the x (input variable) causes a change of β_1 units in the output/response. Also note that ϵ_i is a random error scattered around the regression line and it is assumed that these errors are mutually independent (i.e. $Y_1, Y_2, \ldots, Y_n$ values are also mutually independent) and normally distributed with *mean* zero and *variance* of σ^2.

6.1 Simple Linear Regression for Empirical Models

In the linear regression, a single *independent* variable x and *dependent* output/response variable y is used to form a simple linear regression equation as given in Equation 6.2.

$$y_i(x) = \beta_0 + \beta_1 x_i + \epsilon_i \rightarrow i = 1,2, \dots, n \qquad 6.2$$

Since the relationship is linear, our objective is to find estimates $\hat{\beta}_0$ and $\hat{\beta}_1$ of coefficients corresponding to β_0 and β_1 such that the *best-fitted* line or *"closest to" n* data pairs of (x_i, y_i). Please note that β_0 and β_1 regression coefficients of the Equation 6.2 are unknown and yet to be calculated.

A widely known *method of least squares* can be used to calculate estimate these coefficients of $\hat{\beta}_0$ and $\hat{\beta}_1$. The sum of squares of the deviations of each data points (i.e. outputs/observations) from the fitted regression line can be formulated where we minimize the sum of the squared difference between the experimental (actual) output and estimated/fitted output for *all n* data pairs as stated in Equation 6.3. This can also be shown in Figure 6.3 graphically (i.e. see vertical deviations, in other words, we want to minimize the summed square of *all errors*).

$$L(\beta_0, \beta_1) = \sum_{i=1}^{n} \epsilon_i^2 = \sum_{i=1}^{n} (y_i - \beta_0 - \beta_1 x_i)^2 \qquad 6.3$$

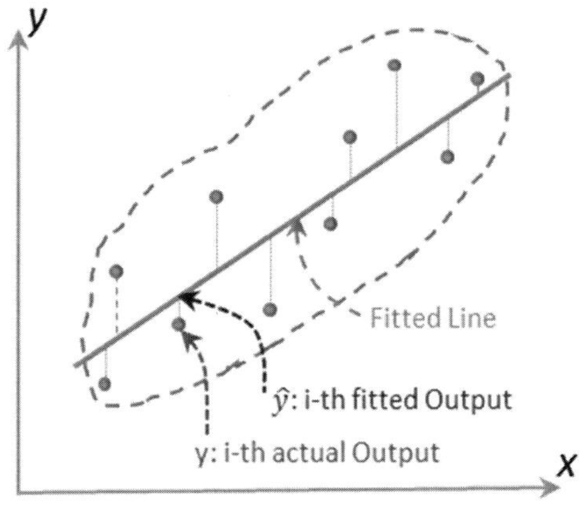

FIGURE 6.3 Deviations of regression - fitted vs. actual output sample data

Note that the sum of squares can be viewed as a measure of the closeness between the data points and the fitted regression line. If the line was going through all the points, it is clear that L would be zero. Find coefficient estimates (or *least-square estimates*) of $\hat{\beta}_0$ and $\hat{\beta}_1$ to satisfy corresponding coefficient derivatives which are set to zero and given in Equation 6.4,

$$\frac{\partial L(\hat{\beta}_0, \hat{\beta}_1)}{\partial \beta_0} = -2\sum_{i=1}^{n}(y_i - \hat{\beta}_0 - \hat{\beta}_1 x_i) = 0 \tag{6.4}$$

$$\frac{\partial L(\hat{\beta}_0, \hat{\beta}_1)}{\partial \beta_1} = -2\sum_{i=1}^{n}(y_i - \hat{\beta}_0 - \hat{\beta}_1 x_i)x_i = 0$$

Results in simpler equations of the forms,

$$n\hat{\beta}_0 + \hat{\beta}_1 \sum_{i=1}^{n} x_i = \sum_{i=1}^{n} y_i \tag{6.5}$$

$$\hat{\beta}_0 \sum_{i=1}^{n} x_i + \hat{\beta}_1 \sum_{i=1}^{n} x_i^2 = \sum_{i=1}^{n} x_i\, y_i$$

Equations 6.5 are called *least-squares normal equations*. After solving these equations we obtain least-square estimator coefficients of $\hat{\beta}_0$ and $\hat{\beta}_1$,

$$\hat{\beta}_0 = \bar{y} - \hat{\beta}_1 \bar{x} \tag{6.6}$$

$$\bar{x} = \frac{1}{n}\sum_{i=1}^{n} x_i \text{ and } \bar{y} = \frac{1}{n}\sum_{i=1}^{n} y_i$$

and

$$\hat{\beta}_1 = \left[\sum_{i=1}^{n} x_i y_i - \frac{(\sum_{i=1}^{n} x_i)(\sum_{i=1}^{n} y_i)}{n}\right] \Bigg/ \left[\sum_{i=1}^{n} x_i^2 - \frac{(\sum_{i=1}^{n} x_i)^2}{n}\right] \tag{6.7}$$

In the following section, more focus is going to be given on the calculation of *fitted regression line coefficient estimates* for Example 6.1.

6.2 Residual and Fitted Values

The least-square estimates, $\hat{\beta}_0$ and $\hat{\beta}_1$, define the fitted regression line estimates as presented in Equation 6.8, i.e. general regression equation form. Whereas, Equation 6.9 defines the fitted regression equation where approximation errors are included in the output estimation,

$$y_i(x) = \hat{\beta}_0 + \hat{\beta}_1 x_i + e_i \rightarrow i = 1,2,\ldots n \qquad 6.8$$

$$\hat{y}(x) = \hat{\beta}_0 + \hat{\beta}_1 x \qquad 6.9$$

where $\hat{y}(x)$ is the estimated/fitted value of the output. Please also note that the differences between actual output $y_1, y_2, \ldots, y_n$ and the fitted/estimated outputs $\hat{y}_1, \hat{y}_2, \ldots, \hat{y}_n$ are called e_i, *residuals* for each, say, $i=1,2,\ldots.n$ data pairs.

$$e_i = y_i - \hat{y}_i \rightarrow i = 1,2,\ldots,n \qquad 6.10$$

Briefly, the residual describes the approximation error made between each actual output/observation and fitted output by the regression model. Residuals are also used for deciding about the *adequacy* of the fitted model, as it would be explained later.

We can go ahead and simplify the $\hat{\beta}_1$ formulation in Equation 6.7 by giving symbols to the numerator (S_{xx}) and denominator (S_{xy}) and keep $\hat{\beta}_0$ same as before,

$$\hat{\beta}_0 = \bar{y} - \hat{\beta}_1 \bar{x} \qquad 6.11$$

$$\bar{x} = \frac{1}{n}\sum_{i=1}^{n} x_i \; and \; \bar{y} = \frac{1}{n}\sum_{i=1}^{n} y_i$$

$$\hat{\beta}_1 = \frac{S_{xy}}{S_{xx}} \qquad 6.12$$

where,

$$S_{xy} = \sum_{i=1}^{n} x_i y_i - \frac{(\sum_{i=1}^{n} x_i)(\sum_{i=1}^{n} y_i)}{n} \qquad S_{xy} = \sum_{i=1}^{n} x_i^2 - \frac{(\sum_{i=1}^{n} x_i)^2}{n} \qquad 6.13$$

Now, let's go ahead and use these equations to fit a simple linear regression model to Example 6.1, %Attendance vs. GPA data, given in Table 6.1. The following format of the table is quite useful to calculate required values from equations as presented in Table 6.2.

TABLE 6.2 Calculation steps of regression coefficients

$n =$	21
$\sum_{i=1}^{n} x_i =$	1345
$\sum_{i=1}^{n} y_i =$	1292
$\bar{x} = \frac{1}{n}\sum_{i=1}^{n} x_i =$	64.0476
$\bar{y} = \frac{1}{n}\sum_{i=1}^{n} y_i =$	61.5238
$\sum_{i=1}^{n} x_i^2 =$	94953.0
$\sum_{i=1}^{n} y^2 =$	93668.0
$\sum_{i=1}^{n} x_i y_i =$	93559
$S_{xx} =$	8808.952
$S_{xy} =$	10809.476
$\boldsymbol{\widehat{\beta}_1 = \frac{S_{xy}}{S_{xx}} =}$	**1.2271**
$\boldsymbol{\widehat{\beta}_0 = \bar{y} - \widehat{\beta}_1 \bar{x} =}$	**-17.0691**
$\boldsymbol{\hat{y} = -17.0691 + 1.2271\, x}$	

So, our linear regression equations is obtained as,

$$\widehat{GPA} = -17.0691 + 1.2271\ \%Attendance \tag{6.14}$$

In addition, SS_E is called the **error sum squares** of residuals and can be computed using Equation 6.15,

$$SS_E = \sum_{i=1}^{n} e_i^2 = \sum_{i=1}^{n} (y_i - \hat{y}_i)^2 = 915 \qquad 6.15$$

Additionally, SS_R is defined as the **regression sum of squares** of the response/output, *y* given in Equation 6.16 as the sum squares of the difference between the estimated/fitted response/output and its mean,

$$SS_R = \sum_{i=1}^{n} (\hat{y}_i - \bar{y})^2 = 13264 \qquad 6.16$$

Further, SS_T is defined as the **total sum of squares** of the response/output, *y* given in Equation 6.17 as the sum squares of the difference between the actual response/output and its mean,

$$SS_T = \sum_{i=1}^{n} (y_i - \bar{y})^2 = 14179 \qquad 6.17$$

In the method of **analyses of variance**, one can run tests to find out the significance of regression by partitioning the total variability in the response/output variable into components which are statistically meaningful as shown in Equation 6.18.

$$\sum_{i=1}^{n} (y_i - \bar{y})^2 = \sum_{i=1}^{n} (\hat{y}_i - \bar{y})^2 + \sum_{i=1}^{n} (y_i - \hat{y}_i)^2 \qquad 6.18$$

For overall representation of analyses of variance, let's use very convenient symbolic abbreviations presented above, which are commonly used in regression analysis as presented in Equation 6.19,

$$SS_T = SS_R + SS_E \qquad 6.19$$

Now, let's continue defining more quantities commonly encountered in regression analysis outputs. For example, **estimated standard error** of the intercept and slope in simple linear regression given in Equation 6.20, can be calculated easily!

$$SE\left(\hat{\beta}_0\right) = \sqrt{\hat{\sigma}^2\left(\frac{1}{n} + \frac{\bar{x}^2}{S_{xx}}\right)} = 4.972 \text{ and } SE\left(\hat{\beta}_1\right) = \sqrt{\frac{\hat{\sigma}^2}{S_{xx}}} = 0.07394 \qquad 6.20$$

where $\hat{\sigma}^2$ is called *estimator of variance, σ^2,* and calculated using Equation 6.21,

$$\hat{\sigma}^2 = \frac{SS_E}{n-2} = 6.939 \qquad 6.21$$

T-statistics for β_0 and β_1 are determined by using the appropriate hypotheses which are known as *null* and *alternate* hypotheses,

$$H_o\colon \beta_i \;=\; \beta_{i,0} \text{ and } H_1\colon \beta_i \;\neq\; \beta_{i,0} \;\rightarrow i = 0{,}1 \qquad 6.22$$

In H_0 Hypothesis, we test the slope equals a constant value $\beta_{0,0}$ or the same thing is tested for constant/intercept $\beta_{1,0}$ as well, using both values in Table 6.2 and in Equation 6.20, in Equation 6.23 as follows;

$$(T_0)_{\text{slope/constant}} = \frac{\hat{\beta}_i - \hat{\beta}_{i,0}}{SE\left(\hat{\beta}_i\right)} \rightarrow i = 0{,}1 \qquad 6.23$$

$$T_0 = \frac{-17.0691 - 0}{4.972} = -3.43 \text{ for Constant}$$

$$T_0 = \frac{1.2271 - 0}{0.07394} = 16.60 \text{ for Slope}$$

This distribution follows the *t-distribution* with α and *n-2* degrees of freedom i.e. dof for $H_o\colon \beta_i \;=\; \beta_{i,0}$ is rejected for both slope and intercept, if

$$|T_0| > t_{\alpha/2\,,n-2} \rightarrow i = 0{,}1 \qquad 6.24$$

For $\alpha = 0.01$ (level of confidence) and *n-2=21-2=19* dof, if we look at the $t_{0.01/2\,,19} =$ 2.861, from *T-table* in Appendix 3, thus,

$$|-3.43| > 2.861 \text{ for constant thus, } \mathrm{H_o}\colon \beta_0 \;=\; \beta_{0,0} \text{ is rejected!}$$

$$|16.60| > 2.861 \text{ for slope thus, } \mathrm{H_o}\colon \beta_1 \;=\; \beta_{1,0} \text{ is rejected!}$$

P values can be calculated using a calculator or *Excel* sheet with statistical command of TDIST() as in the following,

$p = \text{TDIST}(|T_0|_{n-2;2})$ where T_0=*-3.43 or 16.60* with n=*21*

p= TDIST($|-3.43|_{19;2}$) = 0.003 for constant/intercept

p= TDIST($|16.60|_{19;2}$) = 0.000 for slope

and they are almost zero for both cases which support our above judgments for constant and slope of the regression equation.

$SE\left(\hat{\beta}_0\right)$	=	4.972
$SE\left(\hat{\beta}_1\right)$	=	0.07394
$p_{constant}$	=	0.003
p_{slope}	=	0.000

If we continue our exploration by looking at regression analysis outputs, the coefficient of determination i.e. **R^2 Statistics** which is widely used to provide some useful information to judge the adequacy of a regression model (i.e. how well the regression equation fits the measured data. **R^2** is defined as the ratio of sum squares of the residuals (or errors) to total variance as in Equation 6.25. In practice, R^2 *is* deemed as the quality of the regression fit and needs to be further supported with R^2_{adj} for better judgment and conclusion on this aspect.

$$R^2 = \frac{SS_R}{SS_T} = 1 - \frac{SS_E}{SS_T} = 1 - \frac{\sum_{i=1}^{n}(y_i - \hat{y}_i)^2}{\sum_{i=1}^{n}\left(y_i - \bar{y}_i\right)^2} = 1 - \frac{915}{14179} = 0.935 \qquad 6.25$$

$$= 1 - \frac{\text{Sum of aquares of the residuals (or errors)}}{\text{Total Variance}}$$

Ideally, R^2 will have a value of 1.0 (100%), ranging between $0 \le R^2 \le 1$. However, in practice due to random variations in the data, this quantity is expected to be less than 1.

Suppose that we have four inputs and a single output for our data set. We are looking for a 2nd order regression fit, i.e. nonlinear. Now the further exploration of this concept will introduce an R^2 *(adj)* statistics. Such an equation will consist of the following terms as shown in Equation 6.26, totaling $\sum_{i=0}^{14} a_i$= 15 terms and also assume that the regression equation has 15 data pairs/sets for use in the regression fit. Then, we are going to get a perfect fit with R^2=*1.0* with the quadratic equation given below,

$$y = a_0 + a_1x_1 + a_2x_2 + a_3x_3 + a_4x_4 + a_5x_1x_2 + a_6x_1x_3 + a_6x_1x_4 + a_8x_2x_3 + a_9x_2x_4 + a_{10}x_3x_4 + a_{11}{x_1}^2 + a_{12}{x_2}^2 + a_{13}{x_3}^2 + a_{14}{x_4}^2 \quad 5.26$$

Now, we need to question the validity of this model from the above point of view! As expected, in this case, the above model is not going to make a valid prediction for general use, rather it would misguide us by making us think that everything is so good to be true with $R^2=1.0$. Here mimicking the behavior of the data or memorization occurs. Then, how can we avoid such an occurrence? Because, we know that any term added to the regression equation will reduce the sum of squares of the residuals, even these terms do not contribute to the improvement of the fit. Now, an *adjusted-*R^2 statistic can help us to avoid such occurrences by including a number of data points in the following equation for *Adjusted-*R^2, where n=number of design/experiment pairs and p=number of terms in the regression model,

$${R^2}_{Adj} = 1 - \frac{n-1}{n-p}(1 - R^2) = 0.932 \quad 6.27$$

Note that R^2_{adj} shows the better ability for indication of the adequacy of the regression model to give an idea about the quality of the fit. As far as n=15 data pairs and p=15 terms in the regression model, since $n\text{-}p=0$, this will make R^2_{adj} going to infinity.

s, standard deviation, is the square root of the $\hat{\sigma}^2$(an *estimator of variance,* σ^2) and calculated using Equation 6.28,

$$S = \sqrt{\hat{\sigma}^2} = 6.939 \quad 6.28$$

Please note that we have constant and slope as the number of parameters in the linear regression model, thus p=2,

$S =$	6.939
$R^2 =$	93.5%
${R^2}_{Adj} =$	93.2%
$n = 21$ *and* $p = 2$	

Now, let's take a quick look at the *Analysis of Variance* part of the characteristic regression output which is used to *test for significance*. In other words, we would like to see if there is a relation between inputs and output variables or not! Our expectation is to see such a relationship since we are developing the regression model with this intent We can adopt F_0 test for this purpose.

As we have shown, true definitions of SS_T, SS_R, and SSE, represent the total variability in the response variable and can also be used in the calculation of F_0 as,

$$F_0 = \frac{\frac{SS_R}{1}}{\frac{SS_E}{n-2}} = \frac{MS_R}{MS_E} \qquad 6.29$$

If the *null hypothesis* $H_0 : \beta_1 = 0$ is true! then, the statistics F_0 given in Equation 5.29, follows the $F_{1,n-2}$ distribution, and H_0 is rejected when $f_0 > f_{\alpha, 1, n-2}$. MS_R and MS_E quantities are called *mean squares.* The *analysis of variance* outputs in a proposed format for this test procedure is summarized in the following table. Also, note that n is the number of data pairs used in the regression.

Source	**DF**	**SS**	**MS**	**F**	**P**
Regression	*1*	SS_R	MS_R	MSR/MSE	TDIST($\lvert T_0 \rvert$,n-2,2)
Residual Error	*n - 2*	SS_E	MS_E		
Total	*n - 1*	SS_T			

One thing to notice that,

$$T_0^2 = F_0 = {MS_R}/{(\hat{\sigma}^2 = MS_E)} \qquad 6.30$$

As it can be seen from Equation 6.30, we can use *t-test* using T_0, instead of F_0 testing. Since *t-test* using T_0 is easier to do! by comparing T_0 to the *p-value*. The test statistic, F_0 = MSR/MSE = 275.46 and T_0=16.597 using this value, we calculate the *p-value*=0.000 by *Excel* command **TDIST(T_0,n-2,2)**. As far as the hypothesis testing is involved, we confirm that β_1 is not zero! Analysis of variance table is given in the following for the Example 6.1 problem of our interest.

Source	**DF**	**SS**	**MS**	**F**	**P**
Regression	*1*	13264	13264	275.46	TDIST($\lvert T_0 = 16.597 \rvert$,n-2,2) =0.000
Residual Error	*19*	915	48		
Total	*20*	14179			

Now, go ahead and use these equations to fit a simple linear regression model to Example 6.1 - %Attendance vs. GPA data, as given in Table 6.1.

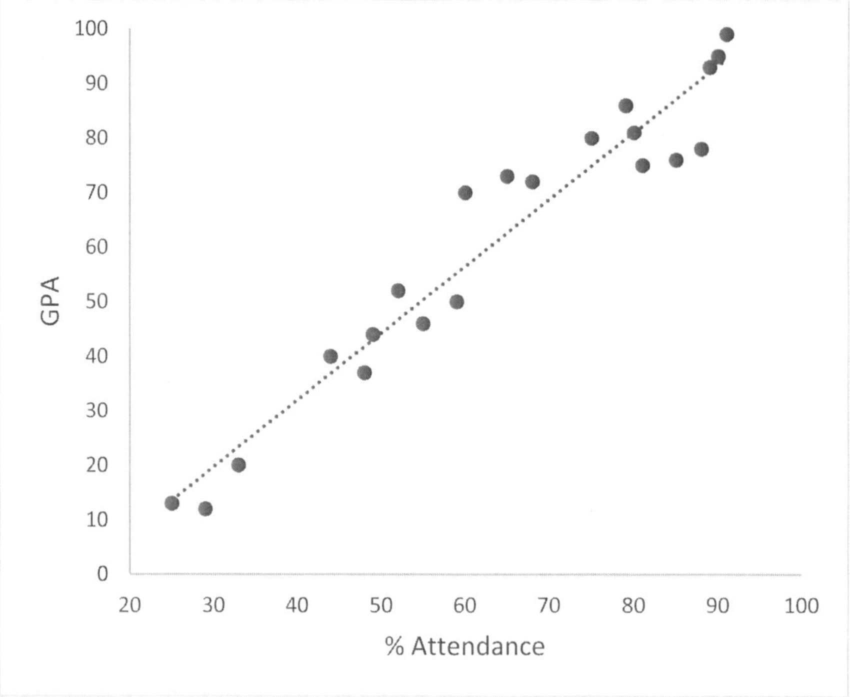

FIGURE 6.4 Scatter plot and regression line of $\hat{y} = -17.0691 + 1.227\ \mathbf{x}$

This fitted regression model line is plotted in Figure 6.4, along with the data used. Note that in this model attendance lower than almost %10 is not permitted! and GPA cannot be taking a negative value since it is not possible!

The following Table 6.3 format is quite useful to calculate required values using the linear regression equation as demonstrated in the next page! I.e. using this regression model, we can calculate the estimated GPA's and residual for each (n=21) data, as presented in Table 6.3.

TABLE 6.3 Example 6.1 %Attendance vs GPA data

n	% Attendance	*Actual* GPA	*Estimated* GPA	e_i (Residual)
1	91	99	94,59	4,40
2	89	93	92,13	0,86
3	90	95	93,36	1,63
4	85	76	87,23	-11,23
5	88	78	90,91	-12,92
6	81	75	82,32	-7,33
7	59	50	55,32	-5,33
8	60	70	56,55	13,44
9	48	37	41,83	-4,83
10	65	73	62,69	10,31
11	79	86	79,86	6,13
12	52	52	46,73	5,26
13	25	13	13,61	-0,61
14	55	46	50,42	-4,42
15	44	40	36,92	3,08
16	80	81	81,09	-0,10
17	29	12	18,51	-6,52
18	68	72	66,37	5,63
19	75	80	74,96	5,04
20	49	44	43,05	0,94
21	33	20	23,42	-3,43

Now, it is time to further explore the simple regression analysis using *Minitab* to equip ourselves with both manual and software-based calculations to build up hands-on experience as presented in the following section.

6.3 Simple Linear Regression with Minitab

Now, let's perform the same simple regression analysis on Example 6.1, using *Minitab*, for the parts/results those have been demonstrated in Section 6.2. Let's start up the **Minitab** software from the Windows Start menu by clicking as in the following order,

> **All Programs>Minitab>Minitab 1X Statistical Software**

1. You can copy or enter your input and output data of $n=21$ pairs in the Worksheet 1 window corresponding to cells C1 and C2 as shown,

Worksheet 1 ***

↓	C1	C2	C3
	% Attendance	GPA	
1	91	99	
2	89	93	
3	90	95	
4	85	76	
5	88	78	
6	81	75	
7	59	50	
8	60	70	
9	48	37	
10	65	73	
11	79	86	
12	52	52	
13	25	13	
14	55	46	
15	44	40	
16	80	81	
17	29	12	
18	68	72	
19	75	80	
20	49	44	
21	33	20	

2. Then, click on **Stat** > **Regression** > **Regression…** tabs in the order as shown below,

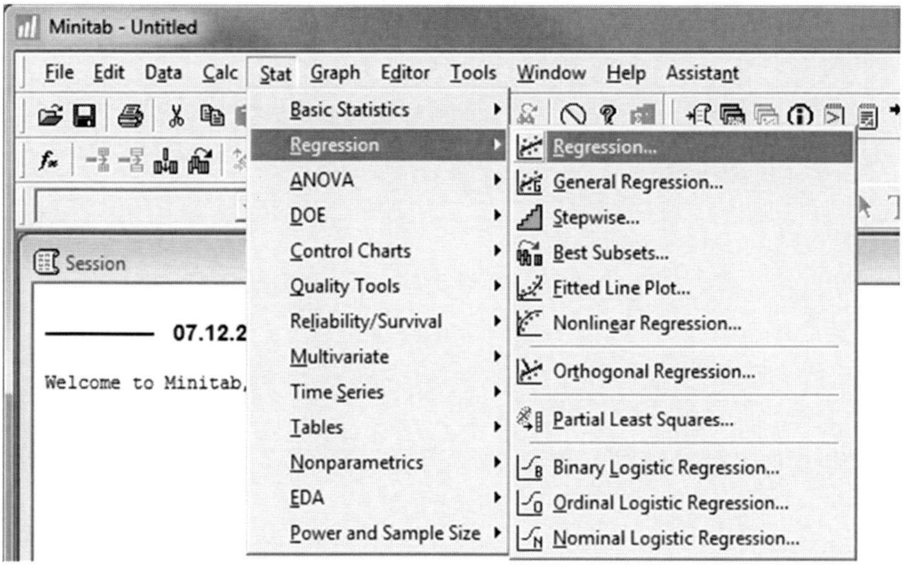

3. **Regression** main window will pop-up, where

- *Specify C1 % **Attendance*** as Predictors/Input
- *Specify C2* ***GPA*** as Response/Output, i.e. first highlight %Attendance next click on **Predictors:** window, then click on the select button, repeat the same thing for GPA, select it to **Response:** window as shown below, click OK to close the **Regression** window.

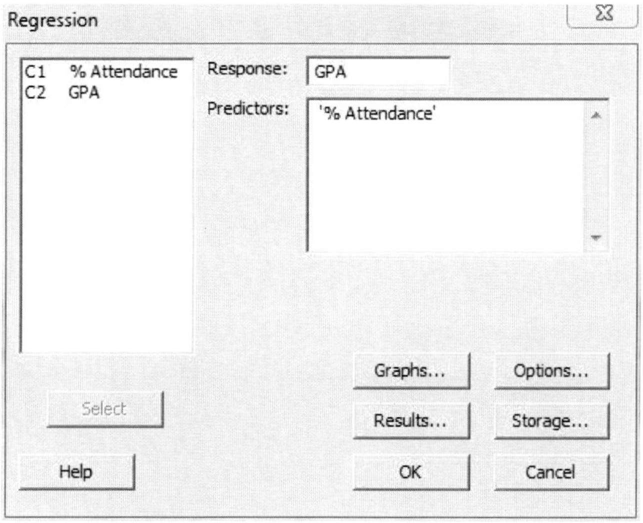

3.1. Under **Regression** main window, click on **Graphs…** tab and click on **Four in one** option as shown below, click OK to close the **Regression-Graphs** window.

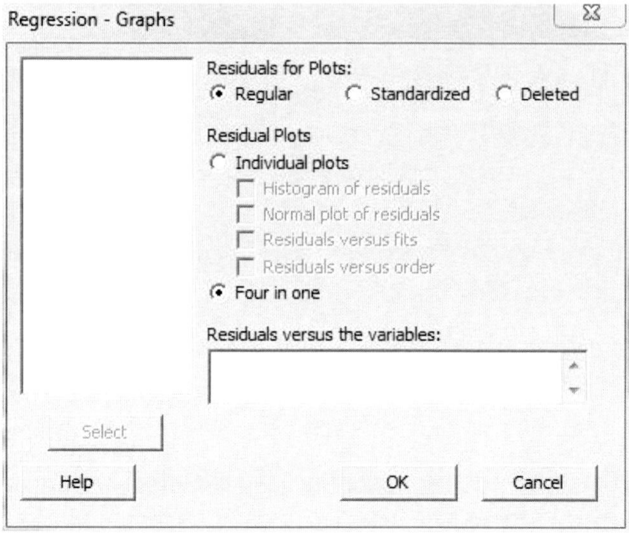

3.2. Under **Regression** main window, click on **Options…** tab some useful statistics option will appear in **Regression-Options** window, at this point we are ok with DEFAULT,

- ✓ Fit intercept and
- ✓ 95% Confidence level settings

as shown below, click OK to close **Regression-Options** window.

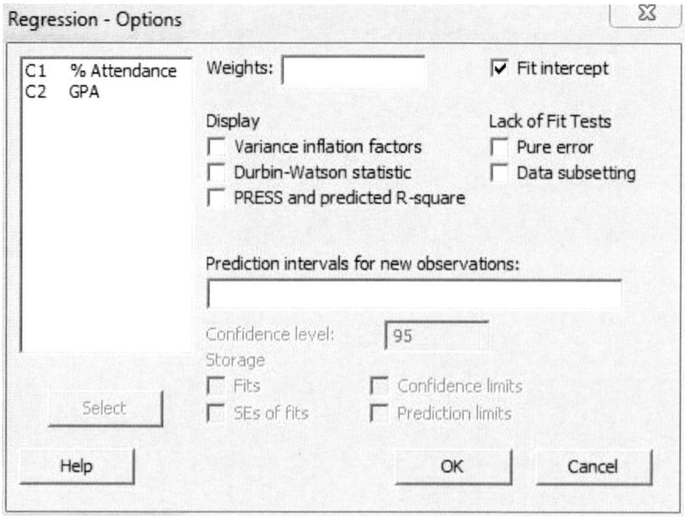

3.3. Under **Regression** main window, click on **Results…** tab you will get **Regression-Results** window, use DEFAULT settings as shown below, click OK to close **Regression-Results** window.

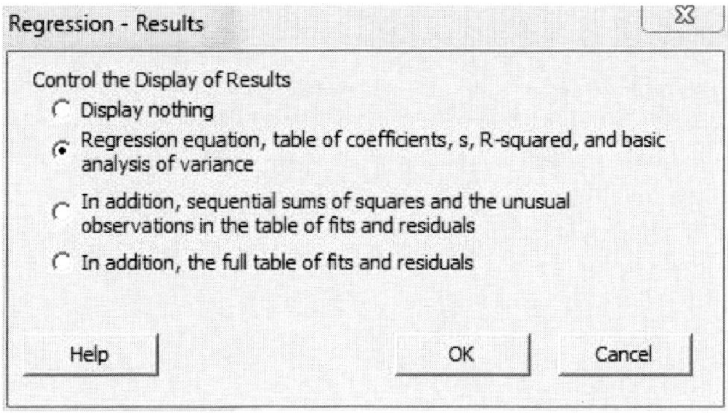

3.4. Under **Regression** main window, click on **Storage…** tab, you will get **Regression-Storage** window where click on,

- ✓ Residuals
- ✓ Coefficients
- ✓ Fits
- ✓ MSE

options as shown below, click OK to close **Regression-Storage** window.

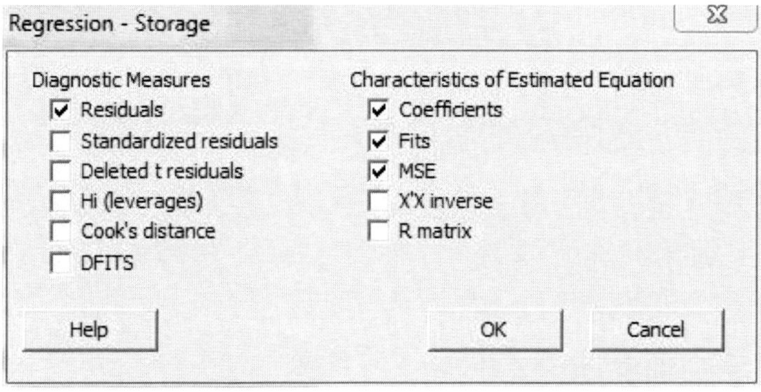

3.5. Now, you are done in **Regression** main window, click on OK to **RUN** the regression. You will get the following Minitab output where all summary of statistic is documented. In the following section, explanations for this information are provided in detail.

In the following;

- RESI2: residuals for each data pair
- COEF2:regression coefficients
- FITS2: estimated fits used regression equation.

Regression Analysis: GPA versus % Attendance

```
The regression equation is
GPA = - 17,1 + 1,23 % Attendance


Predictor        Coef   SE Coef       T      P
Constant      -17,069     4,972   -3,43  0,003
% Attendance  1,22710   0,07394   16,60  0,000


S = 6,93927   R-Sq = 93,5%   R-Sq(adj) = 93,2%


Analysis of Variance

Source          DF     SS     MS       F      P
Regression       1  13264  13264  275,46  0,000
Residual Error  19    915     48
Total           20  14179
```

↓	C1	C2	C3	C4	C5
	% Attendance	GPA	RESI1	COEF1	FITS1
1	91	99	4,4029	-17,0691	94,5971
2	89	93	0,8571	1,2271	92,1429
3	90	95	1,6300		93,3700
4	85	76	-11,2345		87,2345
5	88	78	-12,9158		90,9158
6	81	75	-7,3261		82,3261
7	59	50	-5,3299		55,3299
8	60	70	13,4430		56,5570
9	48	37	-4,8318		41,8318
10	65	73	10,3075		62,6925
11	79	86	6,1281		79,8719
12	52	52	5,2598		46,7402
13	25	13	-0,6084		13,6084
14	55	46	-4,4215		50,4215
15	44	40	3,0766		36,9234
16	80	81	-0,0990		81,0990
17	29	12	-6,5168		18,5168
18	68	72	5,6262		66,3738
19	75	80	5,0365		74,9635
20	49	44	0,9411		43,0589
21	33	20	-3,4252		23,4252

3.6. Finally, you will get **Four-in-one** plot where,

- **Normal Probability Plot** indicates good fit because if we put a thick pen on top of the line, almost all dots will be disappearing
- **Versus Fits** plot also is good, showing fitted value vs. residuals look balanced at both sides
- **Histogram** plots show disperse and distribution frequency of the residuals around target zero value
- **Versus order** plot shows for each observation, a number of residuals disperse around target zero value.

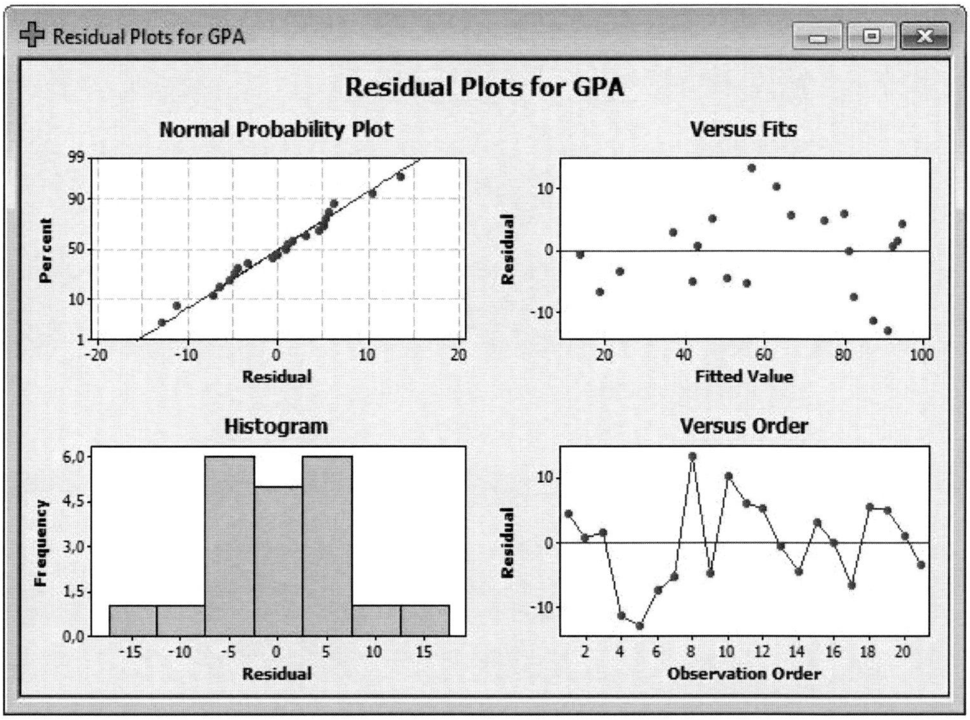

6.4 Understanding Minitab Output

Now, let's go ahead and provide detailed information on the Minitab output for regression analysis as given next.

Regression Analysis: GPA versus % Attendance

```
The regression equation is
GPA = - 17,1 + 1,23 % Attendance


Predictor        Coef  SE Coef      T      P
Constant      -17,069    4,972  -3,43  0,003
% Attendance  1,22710  0,07394  16,60  0,000


S = 6,93927   R-Sq = 93,5%   R-Sq(adj) = 93,2%


Analysis of Variance

Source          DF     SS     MS       F      P
Regression       1  13264  13264  275,46  0,000
Residual Error  19    915     48
Total           20  14179
```

- The first set of output shows clearly regression equation where GPA is defined in terms of *%Attendance* in a linear equation
- The second set of output shows in,
    I. Column 1: Predictors i.e. $\hat{\beta}_0$ as Constant and $\hat{\beta}_1$as % Attendance (Slope)
    II. Column 2: $\hat{\beta}_0$ and $\hat{\beta}_1$ coefficients.
    III. Column 3: Estimated standard error of the intercept and slope in simple linear regression given in Equation 6.20
    IV. Column 4 and 5: T values and corresponding P values for constant and slope, respectively for each column
- S, R^2, and $R^2{}_{Adj}$ values are presented
- Finally, a simple *Analysis of Variance* values is presented.

Please note all of these values and their associated formulations are presented and explained to be consistent with this format in the previous section.

6.5 Simple Linear Regression with *Excel*

In this section, an alternative linear regression analysis tool and its use in statistics are introduced. In this context, the same example problem is modeled and analyzed using *Excel* worksheet.

Let's go ahead and adopt the same Example 6.1, used in the previous section and demonstrate all the steps needed to perform a linear regression analysis in *Excel.* As far as any specific analysis is concerned, in this case, "*Regression*", we need to use **"Data**

analysis" added-in feature in *Excel*, which is not added in the *default* installation. Therefore, before starting to regression analysis, we need to check whether this **"Data analysis"** option/added-in feature is available/added in our present *Excel* program window, first. This check can be done as outlined in the following:

- Open up a new Microsoft *Excel* document and click on the **DATA** tab, look all the way right on the top of the *Excel* page, as also shown i.e. in the first and second pictures.
- If you do not see a **"Data Analysis"** button there in the dashed rectangle area in the second picture on the left, then you should go ahead and add that in there, as explained in the following.

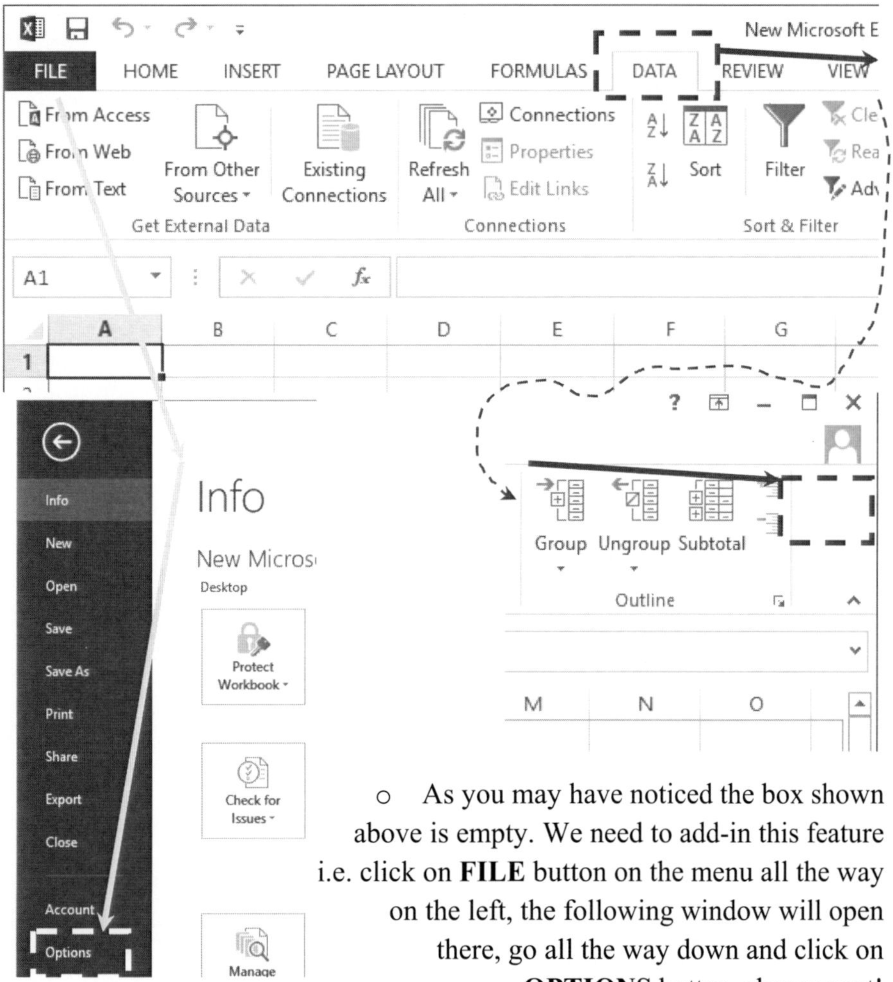

- As you may have noticed the box shown above is empty. We need to add-in this feature i.e. click on **FILE** button on the menu all the way on the left, the following window will open there, go all the way down and click on **OPTIONS** button, shown next!

- ***Excel* Options** menu will open, where you would see **Add-Ins tab** in shaded light green and **Analysis ToolPak** highlighted in grey, where go ahead check out **Analysis ToolPak**,

  Then, click on **Go…** button

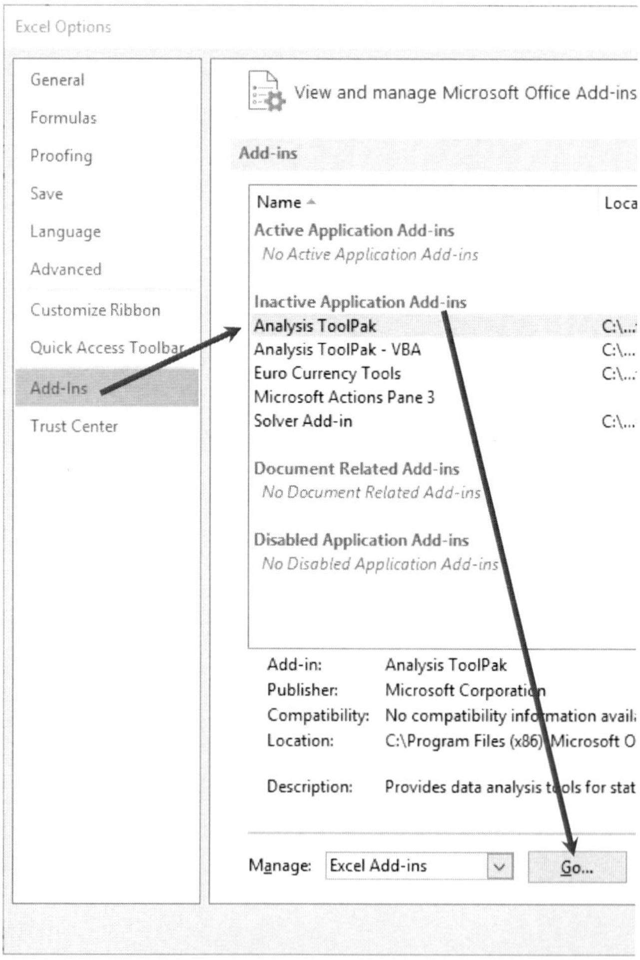

- You can also do the same thing From **Manage** option shown above, and set it to ***Excel* Add-ins** then click **Go…** button, you will also get the following Add-ins window.

- **Add-Ins** window will appear, there go ahead check out **Analysis ToolPak** then click OK!

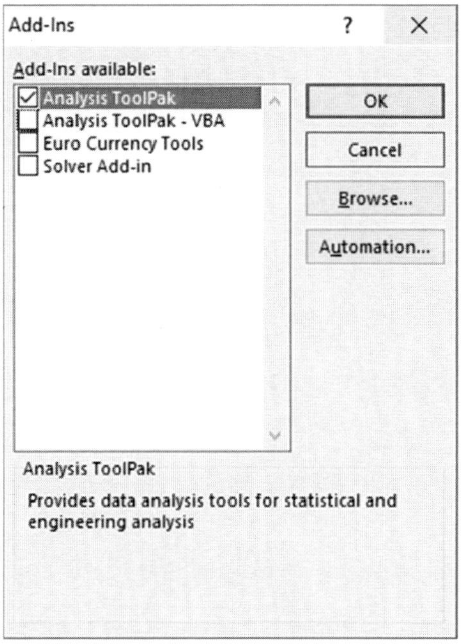

- Now, you are done in adding **Data Analysis** Add-Inn capability to your *Excel* Sheet as shown in the following. Now, You are ready to go ahead and perform regression analysis as demonstrated in the following section.

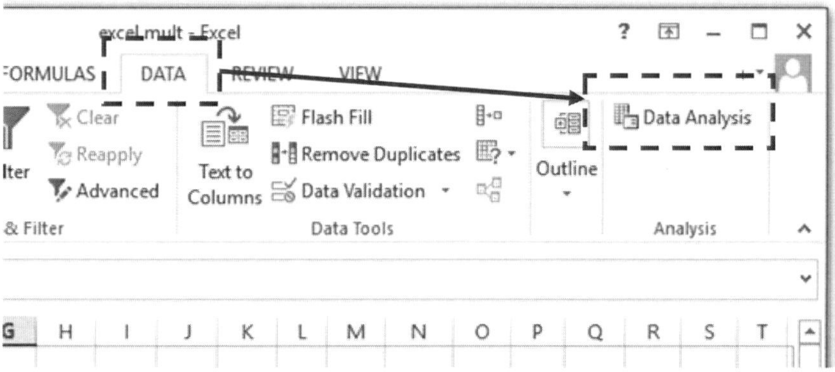

- Using the same example data given below in *Excel* format, Click on **Data Analysis** *button* that you added in.

	A	B	C	D
1				
2	n	X	y	
3	1	91	99	
4	2	89	93	
5	3	90	95	
6	4	85	76	
7	5	88	78	
8	6	81	75	
9	7	59	50	
10	8	60	70	
11	9	48	37	
12	10	65	73	
13	11	79	86	
14	12	52	52	
15	13	25	13	
16	14	55	46	
17	15	44	40	
18	16	80	81	
19	17	29	12	
20	18	68	72	
21	19	75	80	
22	20	49	44	
23	21	33	20	
24				

- You will get the following **Data Analysis** window as shown below. Scroll down all the way down and once you see the **Regression**, highlight that tool by clicking on it, then Hit OK!

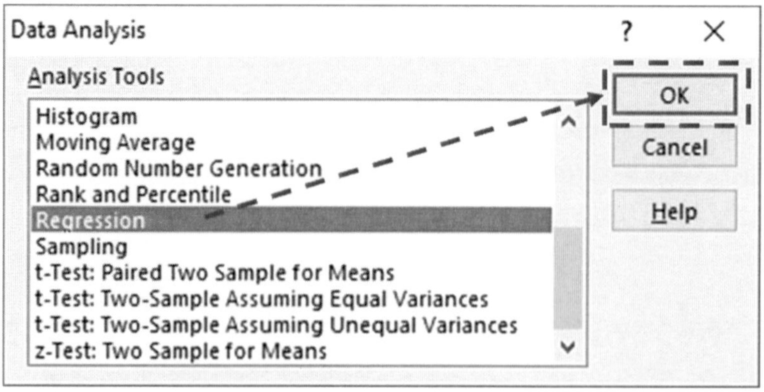

- **Regression** window will pop-up! where you will have to select specific Ranges for input and output data.
  - First, select Output y range between **=C2:C23** with ***y*-Label,** hit Enter!
  - Second, select Input X range between **=B2:B23** with ***X*-Label**, hit Enter!
  - Since you selected the data pairs with their *Labels*, check *Labels* box, and let the program know about that.
  - In the **Output options**, select **New Worksheet Ply:** option so that regression results are stored in a new worksheet in your already open *Excel* file!
  - You may print the **Residuals** and **Line Fit Plots**, just click the appropriates boxes as shown.
  - Optional, you may want to see **Normal Probability Plots** by clicking the associated box.
  - Hit **OK** to close the window!

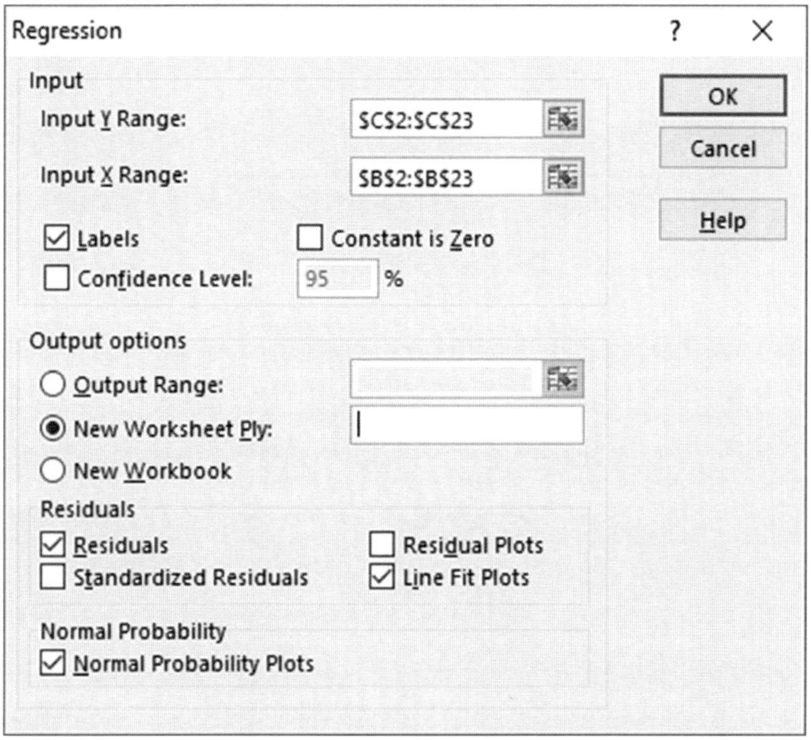

- Next, you will get the regression results in open *Excel* Sheet in a new Worksheet Ply as shown in the SUMMARY OUTPUT below, where you will have a similar output that you got in the previous chapter.

SUMMARY OUTPUT

Regression Statistics	
Multiple R	0.96719952
R Square	0.9354749
Adjusted R Square	0.93207885
Standard Error	6.93927306
Observations	21

ANOVA

	df	*SS*	*MS*	*F*	*Significance F*
Regression	1	13264.32139	13264.3214	275.45907	9.16941E-13
Residual	19	914.9167027	48.1535107		
Total	20	14179.2381			

	Coefficients	*Standard Error*	*t Stat*	*P-value*	*Lower 95%*	*Upper 95%*	*Lower 95.0%*	*Upper 95.0%*
Intercept	-17.069102	4.971604179	-3.4333187	0.0027862	-27.47478892	-6.6634146	-27.4747889	-6.663414644
X	1.22710122	0.073935301	16.5969597	9.169E-13	1.072352853	1.38184958	1.072352853	1.381849581

- Further, you will get the RESIDUAL and PROBABILITY OUTPUTs for your evaluation of the results as shown in the following.

RESIDUAL OUTPUT

Observation	*Predicted y*	*Residuals*
1	94.597109	4.402890998
2	92.1429066	0.857093433
3	93.3700078	1.629992216
4	87.2345017	-11.2345017
5	90.9158053	-12.91580535
6	82.3260968	-7.326096828
7	55.32987	-5.329870046
8	56.5569713	13.44302874
9	41.8317567	-4.831756654
10	62.6924773	10.30752265
11	79.8718944	6.128105607
12	46.7401615	5.259838476
13	13.6084287	-0.608428655
14	50.4214652	-4.421465176
15	36.9233518	3.076648215
16	81.0989956	-0.098995611
17	18.5168335	-6.516833524
18	66.373781	5.626218998
19	74.9634895	5.036510476
20	43.0588579	0.941142128
21	23.4252384	-3.425238394

PROBABILITY OUTPUT

Percentile	*y*
2.3809524	12
7.1428571	13
11.904762	20
16.666667	37
21.428571	40
26.190476	44
30.952381	46
35.714286	50
40.47619	52
45.238095	70
50	72
54.761905	73
59.52381	75
64.285714	76
69.047619	78
73.809524	80
78.571429	81
83.333333	86
88.095238	93
92.857143	95
97.619048	99

- Finally, you will end up with *Excel* regression plots for the above data for ***X*-Line Fit Plot** and Normal Probability Plot.

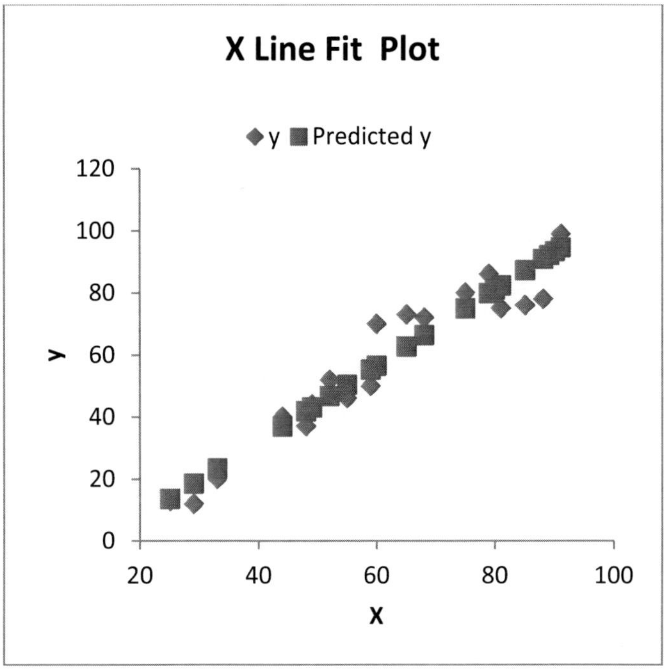

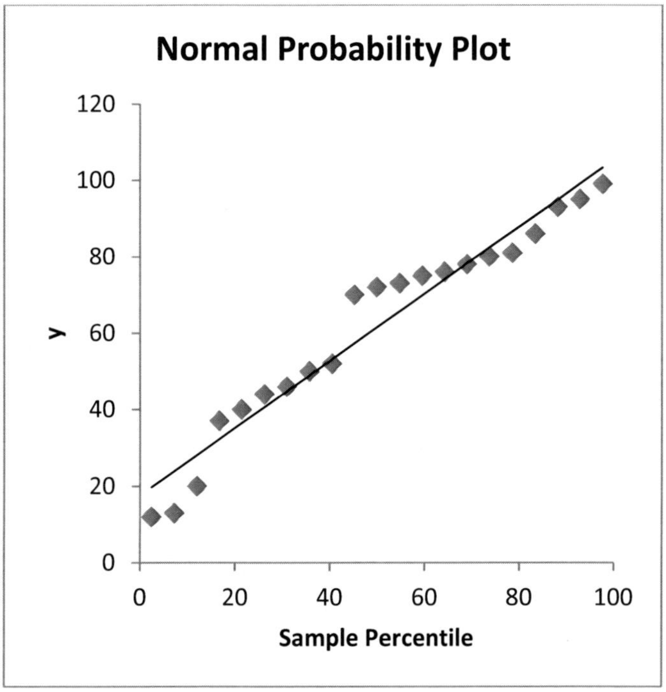

6.6 Simple Linear Regression (SLR) using Matrix Notations

In this section, an alternative way of the same linear regression analysis is performed using a mathematical *matrix notation* and its associated operations. Note that, this method is really useful for the solution of such kind of problems, where we will perform all basic matrix operations i.e. taking the transpose of a matrix, matrix multiplication, taking an inverse of a matrix etc. and all these matrix operations are all going to be done in Microsoft *Excel* program. Let's use the general form of the regression model in which we relate/map the k number of regressor variables or inputs to the response/output by the following first order equation as stated in Equation 6.31.

$$y_i(x) = \beta_0 + \beta_1 x_{i1} + \beta_2 x_{i2} + \cdots + \beta_k x_{ik} + \epsilon_i \qquad 6.31$$

$$\text{where } \epsilon_i \sim N(0, \sigma^2) \text{ and } i = 1,2,\dots,n$$

where, please note that $p=k+1$ (total number of β terms in the regression equation). Now, in an open equation form, let's consider writing associated equations of 6.31 for each output/observation, i.e. $i = 1,2,\dots,n$ as presented in the following equations,

$$\begin{aligned} y_1 &= \beta_0 + \beta_1 x_{11} + \beta_2 x_{12} + \cdots + \beta_k x_{1k} + \epsilon_1 \\ y_2 &= \beta_0 + \beta_1 x_{21} + \beta_2 x_{22} + \cdots + \beta_k x_{2k} + \epsilon_2 \\ &\dots \dots \dots \dots \dots \dots \dots \dots \dots \dots \\ y_n &= \beta_0 + \beta_1 x_{n1} + \beta_2 x_{n2} + \cdots + \beta_k x_{nk} + \epsilon_n \end{aligned} \qquad 6.32$$

So, in matrix form, this equation or model can be stated as in Equation 6.33, (Please note that for convenience, the matrix notation is shown in bold case),

$$\boldsymbol{y} = \boldsymbol{X\beta} + \boldsymbol{\epsilon} \qquad 6.33$$

Now, let's re-write above equations for our **SLR** case, so *Single Linear Regression* model in general equation format reduces to,

$$y_i = \beta_0 + \beta_1 X_i + \epsilon_i \qquad 6.34$$

$$\epsilon_i \sim N(0, \sigma^2)\ , i = 1,2,\dots,n$$

and open form of the equations can be written as,

$$y_1 = \beta_0 + \beta_1 x_1 + \epsilon_1$$ 6.35

$$y_2 = \beta_0 + \beta_1 x_2 + \epsilon_2$$

$$\dots\dots\dots\dots\dots\dots\dots\dots\dots$$

$$y_n = \beta_0 + \beta_1 x_n + \epsilon_n$$

Further, the **SLR** model in matrix form can also be written as,

$$\begin{bmatrix} y_1 \\ y_2 \\ \vdots \\ y_n \end{bmatrix} = \begin{bmatrix} \beta_0 + \beta_1 x_1 \\ \beta_0 + \beta_1 x_2 \\ \vdots \\ \beta_0 + \beta_1 x_n \end{bmatrix} + \begin{bmatrix} \epsilon_1 \\ \epsilon_2 \\ \vdots \\ \epsilon_n \end{bmatrix}$$ 6.36

$$\begin{bmatrix} y_1 \\ y_2 \\ \vdots \\ y_n \end{bmatrix} = \begin{bmatrix} 1 & x_1 \\ 1 & x_2 \\ \vdots & \vdots \\ 1 & x_n \end{bmatrix} \begin{bmatrix} \beta_0 \\ \beta_1 \end{bmatrix} + \begin{bmatrix} \epsilon_1 \\ \epsilon_2 \\ \vdots \\ \epsilon_n \end{bmatrix}$$

According to this matrix form,

- X is called the design vector or in general *"design matrix"* with n × 2 matrix size

$$x_{n\times 2} = \begin{bmatrix} 1 & x_1 \\ 1 & x_2 \\ \vdots & \vdots \\ 1 & x_n \end{bmatrix}$$

- β is the vector of the coefficient of linear regression parameters with 2 × 1 matrix size

$$\beta_{2\times 1} = \begin{bmatrix} \beta_0 \\ \beta_1 \end{bmatrix}$$

- ε is the error vector with n × 1 matrix size

$$\epsilon_{\ n\times 1} = \begin{bmatrix} \epsilon_1 \\ \epsilon_2 \\ \vdots \\ \epsilon_n \end{bmatrix}$$

- y is the output/response vector with an n × 1 matrix size

$$y_{n\times 1} = \begin{bmatrix} y_1 \\ y_2 \\ \vdots \\ y_n \end{bmatrix}$$

Thus, in matrix form,

$$\boldsymbol{y}_{n\times 1} = \boldsymbol{X}_{n\times 2}\,\boldsymbol{\beta}_{2\times 1} + \boldsymbol{\epsilon}_{n\times 1} \qquad 6.37$$

Again, remember that one can summarize these *n-equations,* in a general format of matrix form, for general multiple linear analysis cases as;

$$y = \begin{bmatrix} y_1 \\ y_2 \\ \vdots \\ y_n \end{bmatrix}, X = \begin{bmatrix} 1 & x_{11} & x_{12} & \cdots & x_{1k} \\ 1 & x_{21} & x_{22} & & x_{2k} \\ \vdots & \vdots & \vdots & \ddots & \vdots \\ 1 & x_{n1} & x_{n2} & \cdots & x_{nk} \end{bmatrix}, \beta = \begin{bmatrix} \beta_0 \\ \beta_1 \\ \vdots \\ \beta_k \end{bmatrix} \; and \; \varepsilon = \begin{bmatrix} \epsilon_1 \\ \epsilon_2 \\ \vdots \\ \epsilon_n \end{bmatrix} \qquad 6.38$$

Note that, $\boldsymbol{y}$ is an (n × 1) *vector* dependent variable of outputs/observations, $\boldsymbol{X}$ is an (n × (p = k + 1)) *matrix* of the independent/input variables, $\boldsymbol{\beta}$ is a (p × 1) *vector* of the coefficient of regression and ϵ is an (n × 1) error vector of the regression coefficients. Also, note that the number of terms in regression equation is controlled by k, i.e. if $k=1$, it means that we end up with totaling (p = $k + 1 = 2$), i.e. two terms β_0 and β_1, in simple linear regression equation as it was the case in our example. Further note that, the first column of 1's are basically due to the coefficient of the β_0 in the simple linear regression equation.

6.7 Parameter Estimation using Least Squares

We could calculate the vector of least squares estimators/coefficients that minimizes, L, (Lagrange function), in other words, let's find $\boldsymbol{\beta}$ vector so that, we can minimize the sum squared residuals as follows,

$$L(\beta_0, \beta_1, \dots, \beta_p) = \sum_{i=1}^{n} \epsilon_i^2 = \boldsymbol{\epsilon}'\boldsymbol{\epsilon} = (\boldsymbol{y} - \boldsymbol{X\beta})'(\boldsymbol{y} - \boldsymbol{X\beta}) \qquad 6.39$$

Let's take the derivative of this quadratic equation, with respect to $\boldsymbol{\beta}$ and set out coming quantity to zero (in matrix sense), and solve for β,

$$\frac{\boldsymbol{d}}{\boldsymbol{d}\beta}\{(\boldsymbol{y}-\boldsymbol{X\beta})'(\boldsymbol{y}-\boldsymbol{X\beta})\} = -2\boldsymbol{X}'(\boldsymbol{y}-\boldsymbol{X\beta}) = 0$$

6.40

$$\boldsymbol{X}'\boldsymbol{y} = (\boldsymbol{X}'\boldsymbol{X})\,\boldsymbol{\beta}$$

Multiply on the left by the inverse of the $(\boldsymbol{X}'\boldsymbol{X})$. The resulting least squares estimate vector of (p×1) dimension matrix is obtained as, FITTED values, (remember that fitted quantities are shown with $\widehat{\ }$, i.e. *hat notation*).

$$\widehat{\boldsymbol{\beta}} = (\boldsymbol{X}'\mathbf{X})^{-1}\boldsymbol{X}'\mathbf{y} \qquad 6.41$$

Please remember that what we have obtained so far using matrix notation is nothing but the same formulas that we presented earlier, i.e. Equations 6.6 and 6.7 for β_0 and β_1. Now, we can re-write the same expressions for the fitted regression line's general equation is defined as,

$$\hat{y}_i = \hat{\beta}_0 + \sum_{j=1}^{p+1} \hat{\beta}_j x_{ij} \quad i = 1, \dots, n \qquad 6.42$$

In matrix form, the same statement is written as,

$$\widehat{\boldsymbol{y}} = \boldsymbol{X}\widehat{\boldsymbol{\beta}} \qquad 6.43$$

Likewise, $\boldsymbol{e} = \boldsymbol{y} - \widehat{\boldsymbol{y}}$, corresponding residual vector (n × 1) size is calculated as a difference between both actual and fitted output values. If we perform the following matrix operations we end up with fitted $\hat{\beta}_0$ and $\hat{\beta}_1$, as to be the outcome of series matrix operation as outlined in detail in the following section.

6.8 Matrix Notations

By examining the example data, the ***X*** and ***y*** matrices can be both shown in matrix format as follows,

$$\boldsymbol{X} = \begin{bmatrix} 1 & 91 \\ 1 & 89 \\ 1 & 90 \\ 1 & 85 \\ 1 & 88 \\ 1 & 81 \\ 1 & 59 \\ 1 & 60 \\ 1 & 48 \\ 1 & 65 \\ 1 & 79 \\ 1 & 52 \\ 1 & 25 \\ 1 & 55 \\ 1 & 44 \\ 1 & 80 \\ 1 & 29 \\ 1 & 68 \\ 1 & 75 \\ 1 & 49 \\ 1 & 33 \end{bmatrix} \quad \text{and} \quad \boldsymbol{y} = \begin{bmatrix} 99 \\ 93 \\ 95 \\ 76 \\ 78 \\ 75 \\ 50 \\ 70 \\ 37 \\ 73 \\ 86 \\ 52 \\ 13 \\ 46 \\ 40 \\ 81 \\ 12 \\ 72 \\ 80 \\ 44 \\ 20 \end{bmatrix}$$

Multiply $\boldsymbol{X}'$ and ***X*** matrices as follows,

$$\boldsymbol{X}'\boldsymbol{X} = \begin{bmatrix} 1 & 1 & \cdot & \cdot & \cdot & 1 & 1 \\ 91 & 89 & \cdot & \cdot & \cdot & 49 & 33 \end{bmatrix} \begin{bmatrix} 1 & 91 \\ 1 & 89 \\ 1 & 90 \\ . & . \\ . & . \\ . & . \\ 1 & 49 \\ 1 & 33 \end{bmatrix} \qquad 6.44$$

$$\boldsymbol{X}'\boldsymbol{X} = \begin{bmatrix} 21 & 1345 \\ 1345 & 94953 \end{bmatrix}$$

Multiply $\boldsymbol{X}'$ and $\boldsymbol{y}$ matrices as follows,

$$\boldsymbol{X}'\boldsymbol{y} = \begin{bmatrix} 1 & 1 & \cdot & \cdot & \cdot & 1 & 1 \\ 91 & 89 & \cdot & \cdot & \cdot & 49 & 33 \end{bmatrix} \begin{bmatrix} 99 \\ 93 \\ 95 \\ \cdot \\ \cdot \\ \cdot \\ 44 \\ 20 \end{bmatrix} \quad 6.45$$

$$\boldsymbol{X}'\boldsymbol{y} = \begin{bmatrix} 1292 \\ 93559 \end{bmatrix}$$

Now, we go ahead and take the inverse of the ***X'X*** matrix as presented in Equation 6.44, note that ***X'X*** multiplied with $(\boldsymbol{X'X})^{-1}$ yield unit matrix of ***I***, as shown in the following

$$(\boldsymbol{X}'\boldsymbol{X})^{-1} = \begin{bmatrix} 0.5132928 & -0.0072707 \\ -0.007271 & 0.00011352 \end{bmatrix} \quad 6.46$$

$$(\boldsymbol{X}'\boldsymbol{X})(\boldsymbol{X}'\boldsymbol{X})^{-1} = \begin{bmatrix} 21 & 1345 \\ 1345 & 94953 \end{bmatrix} \begin{bmatrix} 0.5132928 & -0.0072707 \\ -0.007271 & 0.00011352 \end{bmatrix} = \begin{bmatrix} 1 & 0 \\ 0 & 1 \end{bmatrix} = \boldsymbol{I}$$

Finally, we obtain $\widehat{\boldsymbol{\beta}}$ regression coefficient vector, as also given in Equation 6.11 and 6.12.

$$\widehat{\boldsymbol{\beta}} = (\boldsymbol{X}'\mathbf{X})^{-1}\boldsymbol{X}'\mathbf{y} = \begin{bmatrix} 0.5132928 & -0.0072707 \\ -0.007271 & 0.00011352 \end{bmatrix} \begin{bmatrix} 1292 \\ 93559 \end{bmatrix} \quad 6.47$$

$$\widehat{\boldsymbol{\beta}} = (\boldsymbol{X}'\mathbf{X})^{-1}\boldsymbol{X}'\mathbf{y} = \begin{bmatrix} \hat{\beta}_0 \\ \hat{\beta}_1 \end{bmatrix} = \begin{bmatrix} -17.0691 \\ 1.227101 \end{bmatrix}$$

Briefly, in the following *Excel* spread sheet, $n=21$ data, ***X:*** input matrix with (21 × 2) size and ***y:*** output vector with (21× 1) size, are both placed. Following, both $\boldsymbol{X}'_{21\times 2}$ and $\boldsymbol{X}_{2\times 21}$ matrices multiplied, and $(\boldsymbol{X}'\boldsymbol{X})^{-1}_{2\times 2}$, i.e. (2 × 2) size matrix is obtained.

Next, both $\boldsymbol{X}'_{2\times 21}$ and $\boldsymbol{y}_{21\times 1}$ matrices multiplied, and $(\boldsymbol{X}'\boldsymbol{y})_{2\times 1}$, i.e. (2 × 1) size vector is obtained.

Finally, the inverse of $(\boldsymbol{X}'\boldsymbol{X})_{2\times 2}$ matrix is calculated and shown to be $(\boldsymbol{X}'\boldsymbol{X})^{-1}_{2\times 2}$. Outcoming matrices of both $(\boldsymbol{X}'\boldsymbol{X})^{-1}_{2\times 2}$ and $(\boldsymbol{X}'\boldsymbol{y})_{2\times 1}$ now are multiplied and $\widehat{\boldsymbol{\beta}}_{2\times 1}$ vector is calculated.

6.9 *Excel* Workout

Now, we can perform all of the calculations introduced above by coding appropriate equations and performing matrix manipulations as shown in the following *Excel* sheet in Table 6.4,

TABLE 6.4 Example 6.1 %Attendance vs. GPA data

	A	B	C	D	E	F	G
1							
2	n		X	y			
3	1	1	91	99	$X'X =$	21	1345
4	2	1	89	93		1345	94953
5	3	1	90	95			
6	4	1	85	76	$X'y =$	1292	
7	5	1	88	78		93559	
8	6	1	81	75			
9	7	1	59	50	$\hat{\beta} = (X'X)^{-1}X'y =$	-17.0691	$\hat{\beta}_0$
10	8	1	60	70		1.2271	$\hat{\beta}_1$
11	9	1	48	37			
12	10	1	65	73			
13	11	1	79	86			
14	12	1	52	52			
15	13	1	25	13			
16	14	1	55	46			
17	15	1	44	40			
18	16	1	80	81			
19	17	1	29	12			
20	18	1	68	72			
21	19	1	75	80			
22	20	1	49	44			
23	21	1	33	20			

We could also show corresponding matrix operations needed to perform these operations in *Excel* in terms of *Excel* matrix operation commands are shown/summarized in the following Table, where, (B3,C23) is $\boldsymbol{X'}$ matrix - (D3:D23) is $\boldsymbol{y}$ matrix – $\boldsymbol{X'X}$ is (F3:G4) matrix – (F6:F7) is $\boldsymbol{X'y}$ matrix. You can check everything out from *Excel* matrix command table and *Excel* Sheet given in the following tables.

EXCEL COMMAND	EXPLANATION
=MMULT( )	*Excel* command for matrix multiplication
=TRANSPOSE( )	*Excel* command to take a transpose a matrix
=MINVERSE( )	*Excel* command for taking an inverse of the matrix

Matrix	*Excel* Matrix Def.
$\mathbf{X} =$	= B3:C23
$\boldsymbol{X}' =$	=TRANSPOSE(B3:C23)
$\mathbf{y} =$	=D3:D23
$(\boldsymbol{X}'\mathbf{X}) =$	=MMULT(TRANSPOSE(B3:C23),B3:C23) **= F3:G4**
$(\boldsymbol{X}'\mathbf{X})^{-1} =$	=MINVERSE(F3:G4)

Matrix Oper.	Result	*Excel* Format for Matrix Operations
$\boldsymbol{X}'\mathbf{X} =$	$\begin{vmatrix} 21 & 1345 \\ 1345 & 94953 \end{vmatrix}$	=MMULT(TRANSPOSE(B3:C23),B3:C23)
$\boldsymbol{X}'\mathbf{y} =$	$\begin{vmatrix} 1292 \\ 93559 \end{vmatrix}$	=MMULT(TRANSPOSE(B3:C23),D3:D23)
$\widehat{\boldsymbol{\beta}} = (\boldsymbol{X}'\mathbf{X})^{-1}\boldsymbol{X}'\mathbf{y} =$	$\begin{vmatrix} -17.0691 \\ 1.227101 \end{vmatrix}$	=MMULT((MINVERSE(F3:G4)),F6:F7)

There is one important point which needs to be kept in mind, when executing matrix operations, user needs to know and select the size of the outcoming matrix and then enter the *Excel* matrix operation commands presented in these tables, while out coming matrix cells are **highlighted**, then you need to press **CTRL+SHIFT** at the same time and hit **ENTER** while these two buttons are pressed! These important calculation steps are also summarized in the following Table.

1) First Select/highlight the corresponding cells matching the matrix size to be calculated after matrix operations!	To calculate the result of $\boldsymbol{X'}_{2\times21}\mathbf{X}_{21\times2}$ matrix multiplication, you need to highlight any **2 × 2** cells (i.e. 2 rows and 2 columns) in the *Excel* sheet, first!
2) Then, enter any Excel formulas as given before.	Simultaneously, Type in the corresponding formula to perform multiplication of two matrices (Note, the transpose of matrix ***X*** and itself is multiplied) in *Excel* command window as, =MMULT(TRANSPOSE(B3:C23),B3:C23)
*3) Press **CTRL+SHIFT** and hit **ENTER***	Then, Press **CTRL+SHIFT** and hit **ENTER** You will see 2 × 2 cells matrix for $\boldsymbol{X'}\mathbf{X}$ will be formed as, $\begin{vmatrix} 21 & 1345 \\ 1345 & 94953 \end{vmatrix}$

In this part, matrix notation based calculation of simple linear regression coefficients is demonstrated on an example problem in an *Excel* spreadsheet. The outlined matrix operations capabilities of *Excel* is not even heard by many students, whereas *Excel* provides powerful capabilities through which students can learn all aspects of SLR analysis by doing it for themselves so that each student can develop clear and concise understanding/control of this approach.

6.10 Exercises

1) Explain, when and why would you prefer the simple linear regression model? To what extent you would use such models, please elaborate on this.

2) We are performing a hardness test on titanium carbide cutting tools, which are coated with special coating material. In this test, our objective is to explore the hardness (Rockwell A) vs. coating depth (μm) change as presented in Table 6.5 below.

(In definition, the hardness is the ability of a material to resist Indentation through plastic deformation).

TABLE 6.5 Test data for hardness test

Depth (μm)	Hardness (Rockwell A)
0.000	93
0.004	86
0.006	92
0.008	88
0.010	84
0.012	85
0.014	80
0.016	69
0.018	66
0.020	60

- ✓ Dot plot the data
- ✓ Fit a simple regression line to the data using hand formulations and Tables.
- ✓ Interpret your results and tell us about what can be done next?

3) Repeat the same regression analysis using *Excel* and compare your results to the one, that is obtained in Exercise 2.

4) Repeat the same regression analysis using *Minitab and* compare your results to the ones, those are obtained in Exercise 2 and 3.

Chapter 7

Multiple Regression

The multiple regression (*MR*) analysis, which is basically an extension of simple linear regression analysis, but now we have more than one input/regressor variables. Likewise, MR models are more frequently used to explore the relationship/dependency between several independent variables i.e. inputs/regressor and a dependent variable i.e. output/response. For example, the geometric quality and tolerance conformity of the tile to be produced, highly depend on several input variables used in its composition and process parameters during production, such as the amount of limestone, fineness of additives and amount of Agalmatolite etc. We are interested in assessing the mathematical relationship/equation between the inputs vs. output data which are obtained from experiments/simulations. Once this equation is established, we would be able to see the effects of each input on the output, further, we could perform optimization for the *"best output"* by finding optimal values of inputs.

Let's start with looking at a multiple linear regression model as being the simplest form of *MR* models. In a multiple linear regression analysis, the dependent output/response *Y*, can be mapped/related to *k* independent variables $x_1, x_2, \ldots, x_k$, i.e. inputs/regressors by using the regression model of Equation 7.1 as follows, (where, we have p number of terms, totaling $p = k + 1$ terms)

$$Y = \beta_0 + \beta_1 x_1 + \beta_2 x_2 + \cdots + \beta_k x_k + \epsilon \qquad 7.1$$

But unfortunately, the most of the models encountered in real-life problems are far more complicated than this model as presented in Equation 7.1, i.e. this model is good for modeling the data with linear/well-behaved nature. Therefore, we need a higher degree of representation/complexity introduced in this equation, i.e. we need nonlinear models to accurately account for this relationship. This can be accomplished by introducing more terms on top of the multilinear regression model (shown in bold letters) with a combination of higher degree of freedom terms, i.e. interaction terms (i.e. x_1x_2) and quadratic terms (i.e. $x_1{}^2, x_2{}^2$) etc.

Let's consider a ***quadratic model*** with two inputs and *Y* output, as shown in Equation 7.2 as follows, (totaling *p=5+1* terms)

$$Y = \boldsymbol{\beta_0} + \boldsymbol{\beta_1 x_1} + \boldsymbol{\beta_2 x_2} + \beta_3 x_1 x_2 + \beta_4 x_1{}^2 + \beta_5 x_2{}^2 + \epsilon \qquad 7.2$$

Next, introduce a ***quadratic model*** with three inputs and *Y* output, as shown in Equation 7.3 as follows, (totaling *p=10* terms)

$$Y = \boldsymbol{\beta_0} + \boldsymbol{\beta_1 x_1} + \boldsymbol{\beta_2 x_2} + \boldsymbol{\beta_3 x_3} + \beta_4 x_1 x_2 + \beta_5 x_1 x_3 + \beta_6 x_2 x_3 + \beta_7 x_1{}^2 + \beta_8 x_2{}^2 + \beta_9 x_3{}^2 + \epsilon \qquad 7.3$$

Further, introduce a ***cubic model*** with two inputs and *Y* output, as shown in Equation 7.4 as follows, (totaling *p=10* terms)

$$Y = \boldsymbol{\beta_0} + \boldsymbol{\beta_1 x_1} + \boldsymbol{\beta_2 x_2} + \beta_3 x_1 x_2 + \beta_4 x_1{}^2 + \beta_5 x_2{}^2 + \beta_6 x_1{}^2 x_2 + \beta_7 x_1 x_2{}^2 + \beta_8 x_1{}^3 + \beta_9 x_2{}^3 + \epsilon \qquad 7.4$$

In this context, we follow similar steps that we presented in early chapters, i.e. the method of *least squares* again can be used to calculate regression coefficients, now we are going to end up with more coefficients of β. We could adopt the same methodology as in Chapter 6.1. The only difference that we are going to have is that we are going to need more equations/terms, but keep in mind that the overall strategy is the same.

Example 7.1

Let's focus on a new example problem, where the behavior of chemical composition of the certain base material is investigated for characteristics changes on mechanical properties in *electrical conductivity* by adding *two additives*, *A* and *B* in varying weights to the 500 grams of the molten base material as shown Figure 7.1. Please also refer to Table 7.1 for different composition levels, each designed with certain amounts. For example, inputs design variables for *additive A* and *additive B*, which have four-levels (i.e. 2,4,6, and 8 grams) and 3 levels (3,6, and 9 grams), respectively. These levels are used to set up the overall DOE design input matrix/table by using Minitab for obtaining multi-regression analysis model. In the next page, all required Minitab steps to set up this matrix/table is introduced in detail.

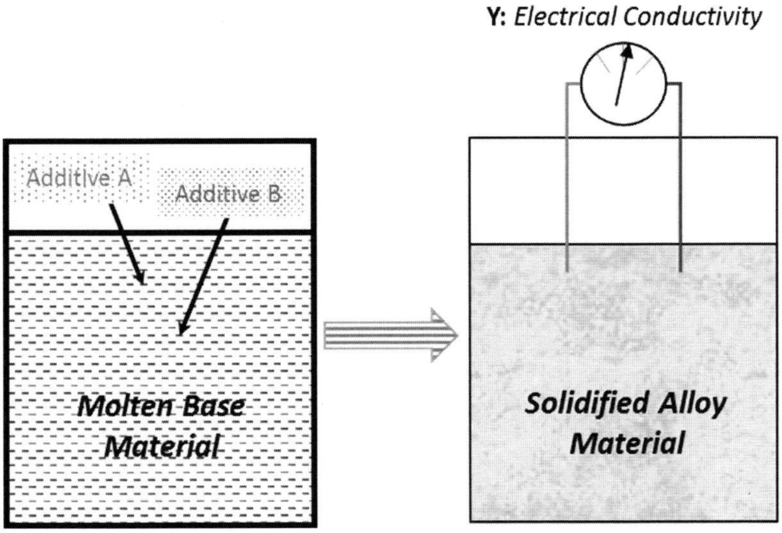

FIGURE 7.1 3D Scatter plot of Example 7.1

TABLE 7.1 DOE Levels for additive A and B in Example 7.1

4 LEVELS	Additive A (grams)	3 LEVELS	Additive B (grams)
Low	2	Low	3
Mid-Low	4	*Mid*	6
Mid-High	6	High	9
High	8		

7.1 DOE Matrix/Table Generation with Minitab

Since we have decided on additive A and B as inputs, the lower and upper bounds and corresponding design levels can be used to set up the DOE matrix. We can easily utilize *Minitab* for this task. Now, go ahead and start up the **Minitab** software from the Windows Start menu by clicking as in the following order, and follow the GUI Flow of instructions as presented next!

> **All Programs>Minitab>Minitab 1X Statistical Software**

4. Using the menu, scroll right to **Stat**, then, click on **Stat** > **DOE** > **Factorial** > **Create Factorial Design…** tabs in the order as shown below,

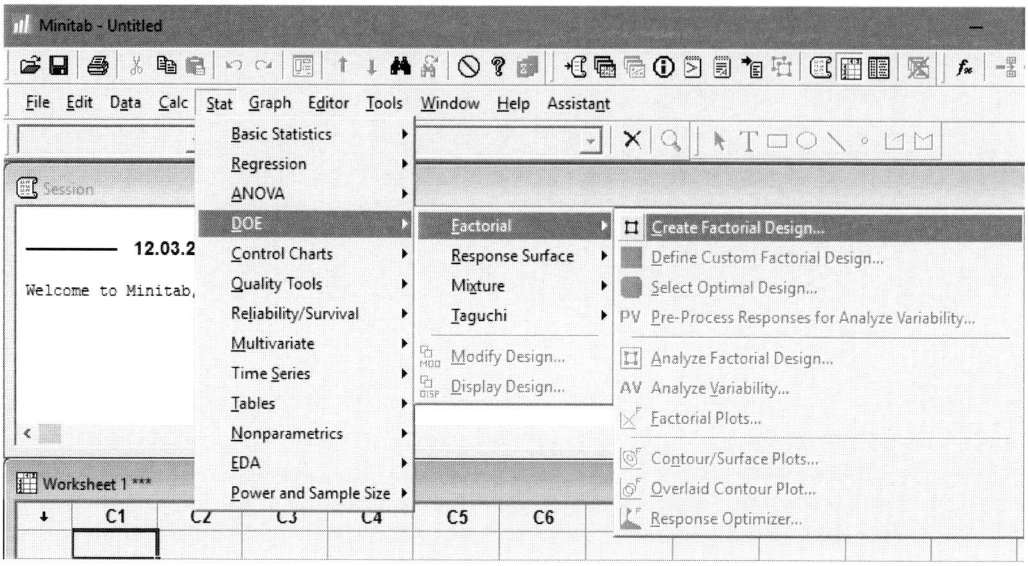

5. **Create Factorial Design** main window will pop-up, where
   5.1. Specify/Click on *Type of Design* menu, **General Full Factorial Design** and be sure **Number of factors:** are set to 2 input variables and then click on **Designs…** tab, **Create Factorial Design** tab will pop up, where you will see input variables A and B, Click on **Number of Levels** window and set the counter to 4 levels for variable A and similarly, click on **Number of Levels** for variable B and set the counter to 3 levels. Make sure **Number of replicates:** is set to 1. Click OK to close **Create Factorial Design – Designs** window.

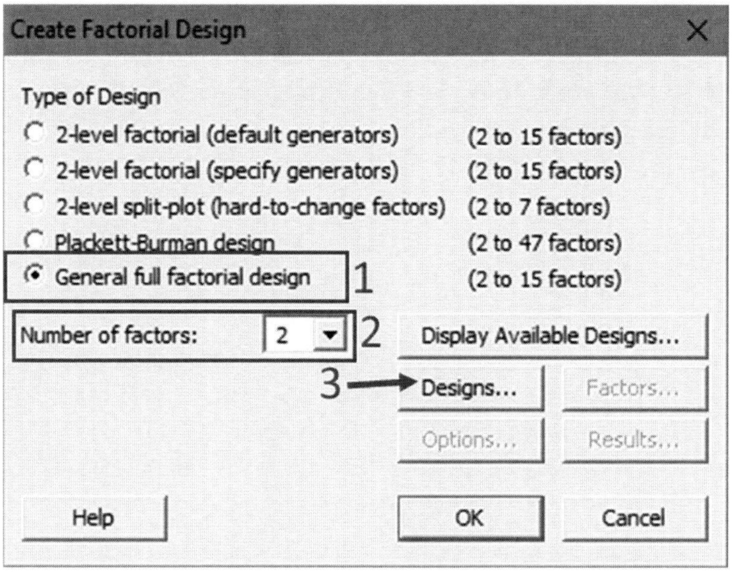
Create Factorial Design
Type of Design
2-level factorial (default generators)
(2 to 15 factors)
2-level factorial (specify generators)
(2 to 15 factors)
2-level split-plot (hard-to-change factors)
(2 to 7 factors)
Plackett-Burman design
(2 to 47 factors)
General full factorial design
1
(2 to 15 factors)
Number of factors:
2
2
Display Available Designs...
3
Designs...
Factors...
Options...
Results...
Help
OK
Cancel

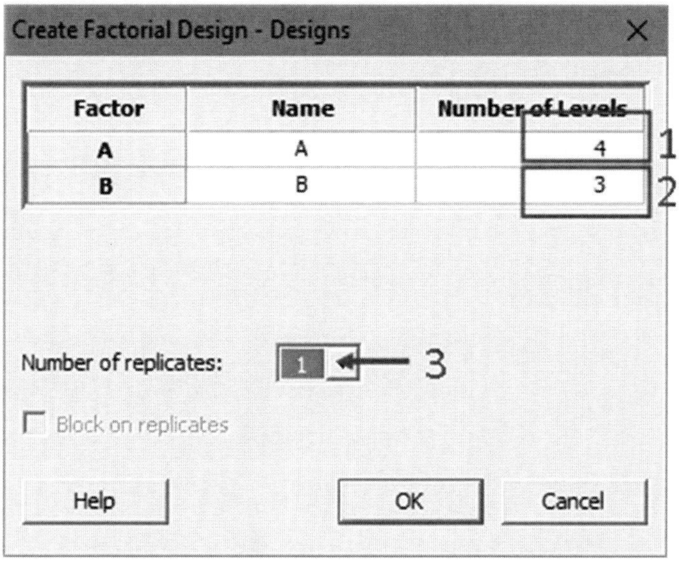
Create Factorial Design - Designs
Factor
Name
Number of Levels
A
A
4
1
B
B
3
2
Number of replicates:
1
3
Block on replicates
Help
OK
Cancel

5.2. **Create Factorial Design** window still remains open, where go ahead and click on, **Factors…** tab, C**reate Factorial Design - Factors** window will pop up, where you are going to *enter* 4 levels of A values and 3 levels of B values, as already given in Table 6.1 to the following **Level Values** cells, manually. Click OK to close **Create Factorial Design – Factors** window.

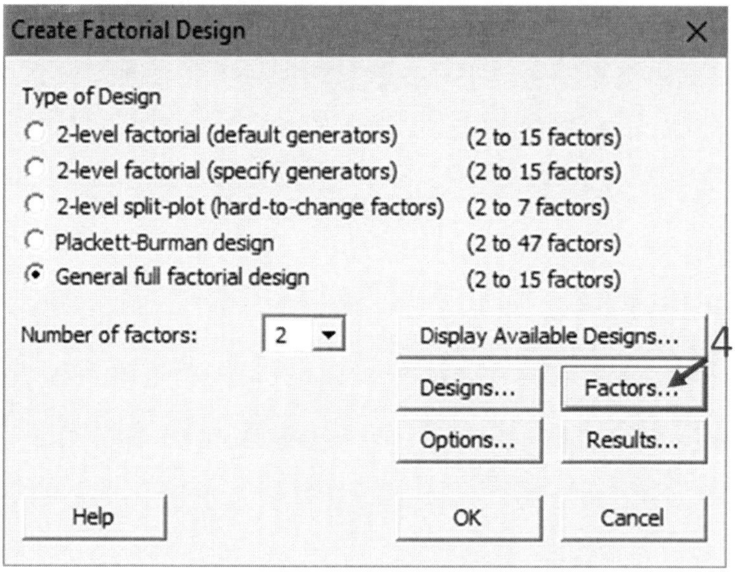

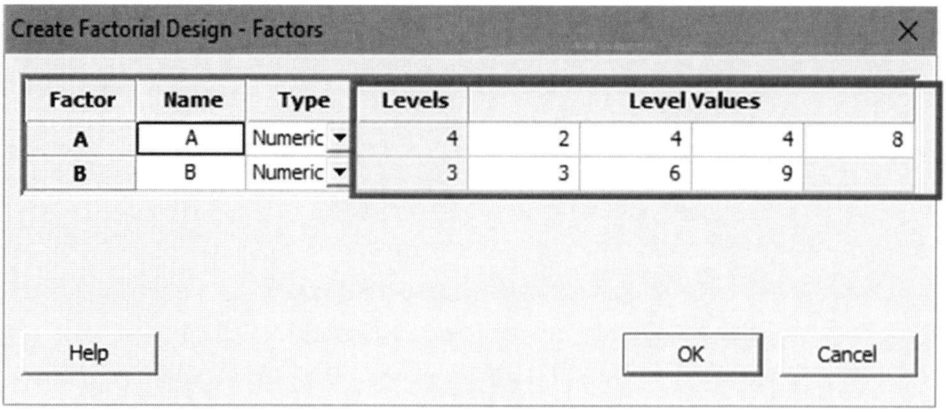

5.3. Under **Create Factorial Design** window again, click on **Options…** tab, **Create Factorial Design – Options** window will pop-up, at this point we are to input i.e. SEED=123 value for random number generation. Click OK to close **Create Factorial Design – Options** window.

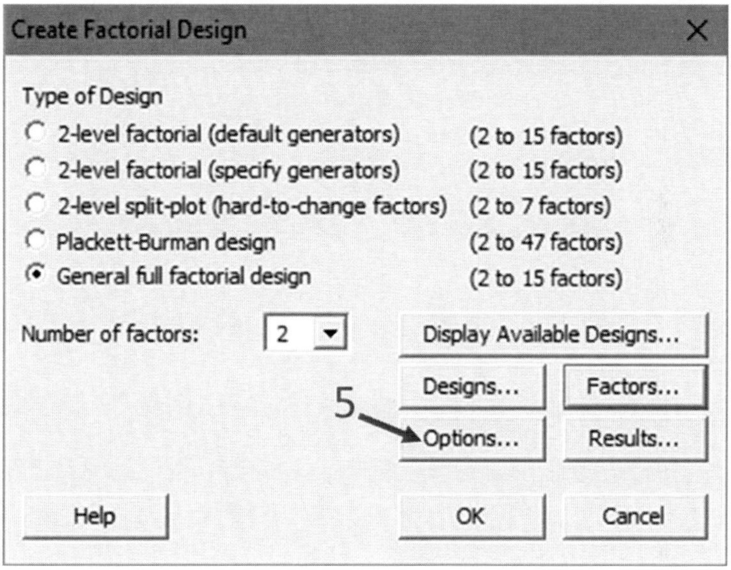

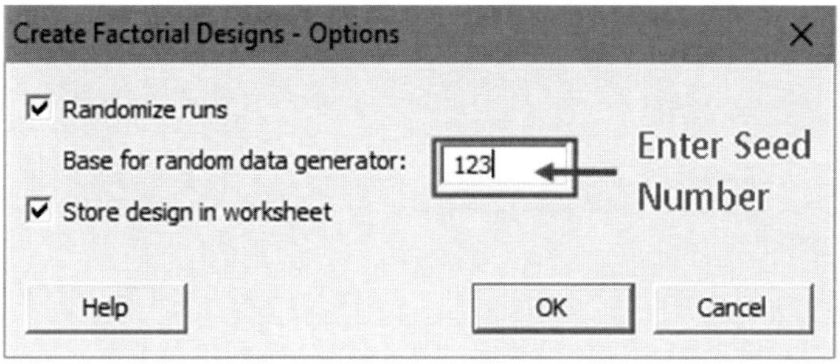

5.4. Again, under **Create Factorial Design** window, click on **Results…** tab, **Create Factorial Design – Results** window will pop-up, at this point use DEFAULT options as shown.

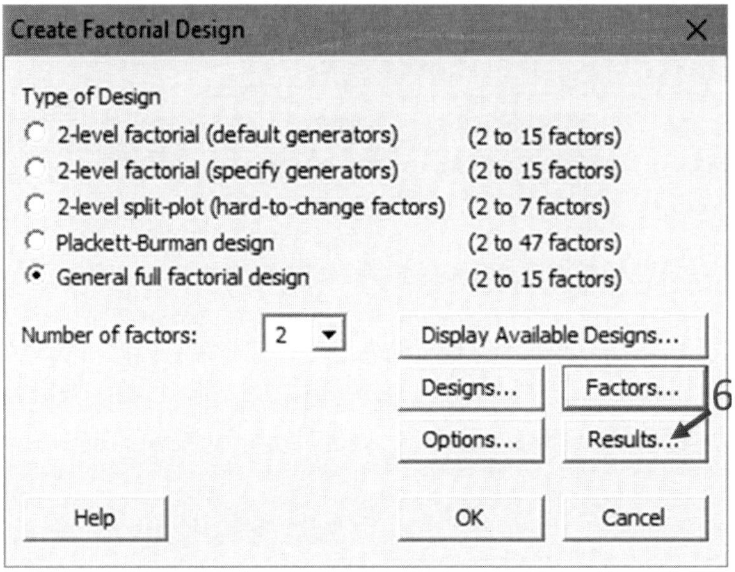

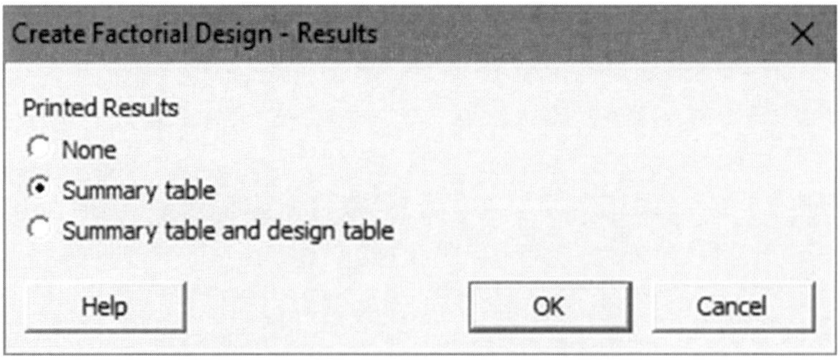

6. Finally, we will end up with **Multilevel Factorial Design**, with 2 factors/inputs, 12 base, and total runs, plus base and total blocks all as 1, with single replicates. As stated earlier, we are more interested in **C5** and **C6** columns, our primary objective is to come up with DOE input matrix/table, the other information is also vital but could easily be utilized when using more than one replicates. It also helps determine random experiment orders etc. so that our experiment would be unbiased and more dependable in terms of minimized experimental errors.

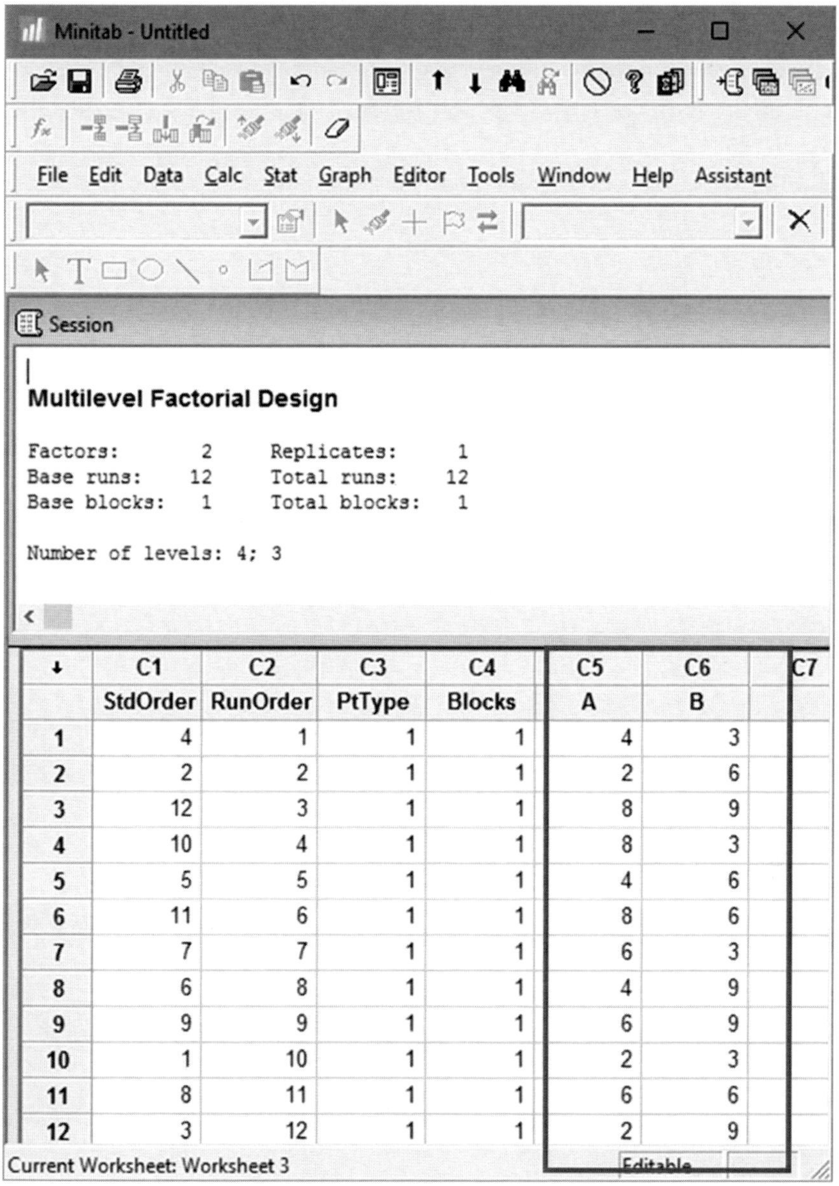

Multilevel Factorial Design

```
Factors:        2    Replicates:     1
Base runs:     12    Total runs:    12
Base blocks:    1    Total blocks:   1

Number of levels: 4; 3
```

↓	C1 StdOrder	C2 RunOrder	C3 PtType	C4 Blocks	C5 A	C6 B	C7
1	4	1	1	1	4	3	
2	2	2	1	1	2	6	
3	12	3	1	1	8	9	
4	10	4	1	1	8	3	
5	5	5	1	1	4	6	
6	11	6	1	1	8	6	
7	7	7	1	1	6	3	
8	6	8	1	1	4	9	
9	9	9	1	1	6	9	
10	1	10	1	1	2	3	
11	8	11	1	1	6	6	
12	3	12	1	1	2	9	

After carrying on all experiments by using the DOE matrix/table formed above, we end up with Y's, i.e. the observed out coming material's electrical conductivity increase or decrease in % and all pertinent data are combined and presented in Table 7.2 to use in multiple regression analysis. In this table, x_1 is the additive-A amount and x_2 is the additive-B amount in *grams* and Y is the output/response in terms of % improvement in the electrical conductivity, i.e. 1, 1.11 and 0.55 refer to %0 no improvement, %11 increase and %45 decrease from base material properties, respectively.

Also, notice that beside inputs columns A and B, we also include an interaction term(s) A*B and quadratic terms of both A*A and B*B. A total number of 12 runs are designed using *Minitab* Full Factorial design with 4 levels of Additive-A and 3 levels of Additive-B, resulting in combinations of 12 runs with a *single* repetition for each run as explained in the previous section.

TABLE 7.2 Example 7.1 DOE inputs/output table

Run Order	A	B	A*B	A*A	B*B	Y
1	4	3	12	16	9	**0,91**
2	2	6	12	4	36	**1,11**
3	8	9	72	64	81	**0,37**
4	8	3	24	64	9	**0,67**
5	4	6	24	16	36	**0,85**
6	8	6	48	64	36	**0,51**
7	6	3	18	36	9	**0,8**
8	4	9	36	16	81	**0,81**
9	6	9	54	36	81	**0,55**
10	2	3	6	4	9	**1,05**
11	6	6	36	36	36	**0,65**
12	2	9	18	4	81	**1,12**

Please note that it is statistically good and sound advice to repeat each experiment more than once (if time permits and feasible) i.e. three times and take the average value for *Y* to use in Table 7.2. Let's assume you would like to run each experiment three times then, it is also advised that 3*12=36 runs are performed in random order for the validity/health of the outcoming results. This design matrix/table shown in Table 7.2 can easily be obtained from *Minitab*.

Since, our primary objective, in this part, is to demonstrate how to perform multiple regression analysis, we are only focusing on these aspects, therefore, we would suffice with this amount of practical information.

Here, we would like to investigate the underlying mathematical formulation thus relationship between the Additives A and B as inputs and *Y* Output by using in multilinear regression analysis. Further, we are also interested in individual effect/importance of each additive on *Y* Output.

7.2 Drawing 3D Scatter Diagram in Minitab

The first step to start out in any statistical analysis is to plot the associated diagram/graph of the data given in Example 7.1. Such a 3D *scatter diagram* of this data is presented in Figure 7.2. Note that, plotting of the regression data is only meaningful in terms of visualization when we are dealing with a maximum of 2 inputs of data. Let's say, in the case of single input and single Output case we have 2D scatter line plot. In the case of two inputs and a single output case, we have 3D scatter plot. Once we go beyond 2 inputs, we cannot visualize the data, since we have a hyperplane. This is true for most of the cases in any real-life data analysis applications.

Now, let's go ahead and use Minitab to draw our 3D scatter diagram as explained in the following steps. Start up the **Minitab** software from the Windows Start menu by clicking as in the following order, and follow the GUI Flow of instructions as presented below.

> **All Programs>Minitab>Minitab 1X Statistical Software**

1. A new worksheet opens up where cut and Paste, copy or manually enter 12 rows of A,B inputs and Y output as given in Table 6.2. As also shown next!

Worksheet 1 ***

↓	C1	C2	C3
	A	B	Y
1	4	3	0,91
2	2	6	1,11
3	8	9	0,37
4	8	3	0,67
5	4	6	0,85
6	8	6	0,51
7	6	3	0,80
8	4	9	0,81
9	6	9	0,55
10	2	3	1,05
11	6	6	0,65
12	2	9	1,12

Current Worksheet: Worksheet 1

2. Using the menu, scroll right to the **Graph**, then, click on **Graph >3D Scatterplot…** tab in the order as shown below. 3D Scatterplots window will open, where choose *Simple* icon. Click OK to close **3D Scatterplots** window as shown below.

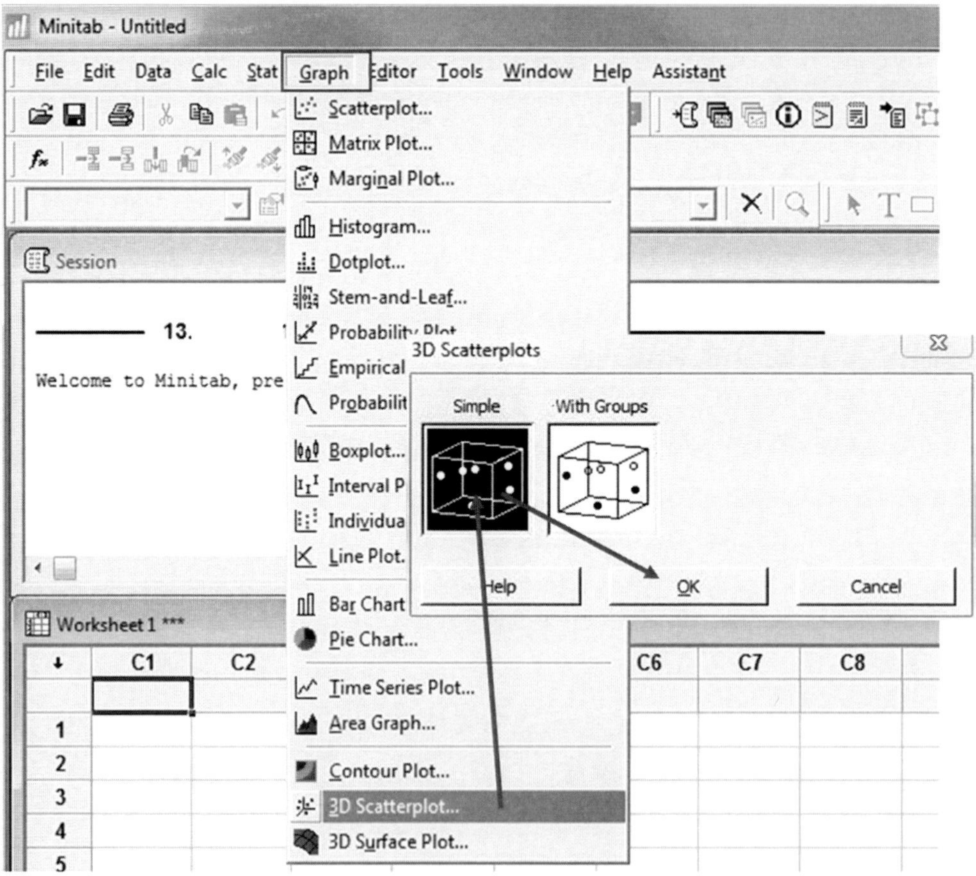

3. **3D Scatterplot - Simple** window will open, where enter/select Y as **Z** variable, B as **Y** variable and A as **X** variable , as shown.

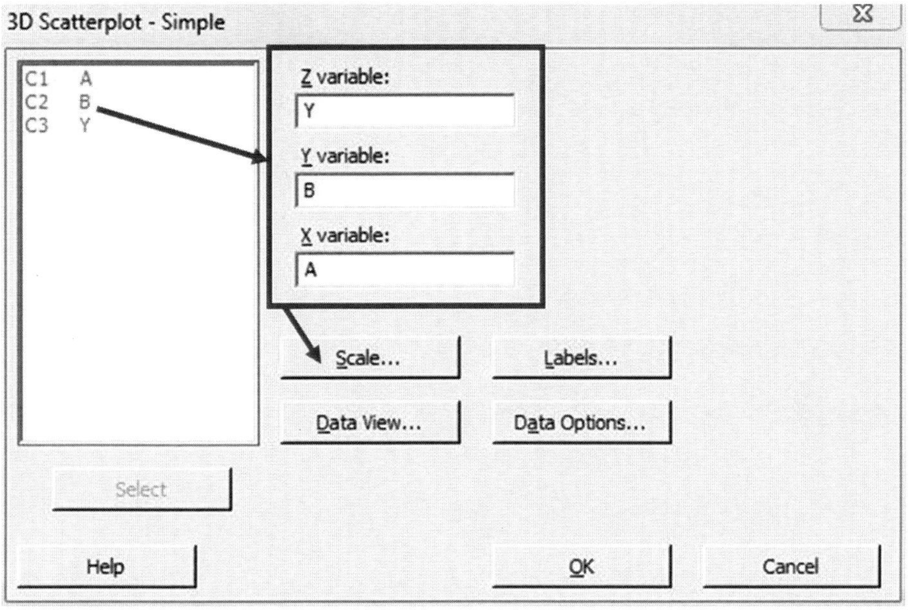

3.1. Click on **Scale...** tab, **3D Scatterplot - Scale** window will open, click on additional options in Minor ticks as shown. Click OK to close this window.

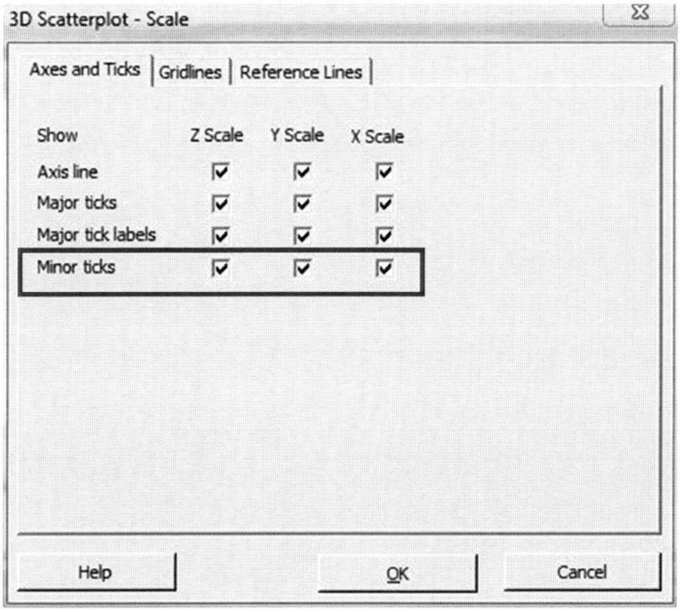

3.2. You may examine the contents of **Labels…** and **Data Options…** tabs, which are left unchanged, as DEFAULT options are used, in this part, therefore no illustration is shown.

3.3. Click on **Data View…** tab, **3D Scatterplot – Data View** window will open, click on additional Project lines option. Click OK to close this window.

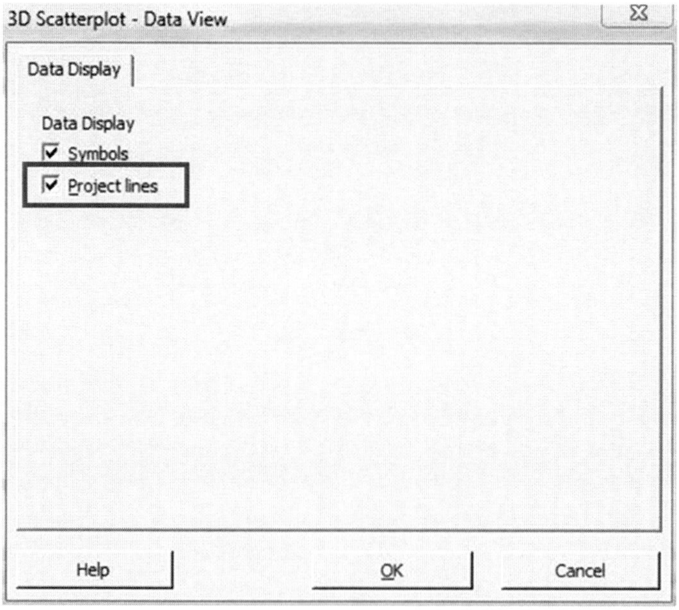

3.4. Now, you may go ahead close main **3D Scatterplot - Simple** window by hitting Ok! Right after that **3D Scatterplot of Y vs B vs A** window with 3D Scatter plot will pop up. Click on this window **3D Graph Tools** touch small window will appear in enabled/highlighted mode using which you may *modify* your plot on the right orientation that you liked.

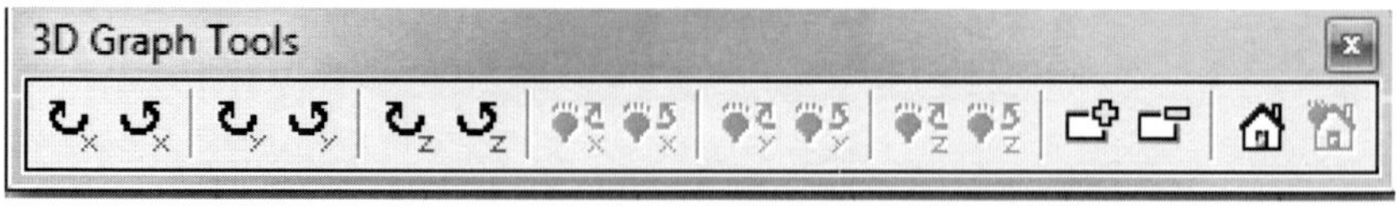

Further, if you double click on **3D Scatterplot of Y vs B vs A** window and click on titles, labels etc. you may change, modify the font sizes and type or edit this information the way you wanted.

In the following we did exactly that, you may notice slight title change and bigger font sizes, as shown in Figure 7.2.

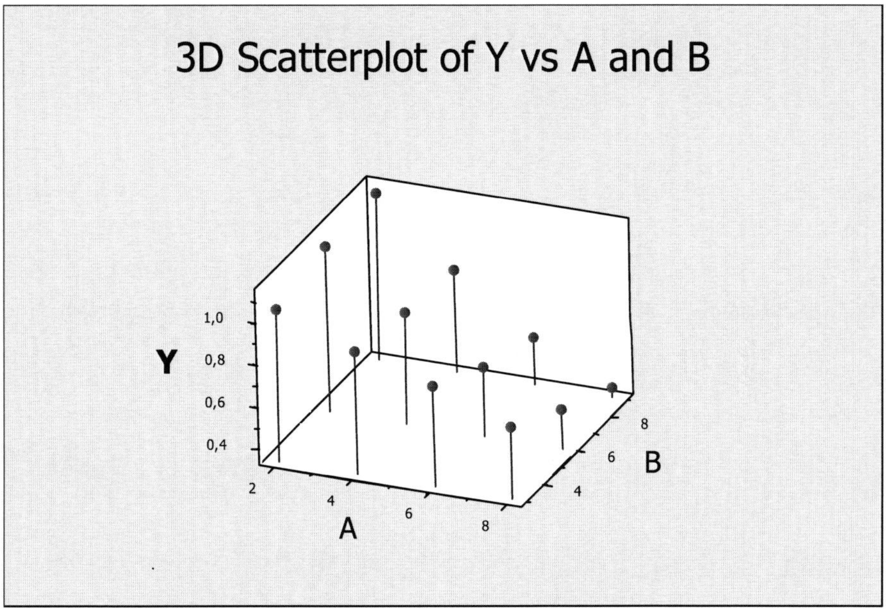

FIGURE 7.2 3D Scatter plot of Example 7.1

In brief, in this graph, each *n* number of (x_{1i}, x_{2i}, Y_i) data are plotted as a point in a 3D Cartesian *X-Y-Z* coordinate system. Also, note that this scatter plot is drawn in *Minitab* as explained and shown in detail.

After careful inspection of this scatter plot, the uniform behavior of the data is observed, from the accuracy standpoint, it is could be easily seen that no simple plane will pass through all the data points, in essence, this is really not we want!

We are more interested in trend rather than having a magical virtual plane passing through all the 12 points of (x_{1i}, x_{2i}, y_i) coordinates of data, but keep in mind that, as far as any further optimization efforts needed, we are to spend more effort on obtaining better/accurate representation of the data by obtaining high R^2 and $R^2_{adj.}$ values.

7.3 Multiple Linear Regression

As a first step, the simplest form of the linear model, characterizing the overall behavior of the scatter plot, is a simple plane (surface), we can go ahead and fit a plane to all $n=12$ pairs of inputs and output data, as given in the following Equation 7.5,

$$Y = \boldsymbol{\beta}_0 + \boldsymbol{\beta}_1 \boldsymbol{x}_1 + \boldsymbol{\beta}_2 \boldsymbol{x}_2 + \epsilon \qquad 7.5$$

where $Y_i(x)$ is the output/response value, and $\beta_0, \beta_1, \beta_2$ are *regression* coefficients, i.e. β_0 is the intercept and, β_1, β_2 the slope of the plane at specific axes. Also, note that ϵ_i is a random error scattered around the regression plane and it is assumed that these errors are mutually independent (so, $Y_1, Y_2, ..., Y_n$ values are also mutually independent) and normally distributed with *mean* zero and *variance* σ^2.

In the multiple linear regression, independent variables $x_1, x_2,$ and Y is the dependent output/response variable *of x's*, are used to form a multiple linear regression equation given in Equation 6.6. Since the relationship is linear, our objective is to find estimates of $\hat{\beta}_0, \hat{\beta}_1$ and $\hat{\beta}_2$ of coefficients corresponding to β_0, β_1 and β_2 such that the *best*-fitted plane or *"close to"* the n data points of (x_{1i}, x_{2i}, Y_i). The fitted regression model is,

$$\hat{\boldsymbol{y}}_i = \widehat{\boldsymbol{\beta}}_0 + \widehat{\boldsymbol{\beta}}_1 \boldsymbol{x}_{1i} + \widehat{\boldsymbol{\beta}}_2 \boldsymbol{x}_{2i} \rightarrow \boldsymbol{i} = \mathbf{1}, \mathbf{2}, \dots, \boldsymbol{n} \qquad 7.6$$

A widely known *method of least squares* can be used to calculate estimated coefficients of $\hat{\beta}_0, \hat{\beta}_1$ and $\hat{\beta}_2$ as presented in previous Chapter 6.

Since our primary objective in this chapter is more to go through practical ways of handling multi-regression analysis problems, the pertinent formulation to obtain multiple regression coefficients are left out of the scope of this chapter. Rather, we are going to demonstrate how to construct multi-regression models with different levels of linearity/nonlinearity as hands-on experience in nature.

7.4 Multiple Linear Regression with Minitab

Now, let's perform the multiple linear regression analysis, as the simplest form of multi-regression analysis, with the data given in Example 7.1 using *Minitab*. Let's remember, so far, we have formed our both DOE matrix/table and 3D data Scatterplot using *Minitab*. Please also note that, for simplification purposes, we are only going to have the data given in Table 7.2 in our new *Minitab* sheet, i.e. as far as multiple linear regression analysis is concerned, we will be only using inputs A,B, and output Y, the other interaction and high order terms will be used in the following general case of multiple regression models.

Let's start up the **Minitab** software from Windows start menu by clicking as in the following order,

> **All Programs>Minitab>Minitab 1X Statistical Software**

1. Now, go ahead and copy/input your inputs and output data of *n=12* pairs in the *Worksheet 1* window corresponding columns of C1,C2,C3,C4,C5 and C6 as shown, please note that although we have copied all columns for use in a general multiple regression analysis, in this section, we are going to need only Columns C1, C2 and C6 for Multiple Linear Regression analysis, as shown below.

Worksheet 1 ***

↓	C1	C2	C3	C4	C5	C6
	A	B	A*B	A*A	B*B	Y
1	4	3	12	16	9	0,91
2	2	6	12	4	36	1,11
3	8	9	72	64	81	0,37
4	8	3	24	64	9	0,67
5	4	6	24	16	36	0,85
6	8	6	48	64	36	0,51
7	6	3	18	36	9	0,80
8	4	9	36	16	81	0,81
9	6	9	54	36	81	0,55
10	2	3	6	4	9	1,05
11	6	6	36	36	36	0,65
12	2	9	18	4	81	1,12

2. Then, click on **Stat** > **Regression** > **Regression…** tabs in the order as shown below,

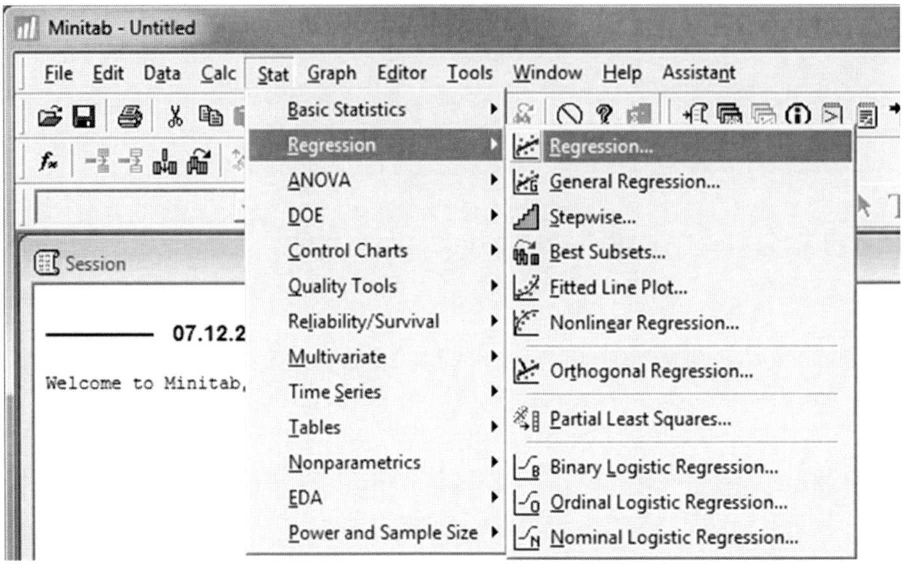

3. **Regression** main window will pop-up, where

3.1. Specify C1 and C2 *i.e.* ***Additive A and B*** as Predictors/Inputs and Specify C6 ***Y*** as Response/Output, i.e. first highlight A and B next click on **Predictors:** window, then click on the select button, repeat the same thing for Y, select it to **Response:** window as shown below.

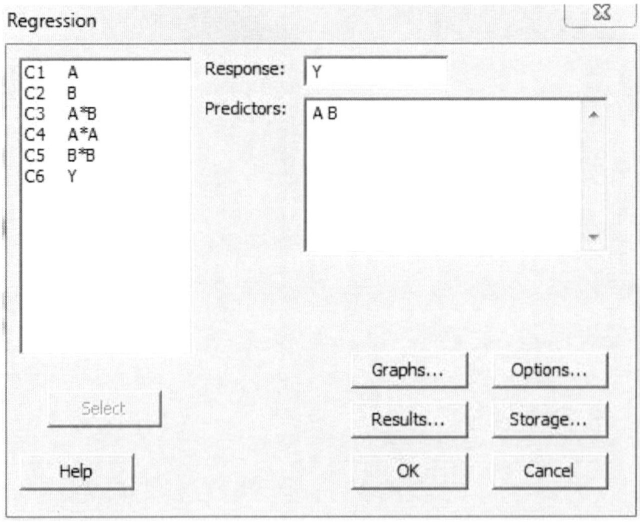

3.2. Under **Regression** main window, click on **Graphs…** tab and click on **Four in one** option as shown below, click OK to close the **Regression-Graphs** window.

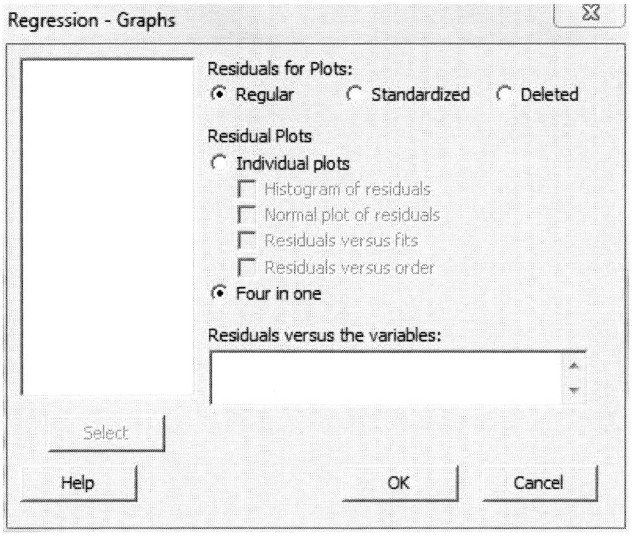

3.3. Still, under **Regression** main window, click on **Options…** tab some useful statistics option will appear in **Regression-Options** window, at this point we are ok with DEFAULT,

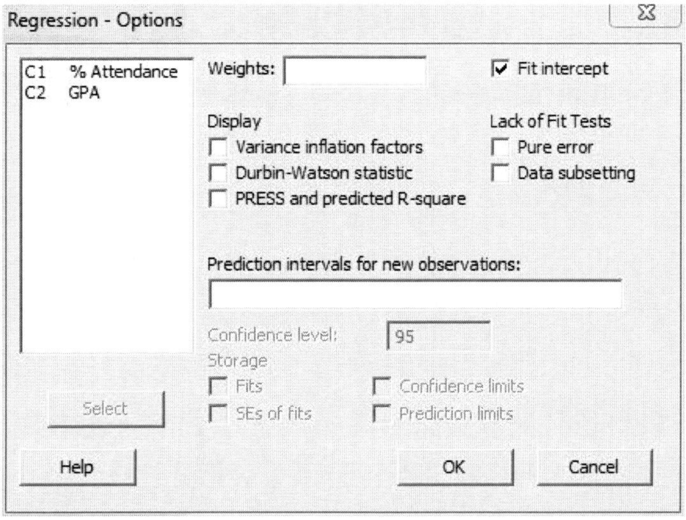

- ✓ Fit intercept and
- ✓ 95% Confidence level settings

are selected in DEFAULT, as shown above, click OK to close **Regression-Options** window.

3.4. Under **Regression** main window, click on **Results…** tab you will get **Regression-Results** window, Choose the settings (2nd option) as shown below, click OK to close **Regression-Results** window.

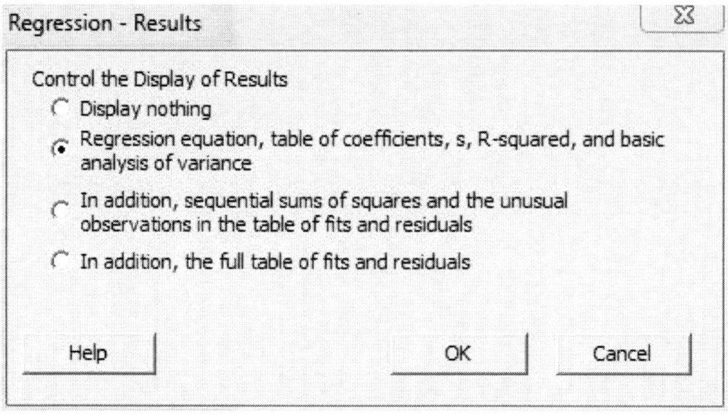

3.5. Under **Regression** main window, click on **Storage…** tab, you will get **Regression-Storage** window where click on,

- ✓ Residuals
- ✓ Coefficients
- ✓ Fits
- ✓ MSE

options as shown below, click OK to close **Regression-Storage** window.

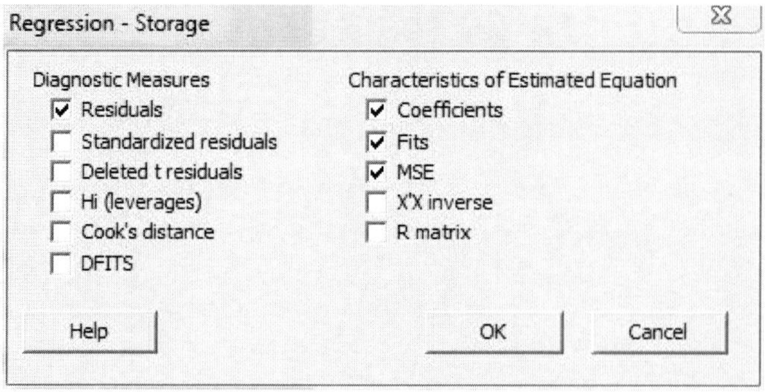

4. Now, you are done in **Regression** main window, click on OK to **RUN** the regression. You will get the following Minitab output where all summary of statistic is documented. In the following section, explanations for this information are provided in detail.

Regression Analysis: Y versus A; B

```
The regression equation is
Y = 1,41 - 0,0960 A - 0,0242 B


Predictor        Coef    SE Coef       T      P
Constant      1,40833    0,07322   19,23  0,000
A           -0,096000   0,009453  -10,16  0,000
B           -0,024167   0,008630   -2,80  0,021


S = 0,0732247   R-Sq = 92,5%   R-Sq(adj) = 90,8%


Analysis of Variance

Source          DF        SS        MS      F      P
Regression       2   0,59501   0,29750  55,49  0,000
Residual Error   9   0,04826   0,00536
Total           11   0,64327
```

and

- RESI2: residuals for each data pair
- COEF2: regression coefficients
- FITS2: estimated fits used regression equation.

↓	C1	C2	C3	C4	C5	C6	C7	C8	C9
	A	B	A*B	A*A	B*B	Y	RESI1	COEF1	FITS1
1	4	3	12	16	9	0,91	-0,041833	1,40833	0,95183
2	2	6	12	4	36	1,11	0,038667	-0,09600	1,07133
3	8	9	72	64	81	0,37	-0,052833	-0,02417	0,42283
4	8	3	24	64	9	0,67	0,102167		0,56783
5	4	6	24	16	36	0,85	-0,029333		0,87933
6	8	6	48	64	36	0,51	0,014667		0,49533
7	6	3	18	36	9	0,80	0,040167		0,75983
8	4	9	36	16	81	0,81	0,003167		0,80683
9	6	9	54	36	81	0,55	-0,064833		0,61483
10	2	3	6	4	9	1,05	-0,093833		1,14383
11	6	6	36	36	36	0,65	-0,037333		0,68733
12	2	9	18	4	81	1,12	0,121167		0,99883

4.1. Finally, you will get **Four-in-one** plot where,

- **Normal Probability Plot** indicates good fit because if we put a thick pen on top of the line, almost all dots will be disappearing
- **Versus Fits** plot also is good, showing fitted value vs. residuals look balanced at both sides
- **Histogram** plots show disperse data and the distribution frequency of the residuals around target zero value
- **Versus order** plot shows for each observation, the residuals disperse around target zero value.

If you examine the *R-Sq* and *R-Sq(adj)* measures, they are both above %90, which is considered to be quite good, but we are still interested in the higher degree of approximation accuracy and representation of the Example 7.1 data, as explained in the following section using **Multiple Regression Analysis.**

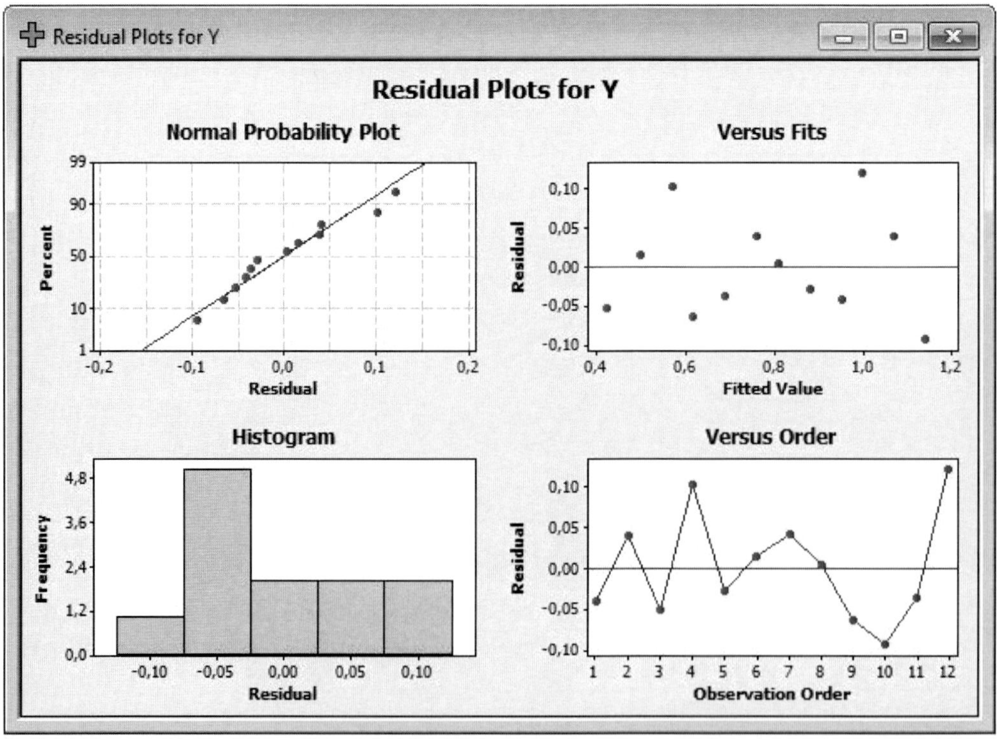

7.5 Multiple Regression with Minitab

At this point, we would like to see/observe the contribution of nonlinear terms in the overall *quadratic* multiple regression models, therefore we are going to repeat the same steps followed in Section 7.4, here as well. Continuing our study with step number 5, as follows,

5. Remember that we have all terms to use in the *quadratic* regression model since our *Minitab* worksheet is already open, we are going to keep using it. In this section, we are going to use all columns C1,C2,C3,C4,C5 and C6 for Multiple Linear Regression analysis as shown in the following.

Worksheet 1 ***

↓	C1	C2	C3	C4	C5	C6
	A	B	A*B	A*A	B*B	Y
1	4	3	12	16	9	0,91
2	2	6	12	4	36	1,11
3	8	9	72	64	81	0,37
4	8	3	24	64	9	0,67
5	4	6	24	16	36	0,85
6	8	6	48	64	36	0,51
7	6	3	18	36	9	0,80
8	4	9	36	16	81	0,81
9	6	9	54	36	81	0,55
10	2	3	6	4	9	1,05
11	6	6	36	36	36	0,65
12	2	9	18	4	81	1,12

6. Again, click on **Stat > Regression > Regression...** tabs in the order shown on the next page.

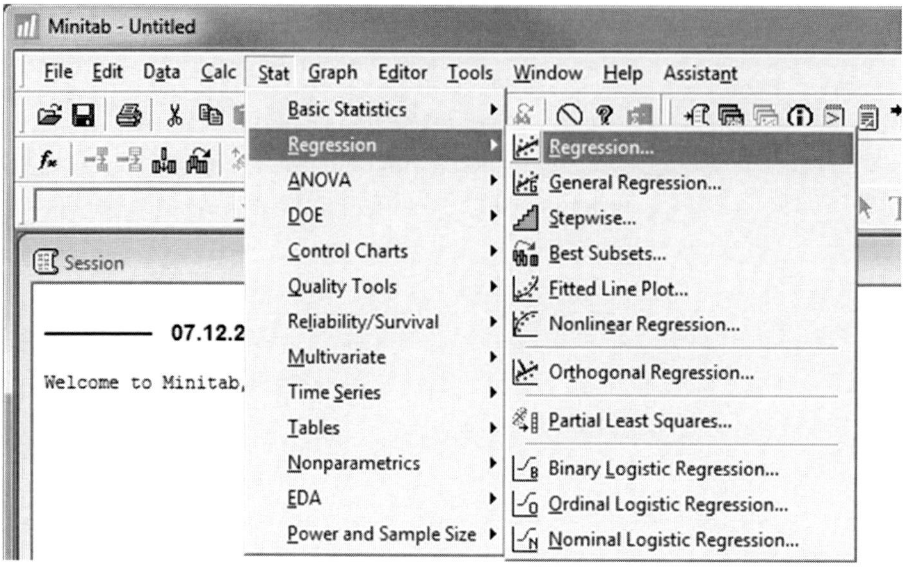

7. **Regression** main window will open, where

7.1. Delete the terms in **Predictors:** window first, then specify C1, C2,C3,C4 and C5 *i.e.* ***Additive A and B and their higher order (quadratic) terms*** as Predictors/Inputs and Specify C6 ***Y*** as Response/Output, i.e. first highlight A,B,A*B,A*A and B*B, next click on **Predictors:** window, then click on select button, Output Y, is already selected in **Response:** window, as shown below.

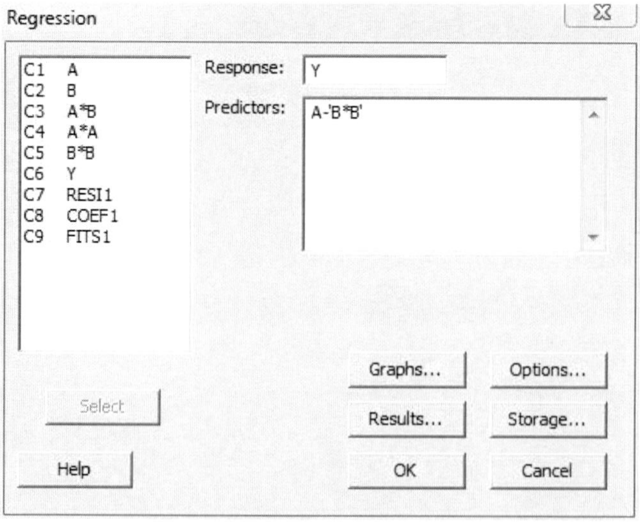

8. You can adopt the same settings that you used, which will not be shown here. Now, you are done in **Regression** main window, click on OK to **RUN** the regression. You will get the following *Minitab* output where all summary of statistics is documented. In the following section, explanations for this information are provided in detail.

Regression Analysis: Y versus A; B; A*B; A*A; B*B

```
The regression equation is
Y = 1,22 - 0,0872 A + 0,0217 B - 0,0105 A*B + 0,00542 A*A + 0,00056 B*B


Predictor       Coef    SE Coef      T      P
Constant     1,21833    0,06825  17,85  0,000
A           -0,08717    0,01745  -4,99  0,002
B            0,02167    0,01893   1,14  0,296
A*B        -0,010500   0,001152  -9,11  0,000
A*A         0,005417   0,001577   3,43  0,014
B*B         0,000556   0,001487   0,37  0,722


S = 0,0218581   R-Sq = 99,6%   R-Sq(adj) = 99,2%


Analysis of Variance

Source          DF       SS       MS       F      P
Regression       5  0,64040  0,12808  268,07  0,000
Residual Error   6  0,00287  0,00048
Total           11  0,64327
```

↓	C1	C2	C3	C4	C5	C6	C7	C8	C9
	A	B	A*B	A*A	B*B	Y	RESI1	COEF1	FITS1
1	4	3	12	16	9	0,91	-0,041833	1,40833	0,95183
2	2	6	12	4	36	1,11	0,038667	-0,09600	1,07133
3	8	9	72	64	81	0,37	-0,052833	-0,02417	0,42283
4	8	3	24	64	9	0,67	0,102167		0,56783
5	4	6	24	16	36	0,85	-0,029333		0,87933
6	8	6	48	64	36	0,51	0,014667		0,49533
7	6	3	18	36	9	0,80	0,040167		0,75983
8	4	9	36	16	81	0,81	0,003167		0,80683
9	6	9	54	36	81	0,55	-0,064833		0,61483
10	2	3	6	4	9	1,05	-0,093833		1,14383
11	6	6	36	36	36	0,65	-0,037333		0,68733
12	2	9	18	4	81	1,12	0,121167		0,99883

Please note that columns of C10,C11, and C12 in the following sheet indicate the latest fit statistics obtained in this section (*quadratic* multi-regression analysis), whereas, the above output table belongs the previous section's multiple linear regression analysis.

C10	C11	C12
RESI2	COEF2	FITS2
0,0096667	1,21833	0,90033
0,0203333	-0,08717	1,08967
0,0183333	0,02167	0,35167
-0,0156667	-0,01050	0,68567
-0,0043333	0,00542	0,85433
-0,0036667	0,00056	0,51367
0,0286667		0,77133
-0,0083333		0,81833
-0,0133333		0,56333
-0,0226667		1,07267
-0,0123333		0,66233
0,0033333		1,11667

8.1. Finally, you will get **Four-in-one** plot as presented in the next page where,

- **Normal Probability Plot** indicates good fit because if we put a thick pen on top of the line, almost all dots will be disappearing
- **Versus Fits** plot also is good, showing fitted value vs. residuals look balanced at both sides
- **Histogram** plots show disperse data and the distribution frequency of the residuals around target zero value
- **Versus order** plot shows for each observation, the residuals disperse around target zero value.

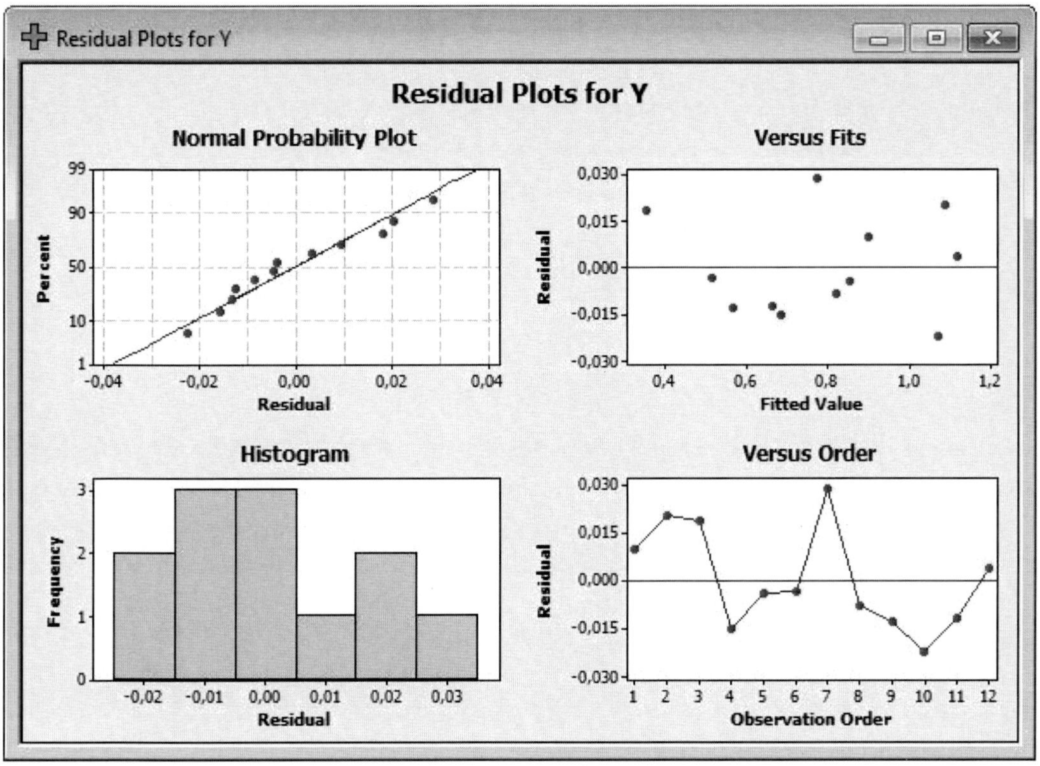

If you examine the *R-Sq* and *R-Sq(adj)* fit measures, they are both above %99, which is considered to be perfect, but we are still interested in simplification in the regression terms by removing insignificant terms from the regression equation. In the following section, a common practice of regression statistics is introduced.

7.5.1 Eliminating Insignificant Terms

Notice that, we tend to have the most basic/simplified model/equation with valid representation with the terms that we have included in our regression model. Then, the question is, how can one do that? According to what metrics/criterions we should decide upon that? The fitted regression model consists of the certain number of β coefficients, that are in fact random variables, with mean and variance(standard deviation). Therefore, each β coefficients, need to be tested to see if it is statistically significant.

7.5.2 P-value of F-test

As it has been introduced in previous chapters, in such situations, we employ hypothesis test for this purpose, as presented in Equation 7.7, where remember in regression Equation 7.1, a total of *k+1* terms exists. In brief, the *p*-value for the *F* significance test tells us one of the following:

$$H_0 : \beta_{k+1} = 0$$
$$H_1 : \beta_{k+1} \neq 0 \qquad 7.7$$

When using *p-statistics*, remember that our DEFAULT confidence level is generally taken as %95 and ($\alpha = 0.05$) and if $p < 0.05$ then we reject the Null Hypothesis, and assert that the β coefficient terms are non-zero, meaning potentially useful and significant for predicting the response in our regression model. Otherwise, the Null Hypothesis is true, and we go ahead and remove that coefficient term from our regression model and conclude that this coefficient is not useful/insignificant.

7.5 Multiple Regression Analysis Continued

Under the light of these very important explanations/highlights as presented in the previous section, now let's go ahead and remove the insignificant term from our previous results. How? By starting with the highest *p-value*!

After the careful examination of this output, we notice that terms for *B* and *B*B* have *p-values* more than 0.05, i.e. 0.296 and 0.722, respectively. But keep in mind that we remove only one term at a time! i.e. B*B term with the highest *p-value*! and re-run the regression analysis and obtain new regression output/results and evaluate them again. Finally, we repeat this procedure until all regression terms are with a *p-value* (preferably much) lower than *p=0.05*.

Regression Analysis: Y versus A; B; A*B; A*A; B*B

```
The regression equation is
Y = 1,22 - 0,0872 A + 0,0217 B - 0,0105 A*B + 0,00542 A*A + 0,00056 B*B


Predictor        Coef   SE Coef      T      P
Constant      1,21833   0,06825  17,85  0,000
A            -0,08717   0,01745  -4,99  0,002
B             0,02167   0,01893   1,14  0,296
A*B         -0,010500  0,001152  -9,11  0,000
A*A          0,005417  0,001577   3,43  0,014
B*B          0,000556  0,001487   0,37  0,722
```

Remove this term first! (B*B)

```
S = 0,0218581   R-Sq = 99,6%   R-Sq(adj) = 99,2%


Analysis of Variance

Source          DF       SS       MS       F      P
Regression       5  0,64040  0,12808  268,07  0,000
Residual Error   6  0,00287  0,00048
Total           11  0,64327
```

9. Remember that we have all our terms used in our quadratic model except B*B term that we identified it as insignificant. In this section, we are only going to need all Columns C1, C2,C3,C4 and C6 for Multiple Linear Regression analysis as shown below.

Worksheet 1 ***

↓	C1	C2	C3	C4	C5	C6
	A	B	A*B	A*A	B*B	Y
1	4	3	12	16	9	0,91
2	2	6	12	4	36	1,11
3	8	9	72	64	81	0,37
4	8	3	24	64	9	0,67
5	4	6	24	16	36	0,85
6	8	6	48	64	36	0,51
7	6	3	18	36	9	0,80
8	4	9	36	16	81	0,81
9	6	9	54	36	81	0,55
10	2	3	6	4	9	1,05
11	6	6	36	36	36	0,65
12	2	9	18	4	81	1,12

10. **Let's go back to Regression** main window, where delete all terms in **Predictors:** window first, then specify C1, C2,C3 and C4 as Predictors/Inputs and Specify C6 ***Y*** as Response/Output, i.e. first highlight A,B,A*B and A*A, next click on **Predictors:** window, then click on select button, Output Y, is already selected in **Response:** window, as shown below.

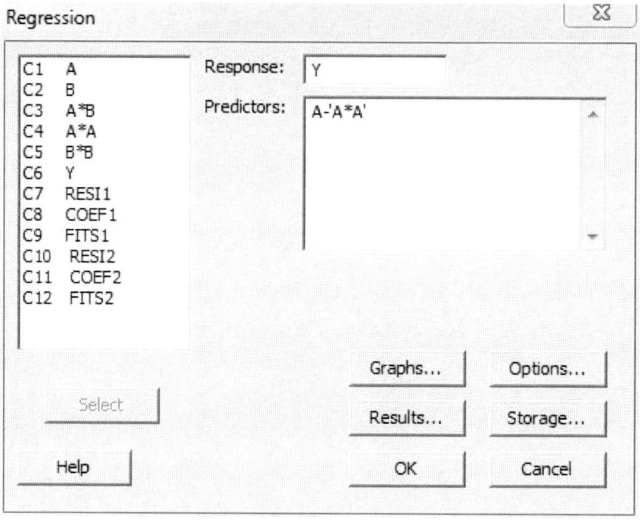

11. You can adopt the same setting that you used which will not be shown here. Now, you are done in **Regression** main window, click on OK to **RUN** the regression. You will get the following Minitab output where all summary of statistic is documented.

Regression Analysis: Y versus A; B; A*B; A*A

```
The regression equation is
Y = 1,20 - 0,0872 A + 0,0283 B - 0,0105 A*B + 0,00542 A*A


Predictor        Coef   SE Coef      T      P
Constant      1,20167   0,04837  24,84  0,000
A            -0,08717   0,01634  -5,33  0,001
B            0,028333  0,005909   4,79  0,002
A*B         -0,010500  0,001079  -9,73  0,000
A*A          0,005417  0,001477   3,67  0,008


S = 0,0204707   R-Sq = 99,5%   R-Sq(adj) = 99,3%


Analysis of Variance

Source          DF        SS        MS       F      P
Regression       4   0,64033   0,16008  382,02  0,000
Residual Error   7   0,00293   0,00042
Total           11   0,64327
```

If we take a look at the above Minitab results, we notice that *R-Sq* and *R-Sq(adj)* remains almost the same but *p-values* are well below 0.05, therefore we conclude that we can stop constructing multiple regression models here, and assert that all coefficients given above are significant for this problem. Because, they are well below the level of confidence, α=0.05.

↓	C1	C2	C3	C4	C5	C6	C7	C8	C9
	A	B	A*B	A*A	B*B	Y	RESI1	COEF1	FITS1
1	4	3	12	16	9	0,91	-0,041833	1,40833	0,95183
2	2	6	12	4	36	1,11	0,038667	-0,09600	1,07133
3	8	9	72	64	81	0,37	-0,052833	-0,02417	0,42283
4	8	3	24	64	9	0,67	0,102167		0,56783
5	4	6	24	16	36	0,85	-0,029333		0,87933
6	8	6	48	64	36	0,51	0,014667		0,49533
7	6	3	18	36	9	0,80	0,040167		0,75983
8	4	9	36	16	81	0,81	0,003167		0,80683
9	6	9	54	36	81	0,55	-0,064833		0,61483
10	2	3	6	4	9	1,05	-0,093833		1,14383
11	6	6	36	36	36	0,65	-0,037333		0,68733
12	2	9	18	4	81	1,12	0,121167		0,99883

Please note that the following columns of C13,C14 and C15 are shown, indicate the latest fit statistics obtained in this section.

C10	C11	C12	C13	C14	C15
RESI2	COEF2	FITS2	RESI3	COEF3	FITS3
0,0096667	1,21833	0,90033	0,0113333	1,20167	0,89867
0,0203333	-0,08717	1,08967	0,0170000	-0,08717	1,09300
0,0183333	0,02167	0,35167	0,0200000	0,02833	0,35000
-0,0156667	-0,01050	0,68567	-0,0140000	-0,01050	0,68400
-0,0043333	0,00542	0,85433	-0,0076667	0,00542	0,85767
-0,0036667	0,00056	0,51367	-0,0070000		0,51700
0,0286667		0,77133	0,0303333		0,76967
-0,0083333		0,81833	-0,0066667		0,81667
-0,0133333		0,56333	-0,0116667		0,56167
-0,0226667		1,07267	-0,0210000		1,07100
-0,0123333		0,66233	-0,0156667		0,66567
0,0033333		1,11667	0,0050000		1,11500

12. Finally, you will get **Four-in-one** plot with valid/similar characteristics as explained in the previous sections.

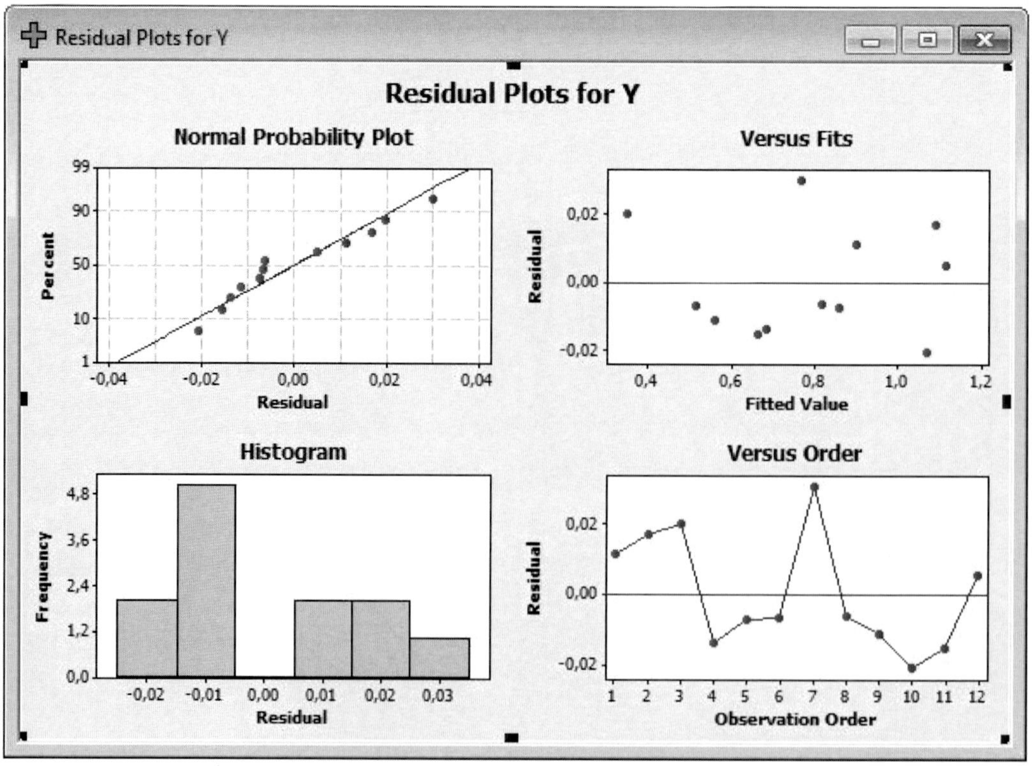

7.6 Multiple Regression with *Excel*

In this section, Example 7.1 can be solved using an alternative statistics tool, i.e. *Excel*, as introduced in the previous section. Now, let's go ahead and demonstrate how to perform multiple regression analysis in *Excel*. The procedure is simple and very similar. Remember, we need to use **"Data analysis"** added-in feature in *Excel*, therefore, make sure this added-in feature is already installed before.

- Click on **Data Analysis** button that you added-in as shown in the following picture!

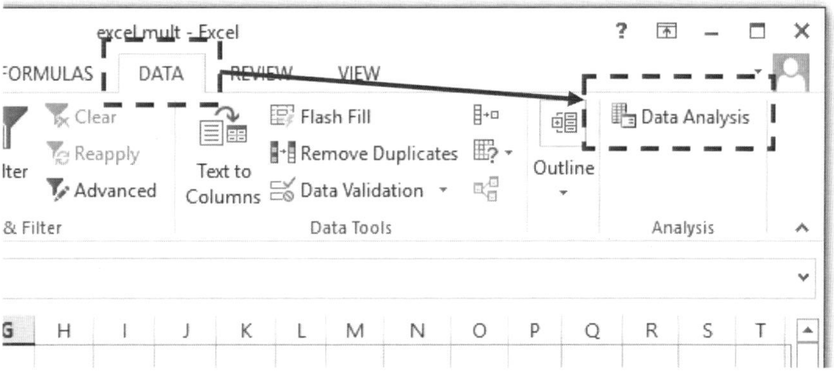

For the sake of demonstration, the last form of the multiple regression models of equation obtained in the previous section is solved/used in this part. Notice that we have all data in the following *Excel* sheet and using the same columns given below,

	A	B	C	D	E	F	G
1							
2	n	A	B	A*B	A*A	B*B	Y
3	1	4	3	12	16	9	0,91
4	2	2	6	12	4	36	1,11
5	3	8	9	72	64	81	0,37
6	4	8	3	24	64	9	0,67
7	5	4	6	24	16	36	0,85
8	6	8	6	48	64	36	0,51
9	7	6	3	18	36	9	0,8
10	8	4	9	36	16	81	0,81
11	9	6	9	54	36	81	0,55
12	10	2	3	6	4	9	1,05
13	11	6	6	36	36	36	0,65
14	12	2	9	18	4	81	1,12

- You will get the following **Data Analysis** window as shown below. Scroll down all the way down and once you see the **Regression**, highlight that tool by clicking on it, then Hit OK!

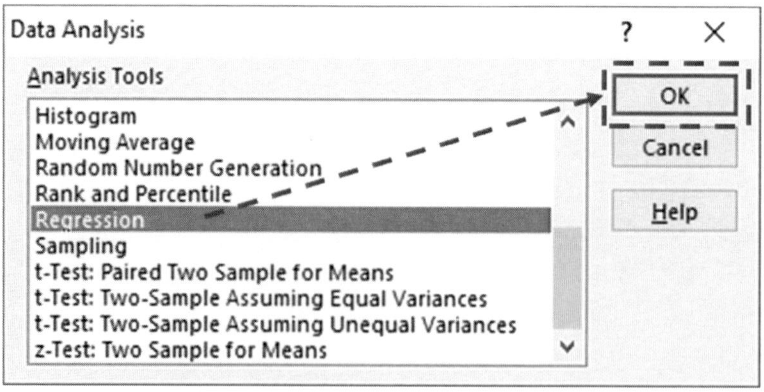

- **Regression** window will pop-up! where you will have to select specific Ranges for input and output data.
  - First, select Output ***y*** range between **=G2:G14** with ***Y*-Label,** hit Enter!
  - Second, select Input ***X*** range between **=B2:E14** with **A,B,A*B and A*A Labels**, hit Enter!
  - Since you selected the data pairs with their labels, check labels box, and let the program know about that.

- In the **Output options**, select **New Worksheet Ply:** option so that regression results are stored in a new worksheet in your already open *Excel* file!
- You may print the **Residuals**, just click the appropriates boxes as shown.
- Optional, you may want to see **Normal Probability Plots** by clicking the associated box.
- Hit **OK** to close the window!

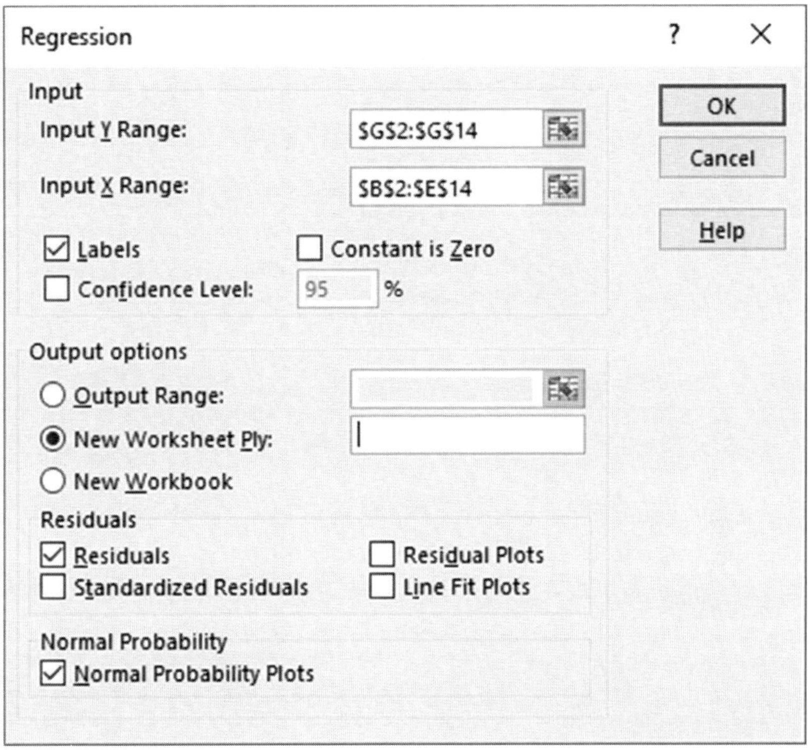

- Finally, you will get the regression results in open *Excel* Sheet in a new Worksheet as shown in the SUMMARY OUTPUT below, where you will have routine/similar outputs that you got in the previous chapter.

Finally, we can re-write our multiple regression models using these coefficients as in equation form as follows:

$$Y = \mathbf{1.201667 - 0.0871667A + 0.0283333B} - 0.0105AB + 0.00541667A^2 \quad 7.8$$

	A	B	C	D	E	F	G
1	**SUMMARY OUTPUT**						
2							
3	*Regression Statistics*						
4	Multiple R	0.99771737					
5	R Square	0.99543994					
6	Adjusted R Square	0.99283419					
7	Standard Error	0.02047065					
8	Observations	12					
9							
10	ANOVA						
11		*df*	*SS*	*MS*	*F*	*Significance F*	
12	Regression	4	0.64033333	0.16008333	382.01705	2.8712E-08	
13	Residual	7	0.00293333	0.00041905			
14	Total	11	0.64326667				
15							
16		*Coefficients*	*Standard Error*	*t Stat*	*P-value*	*Lower 95%*	*Upper 95%*
17	Intercept	1.20166667	0.04837027	24.8430863	4.366E-08	1.08728916	1.316044
18	A	-0.0871667	0.01634451	-5.3330868	0.0010838	-0.1258153	-0.04852
19	B	0.02833333	0.00590937	4.79464664	0.0019785	0.0143599	0.042307
20	A*B	-0.0105	0.0010789	-9.7321515	2.558E-05	-0.0130512	-0.00795
21	A*A	0.00541667	0.00147734	3.66649449	0.0080013	0.00192331	0.00891

- Further, you will get the RESIDUAL and PROBABILITY OUTPUT for your evaluation of the results as shown in the following.

25	**RESIDUAL OUTPUT**				**PROBABILITY OUTPUT**	
26						
27	*Observation*	*Predicted Y*	*Residuals*		*Percentile*	*Y*
28	1	0.89866667	0.01133333		4.1666667	0.37
29	2	1.093	0.017		12.5	0.51
30	3	0.35	0.02		20.833333	0.55
31	4	0.684	-0.014		29.166667	0.65
32	5	0.85766667	-0.00766667		37.5	0.67
33	6	0.517	-0.007		45.833333	0.8
34	7	0.76966667	0.03033333		54.166667	0.81
35	8	0.81666667	-0.00666667		62.5	0.85
36	9	0.56166667	-0.01166667		70.833333	0.91
37	10	1.071	-0.021		79.166667	1.05
38	11	0.66566667	-0.01566667		87.5	1.11
39	12	1.115	0.005		95.833333	1.12

- Finally, you will end up with *Excel* regression plots for the above data for the normal probability plot proves that normal distribution assumption is valid and works well.

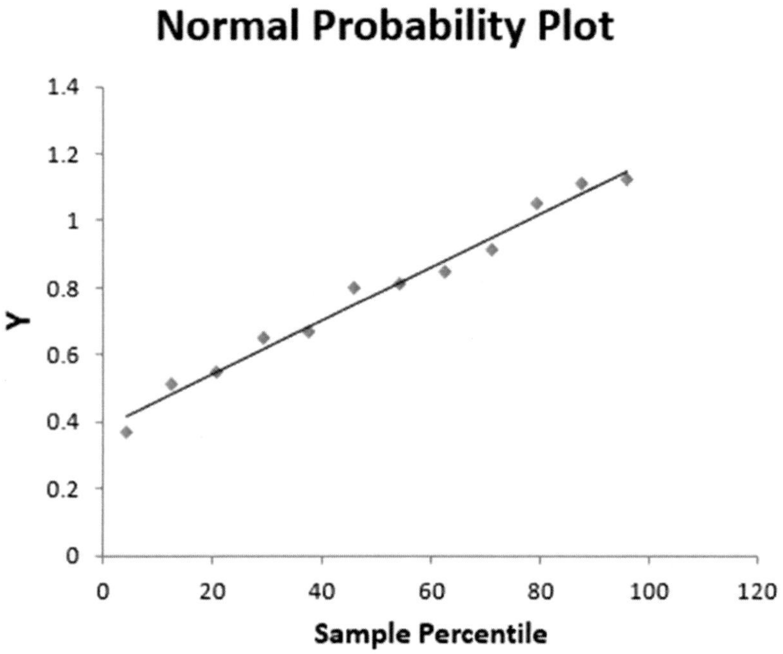

7.7 Multiple Regression with Matrix Notations

In this section, multiple linear regression analysis is performed using a mathematical *matrix notation* and its associated operations, as it is demonstrated in SLR analysis in Chapter 6. Further, for the sake of demonstration, the final form of the regression model obtained in the previous section for Example 7.1 is used in this section to avoid repetitive calculations.

7.7.1 Matrix Notations

Example 7.1, consists of $n=12$ data, where inputs $\boldsymbol{A},\boldsymbol{B}$ and their interaction and quadratic terms are also included as, $\boldsymbol{X} = \{\mathbf{1}, \boldsymbol{A}, \boldsymbol{B}, \boldsymbol{AB}, \boldsymbol{A}^2\}$ and $\boldsymbol{y}$ as output/response, are both shown in matrix format as follows,

$$\boldsymbol{X} = \begin{bmatrix} 1 & 4 & 3 & 12 & 16 \\ 1 & 2 & 6 & 12 & 4 \\ 1 & 8 & 9 & 72 & 64 \\ 1 & 8 & 3 & 24 & 64 \\ 1 & 4 & 6 & 24 & 16 \\ 1 & 8 & 6 & 48 & 64 \\ 1 & 6 & 3 & 18 & 36 \\ 1 & 4 & 9 & 36 & 16 \\ 1 & 6 & 9 & 54 & 36 \\ 1 & 2 & 3 & 6 & 4 \\ 1 & 6 & 6 & 36 & 36 \\ 1 & 2 & 9 & 18 & 4 \end{bmatrix} \quad \text{and} \quad \boldsymbol{y} = \begin{bmatrix} 0.91 \\ 1.11 \\ 0.37 \\ 0.67 \\ 0.85 \\ 0.51 \\ 0.80 \\ 0.81 \\ 0.55 \\ 1.05 \\ 0.65 \\ 1.12 \end{bmatrix}$$

To start with, let's multiply $\boldsymbol{X}'$ and $\boldsymbol{X}$ matrices as follows,

$$\boldsymbol{X}'\boldsymbol{X} = \begin{bmatrix} 1 & 1 & 1 & 1 & 1 & 1 & 1 & 1 & 1 & 1 & 1 & 1 \\ 4 & 2 & 8 & 8 & 4 & 8 & 6 & 4 & 6 & 2 & 6 & 2 \\ 3 & 6 & 9 & 3 & 6 & 6 & 3 & 9 & 9 & 3 & 6 & 9 \\ 12 & 12 & 72 & 24 & 24 & 48 & 18 & 36 & 54 & 6 & 36 & 18 \\ 16 & 4 & 64 & 64 & 16 & 64 & 36 & 16 & 36 & 4 & 36 & 4 \end{bmatrix} \begin{bmatrix} 1 & 4 & 3 & 12 & 16 \\ 1 & 2 & 6 & 12 & 4 \\ 1 & 8 & 9 & 72 & 64 \\ 1 & 8 & 3 & 24 & 64 \\ 1 & 4 & 6 & 24 & 16 \\ 1 & 8 & 6 & 48 & 64 \\ 1 & 6 & 3 & 18 & 36 \\ 1 & 4 & 9 & 36 & 16 \\ 1 & 6 & 9 & 54 & 36 \\ 1 & 2 & 3 & 6 & 4 \\ 1 & 6 & 6 & 36 & 36 \\ 1 & 2 & 9 & 18 & 4 \end{bmatrix} \qquad 7.9$$

$$\boldsymbol{X}'\boldsymbol{X} = \begin{bmatrix} 12 & 60 & 72 & 360 & 360 \\ 60 & 360 & 360 & 2160 & 2400 \\ 72 & 360 & 504 & 2520 & 2160 \\ 360 & 2160 & 2520 & 15120 & 14400 \\ 360 & 2400 & 2160 & 14400 & 16992 \end{bmatrix}$$

Next, multiply $\boldsymbol{X}'$ and $\boldsymbol{y}$ matrices as follows,

$$\boldsymbol{X}'\boldsymbol{y} = \begin{bmatrix} 1 & 1 & 1 & 1 & 1 & 1 & 1 & 1 & 1 & 1 & 1 & 1 \\ 4 & 2 & 8 & 8 & 4 & 8 & 6 & 4 & 6 & 2 & 6 & 2 \\ 3 & 6 & 9 & 3 & 6 & 6 & 3 & 9 & 9 & 3 & 6 & 9 \\ 12 & 12 & 72 & 24 & 24 & 48 & 18 & 36 & 54 & 6 & 36 & 18 \\ 16 & 4 & 64 & 64 & 16 & 64 & 36 & 16 & 36 & 4 & 36 & 4 \end{bmatrix} \begin{bmatrix} 0.91 \\ 1.11 \\ 0.37 \\ 0.67 \\ 0.85 \\ 0.51 \\ 0.80 \\ 0.81 \\ 0.55 \\ 1.05 \\ 0.65 \\ 1.12 \end{bmatrix} \qquad 7.10$$

$$X'y = \begin{bmatrix} 9.4 \\ 41.24 \\ 54.66 \\ 234.96 \\ 225.44 \end{bmatrix}$$

Now, we go ahead and take the inverse of the ***X'X*** matrix as presented in Equation 7.11, note that **X'X** multiplied with **(X'X)**$^{-1}$ yields unit matrix of ***I***, as shown in the following.

$$(X'X)^{-1} = \begin{bmatrix} 5.5833 & -1.625 & -0.5 & 0.083 & 0.104 \\ -1.625 & 0.638 & 0.08 & -0.02 & -0.05 \\ -0.5 & 0.083 & 0.08 & -0.01 & 0 \\ 0.0833 & -0.017 & 0 & 0.003 & 0 \\ 0.104 & -0.052 & 0 & 0 & 0.0052 \end{bmatrix} \quad 7.11$$

$$(X'X)(X'X)^{-1} = \begin{bmatrix} 12 & 60 & 72 & 360 & 360 \\ 60 & 360 & 360 & 2160 & 2400 \\ 72 & 360 & 504 & 2520 & 2160 \\ 360 & 2160 & 2520 & 15120 & 14400 \\ 360 & 2400 & 2160 & 14400 & 16992 \end{bmatrix} \begin{bmatrix} 5.5833 & -1.625 & -0.5 & 0.083 & 0.104 \\ -1.625 & 0.638 & 0.08 & -0.02 & -0.05 \\ -0.5 & 0.083 & 0.08 & -0.01 & 0 \\ 0.0833 & -0.017 & 0 & 0.003 & 0 \\ 0.104 & -0.052 & 0 & 0 & 0.0052 \end{bmatrix}$$

$$= \begin{bmatrix} 1 & 0 & 0 & 0 & 0 \\ 0 & 1 & 0 & 0 & 0 \\ 0 & 0 & 1 & 0 & 0 \\ 0 & 0 & 0 & 1 & 0 \\ 0 & 0 & 0 & 0 & 1 \end{bmatrix} = I$$

Finally, we obtain $\widehat{\boldsymbol{\beta}}$ regression coefficient vector, as given in Equation 7.12.

$$\widehat{\boldsymbol{\beta}} = (X'\mathbf{X})^{-1}X'\mathbf{y} = \begin{bmatrix} 5.5833 & -1.625 & -0.5 & 0.083 & 0.104 \\ -1.625 & 0.638 & 0.08 & -0.02 & -0.05 \\ -0.5 & 0.083 & 0.08 & -0.01 & 0 \\ 0.0833 & -0.017 & 0 & 0.003 & 0 \\ 0.104 & -0.052 & 0 & 0 & 0.0052 \end{bmatrix} \begin{bmatrix} 9.4 \\ 41.24 \\ 54.66 \\ 234.96 \\ 225.44 \end{bmatrix} \quad 7.12$$

$$\widehat{\boldsymbol{\beta}} = (\boldsymbol{X}'\mathbf{X})^{-1}\boldsymbol{X}'\mathbf{y} = \begin{bmatrix} \hat{\beta}_0 \\ \hat{\beta}_1 \\ \hat{\beta}_2 \\ \hat{\beta}_3 \\ \hat{\beta}_4 \end{bmatrix} = \begin{bmatrix} 1.201667 \\ -0.0871667 \\ 0.028333 \\ -0.01050 \\ 0.00541667 \end{bmatrix}$$

Finally, we can re-write our multiple regression models using these coefficients as in equation form as follows:

$$Y = \mathbf{1.201667 - 0.0871667}\boldsymbol{A} + \mathbf{0.028333}\boldsymbol{B} - 0.0105AB + 0.00541667A^2 \quad 7.13$$

Briefly, in the following *Excel* spread shot, *n=12* data, $\boldsymbol{X}$: input matrix with (12× 5) size and $\boldsymbol{y}$**:** output vector with (12× 1) size, are both placed. Following, both $\boldsymbol{X}'_{12 \times 5}$ and $\boldsymbol{X}_{5 \times 12}$ matrices multiplied, and $(\boldsymbol{X}'\boldsymbol{X})^{-1}_{5 \times 5}$, i.e. (5 × 5) size matrix is obtained. Next, both $\boldsymbol{X}'_{5 \times 12}$ and $\boldsymbol{y}_{12 \times 1}$ matrices multiplied, and $(\boldsymbol{X}'\boldsymbol{y})_{5 \times 1}$, i.e. (5 × 1) size vector is obtained.

Finally, the inverse of $(\boldsymbol{X}'\boldsymbol{X})_{5 \times 5}$ matrix is calculated and shown to be $(\boldsymbol{X}'\boldsymbol{X})^{-1}_{5 \times 5}$. Outcoming matrices of both $(\boldsymbol{X}'\boldsymbol{X})^{-1}_{5 \times 5}$ and $(\boldsymbol{X}'\boldsymbol{y})_{5 \times 1}$ now are multiplied and $\widehat{\boldsymbol{\beta}}_{5 \times 1}$ (vector of regression coefficients) is calculated.

7.8 *Excel* Workout

Briefly, in the *Excel* Screenshot Next, total n=12 data are shown where X: input matrix with (12 × 5) size and y: output vector with (12 × 1) size.

	A	B	C	D	E	F	G	H
1								
2	**n**		**A**	**B**	**A*B**	**A*A**	B*B	**Y**
3	1	1	4	3	12	16	9	0.91
4	2	1	2	6	12	4	36	1.11
5	3	1	8	9	72	64	81	0.37
6	4	1	8	3	24	64	9	0.67
7	5	1	4	6	24	16	36	0.85
8	6	1	8	6	48	64	36	0.51
9	7	1	6	3	18	36	9	0.8
10	8	1	4	9	36	16	81	0.81
11	9	1	6	9	54	36	81	0.55
12	10	1	2	3	6	4	9	1.05
13	11	1	6	6	36	36	36	0.65
14	12	1	2	9	18	4	81	1.12

All of the matrix operations are also performed in *Excel* and shown on the next page!

	I	J	K	L	M	N
1						
2		12	60	72	360	360
3		60	360	360	2160	2400
4	$X'X =$	72	360	504	2520	2160
5		360	2160	2520	15120	14400
6		360	2400	2160	14400	16992
7						
8		9.4				
9		41.24				
10	$X'y =$	54.66				
11		234.96				
12		225.44				
13						
14		1.2016667	$\hat{\beta}_0$			
15		-0.087167	$\hat{\beta}_1$			
16	$\hat{\beta} = (X'X)^{-1}X'y =$	0.0283333	$\hat{\beta}_2$			
17		-0.0105	$\hat{\beta}_3$			
18		0.0054167	$\hat{\beta}_4$			

We could also show corresponding matrix operations needed to perform these operations in *Excel* in terms of *Excel* matrix operation commands, are shown/summarized in the following Table X, where, $\boldsymbol{X'}$ and $\boldsymbol{X}$ is (B3,F14) matrix – $\boldsymbol{X'X}$ is (J2:N6) matrix – $\boldsymbol{y}$ is (H3,H14) matrix and (J8:J12) is $\boldsymbol{X'y}$ matrix. You can check everything out from *Excel* matrix command table and *Excel* Sheet given in the following Tables,

Matrix	***Excel* Matrix Definition**
$\mathbf{X} =$	= B3:F14
$\boldsymbol{X'} =$	=TRANSPOSE(B3:F14)
$\mathbf{y} =$	=H3:H14
$(\boldsymbol{X'}\mathbf{X}) =$	= J2:N6
$(\boldsymbol{X'}\mathbf{X})^{-1} =$	=MINVERSE(J2:N6)
$\boldsymbol{X'}\mathbf{y} =$	= J8:J12

Matrix Oper.	*Excel* Format for Matrix Oper.	Result
$X'\mathbf{X} =$	=MMULT(TRANSPOSE(B3:F14),B3:F14)	$\begin{bmatrix} 12 & 60 & 72 & 360 & 360 \\ 60 & 360 & 360 & 2160 & 2400 \\ 72 & 360 & 504 & 2520 & 2160 \\ 360 & 2160 & 2520 & 15120 & 14400 \\ 360 & 2400 & 2160 & 14400 & 16992 \end{bmatrix}$
$X'\mathbf{y} =$	=MMULT(TRANSPOSE(B3:F14),H3:H14)	$\begin{bmatrix} 9.4 \\ 41.24 \\ 54.66 \\ 234.96 \\ 225.44 \end{bmatrix}$
$\hat{\boldsymbol{\beta}} = (X'\mathbf{X})^{-1}X'\mathbf{y} =$	=MMULT((MINVERSE(J2:N6)),J8:J12)	$\begin{bmatrix} 1.201667 \\ -0.0871667 \\ 0.028333 \\ -0.01050 \\ 0.00541667 \end{bmatrix}$

There is one important point which needs to be kept in mind, when executing matrix operations, user needs to know and select the size of the outcoming matrix and then enter the *Excel* matrix operation commands presented in the above tables, while out coming matrix cells are highlighted, then you need to press **CTRL+SHIFT** at the same time and hit ENTER, while these two buttons are pressed!

In this chapter, matrix notation based calculation of multiple regression model coefficients is demonstrated on an example problem in an *Excel* spreadsheet. *Excel* provides powerful capabilities through which students can learn all aspects of MR analysis by doing it for themselves so that each student can develop clear and concise understanding/control of this approach. A summary of the *Excel* Sheet is again given in the following page.

	A	B	C	D	E	F	G	H	I	J	K	L	M	N
1														
2	n		A	B	A*B	A*A	B*B	Y		12	60	72	360	360
3	1	1	4	3	12	16	9	0.91		60	360	360	2160	2400
4	2	1	2	6	12	4	36	1.11	$X'X =$	72	360	504	2520	2160
5	3	1	8	9	72	64	81	0.37		360	2160	2520	15120	14400
6	4	1	8	3	24	64	9	0.67		360	2400	2160	14400	16992
7	5	1	4	6	24	16	36	0.85						
8	6	1	8	6	48	64	36	0.51		9.4				
9	7	1	6	3	18	36	9	0.8		41.24				
10	8	1	4	9	36	16	81	0.81	$X'y =$	54.66				
11	9	1	6	9	54	36	81	0.55		234.96				
12	10	1	2	3	6	4	9	1.05		225.44				
13	11	1	6	6	36	36	36	0.65						
14	12	1	2	9	18	4	81	1.12		1.2016667	$\widehat{\beta}_0$			
15										-0.087167	$\widehat{\beta}_1$			
16									$\hat{\beta} = (X'X)^{-1}X'y =$	0.0283333	$\widehat{\beta}_2$			
17										-0.0105	$\widehat{\beta}_3$			
18										0.0054167	$\widehat{\beta}_4$			

7.9 Exercises

1) Explain, when and why would you prefer a multiple regression model? To what extent you would use such models, please elaborate on this.

2) Assume you have a black-box, and you would like to identify this box. When you input x_1 and x_2 it yields an output y, as presented in Table 7.4 below.

TABLE 7.4 Inputs/output data

x_1	x_2	y
1	3.5	49
2	5	144
3	8	550
4	6	260
5	3	60
6	4	115
7	7	435
8	9	850
9	2	47
10	5	230

- ✓ Do 3D plane surface plot for the data
- ✓ Fit a multiple linear regression model to the data using *Excel*
- ✓ Interpret your results and tell us about what can be done next?

3) Repeat your analysis using *Minitab* and obtain the best multiple regression model with higher order terms as necessary.

Chapter 8

Fundamentals of Designed Experiments

It is no surprise that one of the important characteristics of any living being with "*reasoning*" is to explore causal relationship between input(s) and output of events(experiments, simulations) which are going on around us. Human wisdom is the primary source of this will/urge and needs to be satisfied and utilized. Sometimes, events can take place without our intervention or control, then we do make observations and learn from them, so we get *"experienced"*.

Sometimes, we believe/think, we are in charge, and we may control the things to a certain extent. From the statistics point of view, in such cases, we perform some purposefully designed experiments, also called *trials* or *tests* (sometimes also called *simulations*) to extract any meaningful information about our system or problem (i.e. briefly controlled event). Eventually, we may infer some useful knowledge for the good of our company/society by evaluating these experiments. The latter is directly linked to the main subject of this and the following two chapters.

Keep in mind that all experiments could be treated as "*somewhat*" designed experiments, but the question is then raised, *how well* are they designed? *Poor*, *good* or *well.* The assessment is made by the evaluation of objective characteristic(s) followed by verification. Therefore, designed experiments are a crucial tool to explore the behavior/characteristics of processes or systems so that one can have the ability to improve performance characteristics in the design or development of processes or new products. If we look at designed experiments from this perspective, we need to meet the following objectives,

- Time spent on product/process realization, *Reduce it*!

Introducing new products into the market (or making some modifications in the existing products/parts) is not an easy task from manufacturing perspective with many risks involved i.e. these products are amenable to many faults and troubles, which are not detected/foreseen during design modifications, production, quality control, and sales. Our company may come up with warranty repairs, lawsuits, thus revenue loss etc. From the other side, we need to reduce this time for reducing cost, increasing market share etc.

- Overall performance, *improve it*!
  Product research and development cost has to be further employed and new improvements to the existing products are the only option to be an effective player in the market.
- Quality(reliability, product/process life span etc.), *increase it*!
  Quality is the vital characteristics of products and makes happy or unhappy customers. A happy customer will probably is the best advertisements and sales opportunities and vice versa.
- Robustness, higher is better, *obtain it*!
  Robust products are the one they never *let down* their owner infeasible, meaningful operational limits. For example, a robust car is the one, always starts up no matter what the weather conditions are! i.e. in too cold, freezing weather.
- Miscellaneous improvements supporting the above objectives, *just do it*!
  As an example, provide a friendly working environment and happy workers who do their job with care and commitment.

Make sure these vital issues are taken into consideration well (i.e. especially, in today's rapidly changing and highly competitive global economy and market conditions). In practice, for any product design, development and improvement (as well as process design and improvements) experiments (including close to reality simulations) are heavily employed, in all fields of science, engineering, politics, and social studies that anyone can think of to meet above objectives. In the next section, introduction to designed experiments is briefly introduced.

8.1 Designed Experiments

In today's rapidly changing industrial/social environment, designed experiments play a prominent role in acquiring improved materials, new product/process design, analysis, optimization, and process improvements etc. Now, our challenge is to investigate the full causal relationship between the experiment inputs and output. The system might also be exposed to outer/noise effects as well, which is not in our control (at least complete control) and they are not welcome. Under such complex conditions, we need to identify most of the inherent experiment features and draw useful conclusions, based upon them. In this context, an *experiment* can be defined as the certain number tests/trials by controlling (purposefully choosing/generating/designing) the input variable(s) which are going into the system, where they are being processed and finally the system outputs response (also called output). We can achieve these aforementioned tasks efficiently, only by using both *designed experiments* and corresponding statistical tools, correctly and effectively.

Designed experiments are a very useful method for exploring the targeted nature of many complex systems (i.e. empirical representations). We have a multi-objectives to satisfy in the design, analysis and development of processes/systems or new products and the certain action needs to be taken within this concept.

A good experimental design must ensure that one can separate the effects of all factors/inputs on the problem response (*Orthogonal* design just works great for this cause). The specified values of the factors/inputs used in the experiment are called *factor levels*, these levels define *uppe*r and *lower* bounds on factors.

Typically, we tend to use a small number of levels for each factor, why? because each experiment/simulation is demanding and costly i.e. in terms of CPU, $, time etc. For example, we might select two *high (denoted as 1)* and two *low (denoted as 0 or -1)* levels for each of the factors. This means we will design our experiment using two levels of all participating factors. A powerful experimental design will include all combinations of the factor and their levels to form a series of the experiment. This type of experiment is called a **full factorial experiment.**

8.1.1 Experimental Design Space Resolution

As far as a *full factorial experiment* is concerned, let's start with the easiest case where, assume our experiment has two inputs, *A* and *B* as factors and a single output, *Y* as the response. Using a different number of levels in 2D plane space, assuming the general case of a maximum *four intervals* and *five levels.* Adopt appropriate design levels given below for each alternative design case, i.e. *a, b* and *c* as presented in Figure 8.1,

-1 for "*l o w*"

-1/3 for "*mid-low*"

0 for "*middle*"

1/3 for "*mid-high*"

1 for "*h I g h*"

Using these design points, we can control 2D experimental design space resolution as illustrated in Figure 8.1 where more ***dot*** means more design space resolution or experiments.

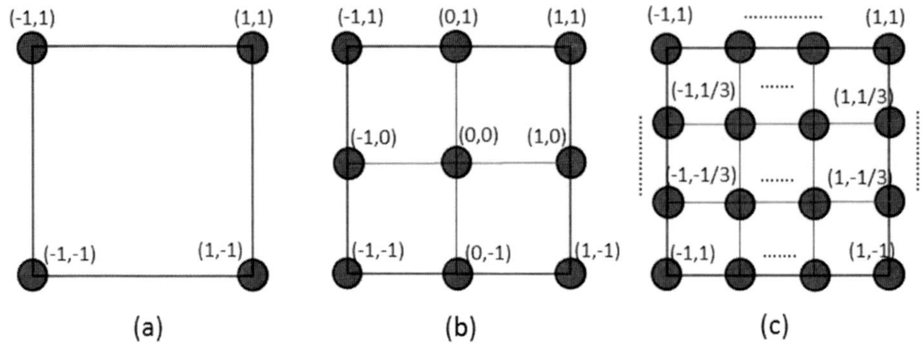

FIGURE 8.1 Different 2D experimental design spaces

- In Figure 8.1a, we have a coarse design space as totaling 4 experiments at *low* and *high* levels as given in Table 8.1,

TABLE 8.1 2^2 Full factorial design inputs

Exp. #	A	B
1	-1	-1
2	1	-1
3	-1	1
4	1	1

- In Figure 8.1b, we have a fine design space as totaling 9 experiments at *low, middle* and *high* levels
- In Figure 8.1c, we have a more refined design space as totaling 16 experiments at *low, middle-low, middle-high* and *high* levels

Further, consider an experiment with three inputs, *A,B* and *C* as factors and a single output, *Y* as a response. Using two design levels, (i.e. now we can use 0 for "*low*" and 1 for "*high*" levels), we end up with eight different possible design combinations for this situation in 3D cubic space.

If we want to use all possible combinations for each factor in our design, we can obtain the following design combinations to be inputted to our system as presented in Table 8.2. Also, notice that in this table shaded rows are part of orthogonal arrays as would be introduced and explained later in this chapter.

Exp. #	A	B	C
1	0	0	0
2	1	0	0
3	0	1	0
4	1	1	0
5	0	0	1
6	1	0	1
7	0	1	1
8	1	1	1

TABLE 8.2 2^3 Full factorial design inputs

In three dimension (3D), we talk about the Cartesian coordinate system. Traditionally, a three-factor design is depicted using a cube, where each design experiment run (shown in Table 8.2) corresponds to eight corner points of the cube. Note that in Figure 8.2, variables *A,B* and *C*, each corresponds to Cartesian axis X,Y, and Z, respectively.

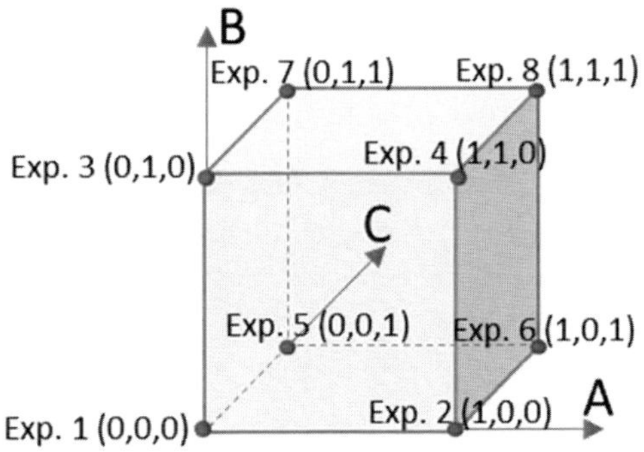

FIGURE 8.2 3D cubic representation of 2^3 full factorial design

Note, that beyond 3D, we talk about hyperplanes and visual representation is not possible other than using may be a very limited 3D trick.

8.2 Objectives in Experiments

It is a good practice not to rush and take some time to plan and use our statistics knowledge in organizing well-designed experiments. By doing so, we can ensure that we would get the right or minimal amount of experiment combinations to obtain the most pertinent characteristics of the experimental system. Keep in mind that, experiments are our valuable means to investigate the behavior of the processes and systems. Every action/input on any system is processed into some form of output/response by the corresponding system.

Now, let's think about the idealized system that we are interested in. The simplified model (slightly different from the one used in literature) to represent the idealized system model is shown in Figure 8.3. This figure looks like an aircraft engine, very simply which

- sucks the air with many variables, i.e. fuel, ambient temperature, pressure, humidity, air amount and its elements can be considered as *inputs*
- outputs push force called thrust which makes the aircraft move on the ground and in the air as *outputs*
- controls flight characteristics pre-adjusted and controlled at optimum levels by the controller as *fix parameters*
- disturbs our aircraft, flow regime i.e. turbulence, ocean air stream characteristics etc. as *uncontrollable (noise) factors*

or in general term, we can call the system as *a black box* as to be identified. There are $\underline{\mathbf{x}}=\{x_{1n},x_{2n},\ldots,x_{kn}\}$ (*k* number of input variables, *n* is the total number of data) inputs, going into the black box and *y* is coming out as a response. The black box simply processes the inputs with an unknown transfer function within, say, *f(x)* that we do not have any idea about.

Further, there are noises affecting our black box, they are, in their nature, uncontrollable and undesirable. We need to deal with them, anyway. Also, we go ahead and fix controllable system parameters to their optimum values that we have comfortable with. Also, note that these parameters are thought/assumed to have secondary effects of importance in the system that we are studying.

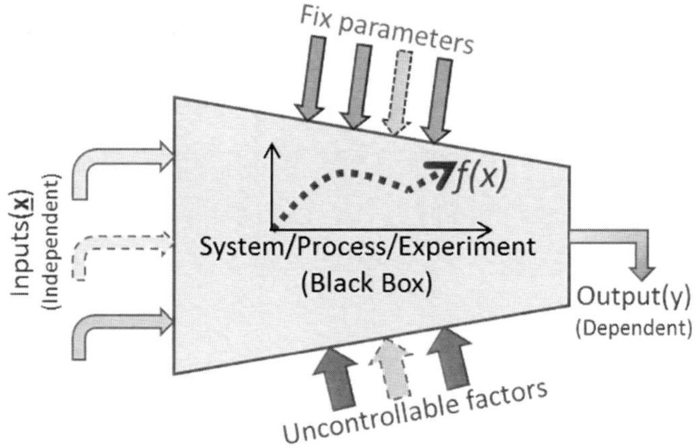

FIGURE 8.3 Schematic model of an idealized system/process

Now in general, we can explain Figure 8.3 as in the following,

- The inputs $\underline{\boldsymbol{x}}=\{x_1,x_2,\ldots,x_k\}$ are controllable factors that we determined as important for a model to be studied, such as temperature, pressures, feed rates, speed, concentration, additive etc.
- A system acts similar to the explicit or implicit transfer function, relating inputs to output, selected input vector $\underline{\boldsymbol{x}}$ consists of input variables, which are pre-determined by design team having the most influences on this system output y.
- The inputs $\underline{\boldsymbol{z}}=\{z_1, z_2\ldots z_q\}$are uncontrollable (or difficult to control) input factors, such as environmental effects, vibration etc.
- Fixed parameters are adjusted to their best levels (static system) and they have a secondary influence on the output for simplification of the system.

In this schematic model of an idealized system/process, we do see things going in and out. For system identification, several objectives need to be satisfied as much as possible. This is easy to say but it may be quite rigorous to do.

Below are the highlights to note for this schematic model;

i. How so, y is dependent on $\underline{\boldsymbol{x}}=\{x_1,x_2,\ldots,x_k\}$? I.e. Assessing the rank of importance of $\underline{\boldsymbol{x}}$ on y.
ii. Making the system *robust (being less sensitive to the outer noises, stay operational)* so y is in safe operating range.
iii. Minimizing variability by setting $\underline{\boldsymbol{x}}$ at an optimal level while satisfying item (ii),
iv. Set the $\underline{\boldsymbol{x}}$ to minimize or eliminate the effects of the uncontrollable variables.

8.3 Steps in Experimental Design

As in all processes, one needs to follow certain steps for successful experimental design implementation. We can generalize and handle any experimental design process by following the steps in the order presented below;

1. Gather the design team and identify/define the system issues which need to be dealt with,
2. Determine experimental objectives, prioritize or combine them in multi-objective form if necessary,
3. Determine important input factors and their levels(upper and lower bounds), i.e. (experimental) design space resolution,
4. Design/generate experiment input table (design combinations),
5. Conduct experiments using input table developed in item (4) above, obtain a response, repeat the experiments as necessary!
6. Analyze the data using statistics,
7. Interpret results and write the report
8. Do predictions using the model obtained in item (6) and verify this prediction experimentally to see how well you identified the system.

In most of the practices in designed experiments studies, we choose our input design table using Taguchi designs, which are based on *orthogonality* concept as described in the next section.

8.4 Orthogonality

In this methodology, we set up our experiments in a way that they would only require a fraction of the full factorial design and the effect of each input can be assessed independently without worrying about the effects of others.

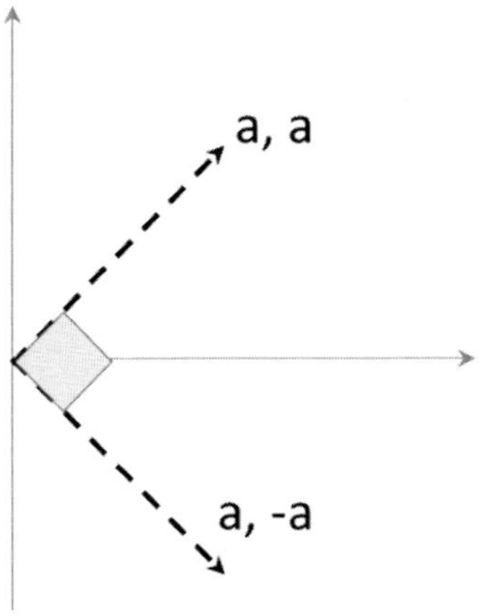

FIGURE 7.4 Orthogonality concept

Two vectors are said to be orthogonal (i.e. the vectors are perpendicular) as shown in Figure 7.4, when their dot product is equal to zero, i.e. the dot product of two vectors **X** and **Y**, is given in Equation 7.1 where θ is the angle between two vectors.

$$X.Y = |X||Y|\cos\theta \qquad 7.1$$

When θ =90°, ***X*** is perpendicular to ***Y***,

$$X.Y = 0 \qquad 7.2$$

where Cos 90 = 0. For example, consider the following vectors **X** and **Y**,

$$\boldsymbol{X} = \begin{bmatrix} -1 \\ -1 \\ 1 \\ 1 \end{bmatrix} \ and\ \boldsymbol{Y} = \begin{bmatrix} -1 \\ 1 \\ -1 \\ 1 \end{bmatrix}$$

If we multiply the corresponding elements of the vectors we get,

X.Y = (-1)(-1)+(-1)(1)+(1)(-1)+(1)(1) = 1-1-1+1 = 0

We show that two vectors are *perpendicular* or *orthogonal.*

In designed experiments, as mentioned before, the effects of one factor can be differentiated from the effect of any other factor(s) by using orthogonal arrays, that's why orthogonality is important in experimental design, therefore we prefer to adopt it. Especially, when design evaluations are difficult or expensive to do!

8.5 Exercises

1) Consider a refrigerator for which you are asked to depict the refrigerator as an idealized system. Identify possible input, output, system parameters and noise variable(s) that you think for essential cooling operation.

2) Consider an experiment with four inputs, *A,B,C and D* as factors and single Output, *Y* as a response. Using two design levels, (i.e. 0 for "*low*" and 1 for "*high*" levels)

- Calculate how many possible *combinations* you would have (i.e. full factorial design)
- Depict the problem using cubes. (Hint: Employ simple trick for this representation in 3D) and show your designs.
- Suggest *orthogonal* design combinations instead of the full factorial design, if possible

3) Consider the following vectors **X** and **Y**,

$$\boldsymbol{X} = \begin{bmatrix} -3-x \\ -1 \\ 1 \\ 1 \end{bmatrix} \ and \ \boldsymbol{Y} = \begin{bmatrix} -1 \\ 1 \\ -1 \\ 1-2x \end{bmatrix}$$

Which value should x take, so that these two vectors are to be *orthogonal*?

Chapter 9

Full Factorial vs. Taguchi Design Experiments

As far as any experimental design is concerned, the full factorial design accounts for all possible combinations of input design variables (or factors). Corresponding response or output are obtained in a predefined design domain, where LOWER and UPPER bounds, i.e. the interval size between LOWER and UPPER bounds are defined or determined by the design team after serious brainstorming. Further note that, the LOWER and UPPER bounds on design (variable) inputs are also called *SIDE constraints* from an optimization point of view. Also notice that, when we divide the design domain into subdomains and make the experiment/simulation evaluations at these locations(i.e. divisions), this means, we can control/define experimental *design space solution resolution*. Therefore, the full factorial experimental design can be viewed as a strength in such sense, if it could practically be done for all design points. Because, in real-life problems, most of the time, it is not always possible to consider/test all of these design combinations, even with coarser design domain resolution. Why? Because each experiment may impose on time/cost limitations in the trial of these experiments.

Keep in mind, we would like to perform a full factorial design of experiments and we are to consider all combinations of design variables combinations as input to our experiment. But, notice that as the number of design variables and their solution resolution (levels) increases, the total number of experiments needed will go up exponentially. Thus, a large number of total experiments would be required, thus making such a design experiment analysis almost impossible. Prove! assume if a single experiment or simulation takes *three days,* then practically we cannot test/run/try all full factorial experiments, as they would be taking so long when you consider all design combinations times *three days,* total duration.

9.1 OFAT Experiments

For the sake of getting good grasp of this concept, let's consider one of the simplest case, i.e. *"One Factor At a Time (OFAT) experiments*" first, with one factor A. Assume A represents baking process temperature levels at 2 (Low, High – *one interval*), 3 (Low, Medium, High – *two intervals*) and 4 (Low, Medium-low, Medium-high High – *three intervals*) temperature settings etc. for obtaining the best tasting cake. Assume, we have used the same ingredients in all cakes. In Table 8.1, sample OFAT experiments at different levels are presented.

TABLE 9.1 Sample OFAT experiments

Tested Levels	Experiments Required
two levels : A_1 A_2	two Experiments
three levels : A_1 A_2 A_3	three Experiments
four levels : A_1 A_2 A_3 A_4	four Experiments
five levels : A_1 A_2 A_3 A_4 A_5	five Experiments
more ...	*more ...*

Since OFAT experimental designs, in their nature, obviously do not account for *interactions*) between the input factors and this procedure is akin to yield premature results. (i.e. an *interaction*, can be stated as, the *adverse change* in the overall system/experiment behavior when one input does not generate the similar effect on the output at different levels of another input).

9.2 Full Factorial Design

Next, consider a *Full Factorial* design to account for all design combinations to overcome this hurdle that was encountered with OFAT. First consider an experiment with two factors A and B, at two levels (1: LOW and 2: HIGH). Two factors, A and B, at two levels (*one interval*) can be tested at 4 combinations of experiments as given in Table 9.2,

TABLE 9.2 Full factorial design for *2^2*

		A	
		A_1	A_2
B	B_1	$A_1 B_1$	$A_2 B_1$
	B_2	$A_1 B_2$	$A_2 B_2$

If three factors, i.e. A,B and C at two levels (*one interval*) are considered, then the total of $2^3 = 8$ experiments are required for *Full Factorial Design* combinations, as given in Table 9.3,

TABLE 9.3 Full factorial design for 2^3

	A , B , C			
$A_{1,2}$	$A_1 B_1 C_1$	$A_1 B_1 C_2$	$A_1 B_2 C_1$	$A_1 B_2 C_2$
$B_{1,2}$	$A_2 B_1 C_1$	$A_2 B_1 C_2$	$A_2 B_2 C_1$	$A_2 B_2 C_2$

By examining Tables 9.2 and 9.3, one can easily come up with the general formulation to find total number of experiments required for full factorial design, using LEVEL (Lower and Upper design variable side constraints or bounds) and FACTORS (# of input variables/factors) in the formulation as given in Equation 9.1,

$$\text{Total Number of Experiments} = \text{LEVEL}^{\text{FACTORS}} \tag{9.1}$$

3 Factors at *2* levels $2^3 = 8$

4 Factors at *2* levels $2^4 = 16$

7 Factors at *2* levels $2^7 = 128$

10 Factors at *2* levels $2^{10} = 1024$

....

We can plot these factors and levels vs. the total number of experiments required as illustrated in Figure 9.1. We quickly notice that the number of experiments required for full factorial design goes up dramatically, as the number of design variables and levels increases.

From the practical point of view, it is almost impossible to do all of these experiments or simulation runs for most of the real-life engineering problems. The remedy is to select the *experimental design* with the least number of experiments and overcome this problem. However, there needs to be solid/fundamental *reasoning* required in such a situation.

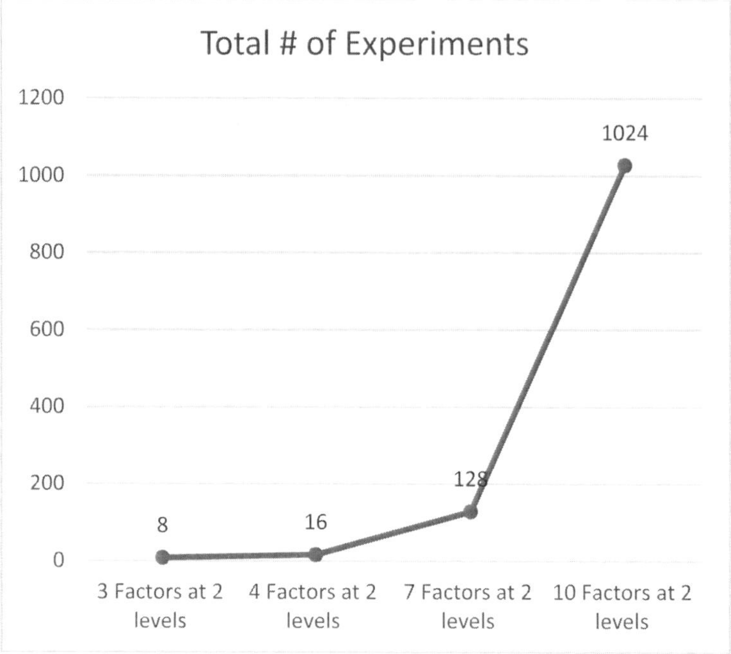

FIGURE 9.1 Total # of experiments needed for full DOE

In addition to our previous hurdles, in the evaluation of any experimental analysis, one also need to identify outside factors/influences, also called "Noise". *Noise* is not desired, when it causes deviations from the desired outputs. Then one can ask - why not eliminate or isolate the *noise*? so that we can determine both each inputs effects on output and interactions among inputs. This is apparently not an easy task, which can be both costly and time-consuming, sometimes impossible from different perspectives. In this respect, Genichi Taguchi (Japanese engineer and statistician) devised a clever/simple way to use orthogonal arrays to isolate the noise factors from the experiment characteristics by proposing the construction of a set of Orthogonal Array (OA) as a solution strategy to overcome these aforementioned hurdles in a cost-effective i.e. much cheaper but effective way. This concept is already introduced in Chapter 8.

In his approach, Taguchi labels the orthogonal array i.e., **L-4** (2^3), **i.e. also labeled as L4**, where **L-4** or **L4** array is used to design an experiment to study 3 Factors at 2 levels case. Both full factorial design and **L-4 array** are shown below in Table 8.4 in bold letters with OA marks. In this table, the first column is shown with EXPT. # represents all required number of 8 trial conditions/experiments and the second column shows the trial combinations needed for full factorial design, i.e. EXPT. 1, consists of inputs values of

A,B,C are all set to their LOW levels, whereas in EXPT. 2, A and B are set to their LOW levels, but C is set to its HIGH level.

TABLE 9.4 Full Factorial Design for 2^3

EXPT. #	Full Fact. 2^3 *A*	*B*	*C*	
1	***1***	***1***	***1***	OA
2	*1*	*1*	*2*	
3	*1*	*2*	*1*	
4	***1***	***2***	***2***	OA
5	*2*	*1*	*1*	
6	***2***	***1***	***2***	OA
7	***2***	***2***	***1***	OA
8	*2*	*2*	*2*	

In Table 9.5, the L-4 array is given which corresponds to full factorial design as shown in Table 9.4. Where Taguchi selects 4 designs out of 8 possible designs, i.e. three columns of an L-4 array are *orthogonal* with respect to each other, i.e.

- Levels appear in equal numbers i.e. *Level-1* and *Level-2*, each six times
- Combination of two designs appear in equal numbers i.e. $A_1B_1, A_1B_2, A_2B_1, A_2B_1$ each appear once in all designs
- Thus, the effect of one factor (i.e. *A*) can be separated from the remaining factor's (i.e. *B* and *C*) effects.

TABLE 9.5 Taguchi L-4 Design instead of 2^3

EXPT. #	***L-4*** *A*	*B*	*C*
1	*1*	*1*	*1*
2	*1*	*2*	*2*
3	*2*	*1*	*2*
4	*2*	*2*	*1*

In summary, our selected input variables (factors) A,B and C all at 2-levels (i.e. LOW, HIGH) produces 8 possible design combinations in FULL FACTORIAL design, yet, Taguchi selects 4 out of 8 designs and sets up an Orthogonal L-4 array as given in Table 9.6, where total number of experiments required are *four* as per Taguchi Orthogonal design. Taguchi array selector as a function of *design levels* and the number of *factors* is given with some examples i.e. L-4, L-8, L-9 and L-12 arrays in Appendix.

TABLE 9.6 Taguchi L-4 experiment inputs

EXPT. #1	A_1 B_1 C_1
EXPT. #2	A_1 B_2 C_2
EXPT. #3	A_2 B_1 C_2
EXPT. #4	A_2 B_2 C_1

Now, let's go ahead and conduct all of these experiments i.e. for 3 *factors* at 2 *levels* and we obtain single OUTPUT, called as a dependent variable, *Y* or *y*, as presented in Table 9.7,

TABLE 9.7 Taguchi L-4 experiment inputs/output

EXPT.	***L-4***			***Y***
#	A	B	C	(OUTPUT)
1	*1*	*1*	*1*	Y_1
2	*1*	*2*	*2*	Y_2
3	*2*	*1*	*2*	Y_3
4	*2*	*2*	*1*	Y_4

Our next step is to show that, by using Taguchi Orthogonal Array, we can get the Full Factorial Design experiment' results by using a simple mathematical formula, without the need for performing all remaining *128 – 8 = 120* experiments. Both results comparisons are given for verification purpose in the next section.

Now, our main question arises, i.e. "how could someone evaluate(formulate) output(s) in Taguchi Experiments in terms of inputs? Further, the number of output can be more than one in most of the situations. In the literature, this calculation/evaluation is achieved mainly in two different ways,

- Using **mean responses analysis** of the output(s), this is a very basic but effective averaging method, most of the time yields good results.
- The second approach is using **Signal to Noise ratio (S/N)** approach, which requires using appropriate formulations for different scenarios for the objective(output), i.e. larger is better, nominal is best or smaller is better. The S/N ratio formulations will be explained in a separate section due to its importance.

9.3 L-8 Taguchi/Full Factorial Experiments by Mean Responses

All of the points highlighted up to this point, can be demonstrated on example easily to assess influences/effects/relationships of each inputs on output, using *mean responses analysis*. For this purpose, consider an idealized input/output system as depicted in Figure 9.2,

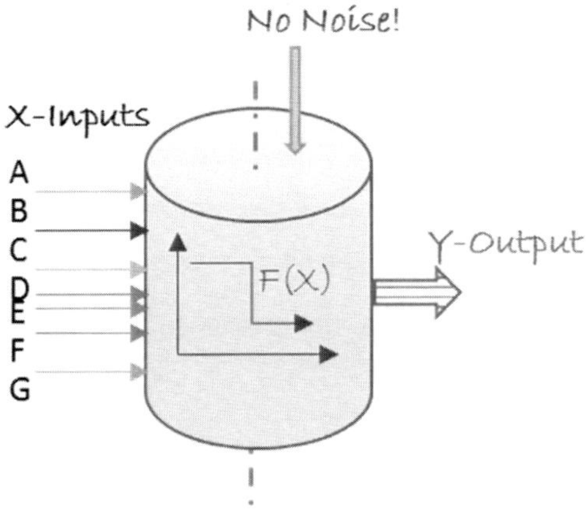

FIGURE 9.2 System inputs/output idealization

where conduct an experiment with 7 inputs/factors each at 2 levels, i.e. *A,B,C,D,E,F* and *G* are as given in the following boundaries (i.e. experiment design space) as LOW and HIGH levels, given in Table 9.8,

TABLE 9.8 Taguchi L-8 Experiment LOW/HIGH Levels

LOW	HIGH
$A_1 = 2$	$A_2 = 5$
$B_1 = 10$	$B_2 = 20$
$C_1 = 0.1$	$C_2 = 0.8$
$D_1 = 50$	$D_2 = 100$
$E_1 = 2$	$E_2 = 4$
$F_1 = 60$	$F_2 = 80$
$G_1 = 3$	$G_2 = 6$

For the sake of demonstration, also assume that the outcome of these experiments takes place according to Equation 9.*2*. Note in this example we assume that we know the transfer function as a function of input variables as given in Equation 9.2, which clearly reveals the relationship between inputs/factors and output (for this experiment, we assume *no noise* affecting our experiment). Therefore, we can verify our results analytically. Keep in mind that, as far as the real-life experiments are involved, we do not have any idea of such a relationship, i.e. in an explicit (closed) form.

$$Y = 2A - 0.1B + 3C - 0.1D + 5E + 1.1F + G \qquad 9.2$$

Using Equation 9.2, we now can go ahead and input *A,B,C,D,E,F* and *G* LOW/HIGH values according to the L-8 design that we have decided to use as given in Table 9.9 and calculate corresponding outputs, Y_i, $i=1,8$ as also presented in Table 9.9,

TABLE 9.9 L-8 Taguchi Design Experiments (with design/experiment evaluations)

EXPT. #	A	B	C	D	E	F	G	Y
1	1	1	1	1	1	1	1	$Y_1 = 2A_1-0.1B_1+3C_1-0.1D_1+5E_1+1.1F_1+G_1 = \mathbf{77.3}$
2	1	1	1	2	2	2	2	$Y_2 = 2A_1-0.1B_1+3C_1-0.1D_2+5E_2+1.1F_2+G_2 = \mathbf{107.3}$
3	1	2	2	1	1	2	2	$Y_3 = 2A_1-0.1B_2+3C_2-0.1D_1+5E_1+1.1F_2+G_2 = \mathbf{103.4}$
4	1	2	2	2	2	1	1	$Y_4 = 2A_1-0.1B_2+3C_2-0.1D_2+5E_2+1.1F_1+G_1 = \mathbf{83.4}$
5	2	1	2	1	2	1	2	$Y_5 = 2A_2-0.1B_1+3C_2-0.1D_1+5E_2+1.1F_1+G_2 = \mathbf{98.4}$
6	2	1	2	2	1	2	1	$Y_6 = 2A_2-0.1B_1+3C_2-0.1D_2+5E_1+1.1F_2+G_1 = \mathbf{102.4}$
7	2	2	1	1	2	2	1	$Y_7 = 2A_2-0.1B_2+3C_1-0.1D_1+5E_2+1.1F_2+G_1 = \mathbf{114.3}$
8	2	2	1	2	1	1	2	$Y_8 = 2A_2-0.1B_2+3C_1-0.1D_2+5E_1+1.1F_1+G_2 = \mathbf{80.3}$

Please also note that using Equation 9.3, we can obtain an average value of outputs as shown with *T*. By calculating the mean(average) value of all calculated outputs i.e. Y_i, $i=1,8$, we calculate *T* as,

$$T = \frac{(Y_1 + Y_2 + Y_3 + Y_4 + Y_5 + Y_6 + Y_7 + Y_8)}{8} \qquad 9.3$$

$$T = \frac{(77.3 + 107.3 + 103.4 + 83.4 + 98.4 + 102.4 + 114.3 + 80.3)}{8} = 95.85$$

Now, using L-8 Taguchi design, we can assess the effects of each input variable at LOW and HIGH values, by averaging the corresponding columns LOW and HIGH output results for each design variable (in column format).

For example, to find A-LOW, consider A column of Table 9.9 and average all outputs values corresponding to LEVEL-1, similarly, for A-HIGH, again in the A column, average all outputs values corresponding to LEVEL-2. All of these calculations are as demonstrated in Table 9.10, for all input variables (see their columns).

TABLE 9.10 L-8 Taguchi design total effects of inputs variables on output at LOW/HIGH values

#	LOW		HIGH	
A	$A_1 = (Y_1+Y_2+Y_3+Y_4) / 4$ $A_1 = (77.3+107.3+103.4+83.4) / 4$	**92.85**	$A_2 = (Y_5+Y_6+Y_7+Y_8)/4$ $A_2 = (98.4+102.4+114.3+80.3) / 4$	**98.85**
B	$B_1 = (Y_1+Y_2+Y_5+Y_6) / 4$ $B_1 = (77.3+107.3+98{,}4+102.4) / 4$	**96.35**	$B_2 = (Y_3+Y_4+Y_7+Y_8) / 4$ $B_2 = (103.4+83.4+114.3+80.3) / 4$	**95.35**
C	$C_1 = (Y_1+Y_2+Y_7+Y_8) / 4$ $C_1 = (77.3+107.3+114.3+80.3) / 4$	**94.8**	$C_2 = (Y_3+Y_4+Y_5+Y_6) / 4$ $C_2 = (103.4+83.4+98{,}4+102.4) / 4$	**96.90**
D	$D_1 = (Y_1+Y_3+Y_5+Y_7) / 4$ $D_1 = (77.3+103.4+98.4+114.3) / 4$	**98.35**	$D_2 = (Y_2+Y_4+Y_6+Y_8) / 4$ $D_2 = (107.3+103.4+98{,}4+114.3) / 4$	**93.35**
E	$E_1 = (Y_1+Y_3+Y_6+Y_8) / 4$ $E_1 = (77.3+103.4+102.4+80.3) / 4$	**90.85**	$E_2 = (Y_2+Y_4+Y_5+Y_7) / 4$ $E_2 = (107.3+83.4+98{,}4+114.3) / 4$	**100.85**
F	$F_1 = (Y_1+Y_4+Y_5+Y_8) / 4$ $F_1 = (77.3+83.4+98.4+80.3) / 4$	**84.85**	$F_2 = (Y_2+Y_3+Y_6+Y_7) / 4$ $F_2 = (107.3+103.4+102.4+114.3) / 4$	**106.85**
G	$G_1 = (Y_1+Y_4+Y_6+Y_7) / 4$ $G_1 = (77.3+83.4+102.4+114.3) / 4$	**94.35**	$G_2 = (Y_2+Y_3+Y_5+Y_8) / 4$ $G_2 = (107.3+103.4+98.4+80.3) / 4$	**97.35**

These results can be re-arranged into the classic form of Response Table for Means format so that the relationship between inputs and output can be seen more clearly in *rank order,* as shown in Table 9.11,

TABLE 9.11 Response table for mean effects

	A	*B*	*C*	*D*	*E*	*F*	*G*
LOW	92.85	**96.35**	94.80	**98.35**	90.85	84.85	94.35
HIGH	**98.85**	95.35	**96.90**	93.35	**100.85**	**106.85**	**97.35**
| Delta |	6	1	2.1	5	10	22	3
%Delta	12.22	2.03	4.27	10.18	20.36	44.81	6.11
Rank	**3**	**7**	**6**	**4**	**2**	**1**	**5**

where *Rank* reveals the most important/effective input variable on output, i.e. 1 is being the most influential input, as 7 shows the least influential input on the *output*. We can also plot **"Main effects graphs for means"** for all input variables, where the higher level of the *slope* indicates the higher level of influence on the output/response.

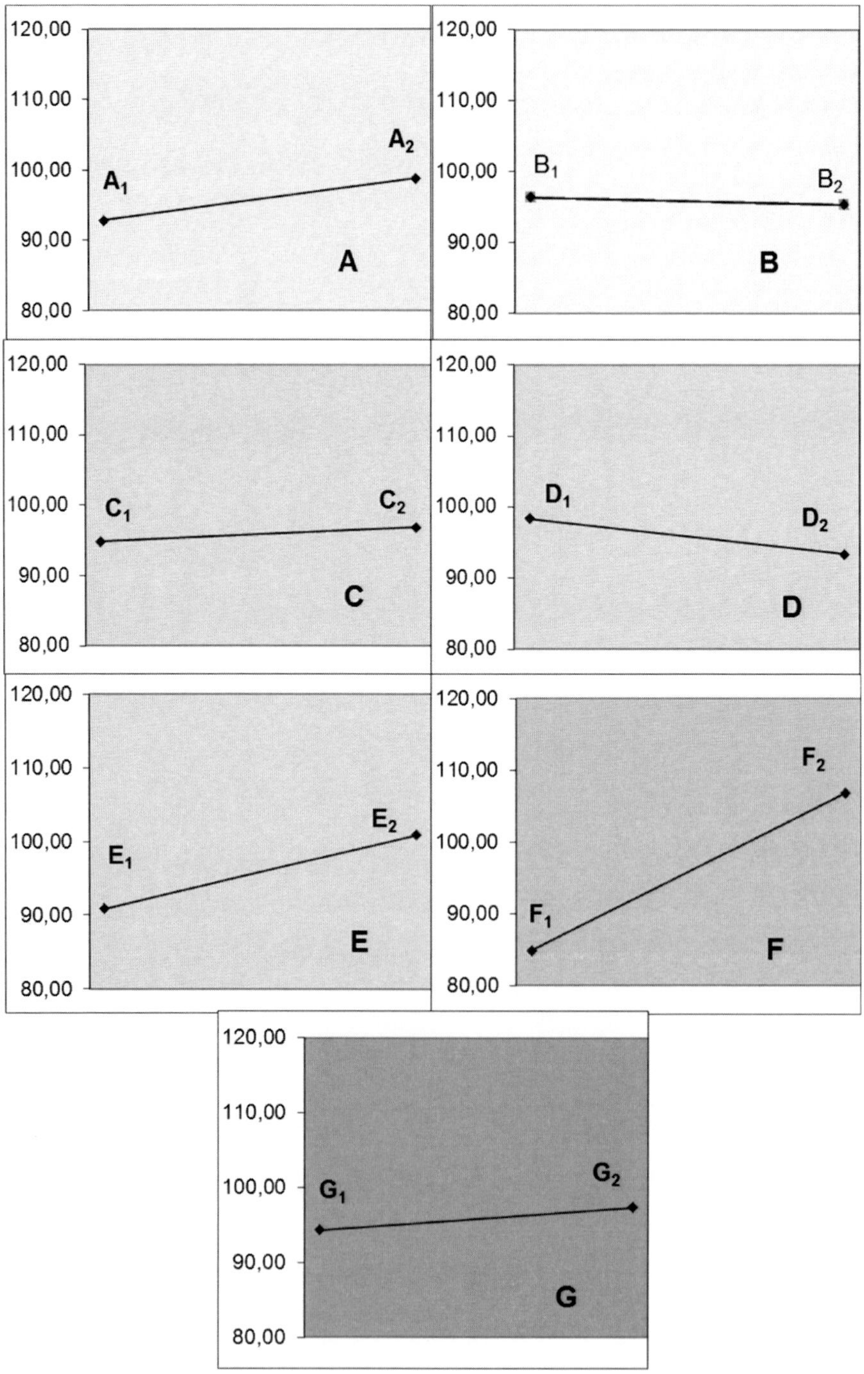

FIGURE 9.3 Mean effects graphs for input variables

Once we obtained this *Response Table for Means*, now, we can proceed with selecting the optimum design variable combinations according to the nature of our optimization's objective (or scenario). Let's assume this is a *maximization* problem, then we can go ahead and choose LOW or HIGH input values as given in Table 9.11 so that each input level contributes to the *maximization of* our objective function (i.e. output).

How? We choose the design variable input level in Figure 9.3 which corresponds to ***maximum*** values in *Means Effects* graph. For example, A_2 produces higher *Mean Effects* and so on

Optimum Design Input Combinations for Y_{max} → A_2, B_1, C_2, D_1, E_2, F_2, G_2

Thus, Y_{max} is given in Equation 9.4,

$$Y_{max} = T + (A_2 - T) + (B_1 - T) + (C_2 - T) + (D_1 - T) + (E_2 - T) + (F_2 - T) + (G_2 - T) \quad 9.4$$

$$Y_{max} = 95{,}85 + (98{,}85 - 95{,}85) + (96{,}35 - 95{,}85) + (96{,}90 - 95{,}85) + (98{,}35 - 95{,}85) + (100{,}85 - 95{,}85) + (106{,}85 - 95{,}85) + (97{,}35 - 95{,}85)$$

$$Y_{max} = 120.4$$

Similarly, if this was a *minimization* problem, then

Optimum Design Input Combinations for Y_{min} → A_1, B_2, C_1, D_2, E_1, F_1, G_1

$$Y_{min} = T + (A_1 - T) + (B_2 - T) + (C_1 - T) + (D_2 - T) + (E_1 - T) + (F_1 - T) + (G_1 - T) \quad 9.5$$

$$Y_{min} = 95{,}85 + (92{,}85 - 95{,}85) + (95{,}35 - 95{,}85) + (94{,}80 - 95{,}85) + (93{,}35 - 95{,}85) + (90{,}85 - 95{,}85) + (84{,}85 - 95{,}85) + (94{,}35 - 95{,}85)$$

$$Y_{min} = 71.3$$

Now, we can *verify* these results by comparing L-8 Taguchi Design, with full factorial design results, where recall that $2^7 = 128$ design combinations are possible and can be calculated easily, using Equation 9.2. Please again note that in real-life most of the time, there is no closed form of the equation like this one. Solely for learning purposes, such a closed equation is formed/supplied.

In Table 9.12, both L-8 Taguchi and Full Factorial Experimental design matrix/tables are presented, with the emphasis of Taguchi design selection out of full factorial design experiments. For example, EXPT. #1 corresponding to A_1,B_1,C_1,D_1,E_1,F_1 and G_1 input variables, Y_1 is calculated using Equation 9.2 as,

$$\boldsymbol{Y_1} = 2A_1 - 0.1B_1 + 3C_1 - 0.1D_1 + 5E_1 + 1.1F_1 + G_1 = \mathbf{77.3}$$

This Y_1 value corresponds to the first shaded area in a 16*8 matrix of full factorial (FF) design in $\mathbf{FF_{1*1}}$ location. Remaining 7 elements belonging to the L-8 array are also shown as shaded areas in the global full factorial design matrix.

If we go ahead and select the *Max.* and *Min Y* values out of full factorial design matrix, we end up with $\mathbf{FF_{8*6}}$,

$$\mathbf{Y_{max} = 120.4} \text{ for } \boldsymbol{A_2, B_1, C_2, D_1, E_2, F_2} \text{ and } \boldsymbol{G_2}$$

and $\mathbf{FF_{9*3}}$,

$$\mathbf{Y_{min} = 71.3} \text{ for } \boldsymbol{A_1, B_2, C_1, D_2, E_1, F_1} \text{ and } \boldsymbol{G_1}$$

as both shown underlined in Table 9.12,

TABLE 9.12 L-8 Taguchi orthogonal array and its corresponding experiments in $2^7 = 128$ full factorial experiments

L-8 / #	**A**	**B**	**C**	**D**	**E**	**F**	**G**
1	1	1	1	1	1	1	1
2	1	1	1	2	2	2	2
3	1	2	2	1	1	2	2
4	1	2	2	2	2	1	1
5	2	1	2	1	2	1	2
6	2	1	2	2	1	2	1
7	2	2	1	1	2	2	1
8	2	2	1	2	1	1	2

FULL FACTORIAL DESIGN EXPTS. **$2^7 = 128$**				A_1				A_2			
				B_1		B_2		B_1		B_2	
				C_1	C_2	C_1	C_2	C_1	C_2	C_1	C_2
D_1	E_1	F_1	G_1	77,3	79.4	76.3	78.4	83,3	85.4	82.3	84.4
			G_2	80.3	82.4	79.3	81.4	86.3	88.4	85.3	87.4
		F_2	G_1	99.3	101.4	98.3	100.4	105.3	107.4	104.3	106.4
			G_2	102.3	104.4	101.3	103.4	108.3	110.4	107.3	109.4
	E_2	F_1	G_1	87.3	89.4	86.3	88.4	93.3	95.4	92.3	94.4
			G_2	90.3	92.4	89.3	91.4	96.3	98.4	95.3	97.4
		F_2	G_1	109.3	111.4	108.3	110.4	115.3	117.4	114.3	116.4
			G_2	112.3	114.4	111.3	113.4	118.3	120.4	117.3	119.4
D_2	E_1	F_1	G_1	72.3	74.4	**71.3**	73.4	78.3	80.4	77.3	79.4
			G_2	75.3	77.4	74.3	76.4	81.3	83.4	80.3	82.4
		F_2	G_1	94.3	96.4	93.3	95.4	100.3	102.4	99.3	101.4
			G_2	97.3	99.4	96.3	98.4	103.3	105.4	102.3	104.4
	E_2	F_1	G_1	82.3	84.4	81.3	83.4	88.3	90.4	87.3	89.4
			G_2	85.3	87.4	84.3	86.4	91.3	93.4	90.3	92.4
		F_2	G_1	104.3	106.4	103.3	105.4	110.3	112.4	109.3	111.4
			G_2	107.3	109.4	106.3	108.4	113.3	115.4	112.3	114.4

Full factorial design combinations are also given in Table 9.13, totaling 128 design combinations. Notice to the shaded L-8 Taguchi designs.

TABLE 9.13 $2^7 = 128$ Full factorial experiments

#	ABCDEFG	#	ABCDEFG	#	ABCDEFG	#	ABCDEFG
1	1111111	33	1111121	65	1111112	97	1111122
2	2111111	34	2111121	66	2111112	98	2111122
3	1211111	35	1211121	67	1211112	99	1211122
4	2211111	36	2211121	68	2211112	100	2211122
5	1121111	37	1121121	69	1121112	101	1121122
6	2121111	38	2121121	70	2121112	102	2121122
7	1221111	39	1221121	71	1221112	103	1221122
8	2221111	40	2221121	72	2221112	104	2221122
9	1112111	41	1112121	73	1112112	105	1112122
10	2112111	42	2112121	74	2112112	106	2112122
11	1212111	43	1212121	75	1212112	107	1212122
12	2212111	44	2212121	76	2212112	108	2212122
13	1122111	45	1122121	77	1122112	109	1122122
14	2122111	46	2122121	78	2122112	110	2122122
15	1222111	47	1222121	79	1222112	111	1222122
16	2222111	48	2222121	80	2222112	112	2222122
17	1111211	49	1111221	81	1111212	113	1111222
18	2111211	50	2111221	82	2111212	114	2111222
19	1211211	51	1211221	83	1211212	115	1211222
20	2211211	52	2211221	84	2211212	116	2211222
21	1121211	53	1121221	85	1121212	117	1121222
22	2121211	54	2121221	86	2121212	118	2121222
23	1221211	55	1221221	87	1221212	119	1221222
24	2221211	56	2221221	88	2221212	120	2221222
25	1112211	57	1112221	89	1112212	121	1112222
26	2112211	58	2112221	90	2112212	122	2112222
27	1212211	59	1212221	91	1212212	123	1212222
28	2212211	60	2212221	92	2212212	124	2212222
29	1122211	61	1122221	93	1122212	125	1122222
30	2122211	62	2122221	94	2122212	126	2122222
31	1222211	63	1222221	95	1222212	127	1222222
32	2222211	64	2222221	96	2222212	128	2222222

9.4 Taguchi Analysis using Signal to Noise Ratio (S/N)

After successfully implementing the *mean response effects* in Taguchi analysis, we have shown that we can assess the effect of each input on the global response (output). It has been proven that we could determine the optimal (critical) or near optimal design point for our objective quantitatively. In robust design, our target is to identify/rank the input control factors and their influences on output, so that we can minimize or control the *variability* of our system (likewise the effects of noise).

Now, we can take a look at an alternate (second) measure called, *Signal to Noise Ratio (S/N),* which is considered a key measure in the robust parameter design (optimization) process which quantifies the level of desired effects of inputs on output variable(s) with minimizing the unwanted noise(s). Using *S/N* ratio has many usage benefits as explained in great details in the literature, therefore we would like to limit our discussion and focus more on its application.

The *S/N* ratio can be developed by accounting for both mathematical aspects and human judgment/intervention. By combining these two factors, three possible scenarios in any experimental design (optimization) are encountered, as presented in the following;

- **Larger is better**, in these cases, the output of the experimental design or objective function is to be maximized, i.e. gas mileage of the automotive, distance traveled, profit, battery life, customer satisfaction etc. as presented in equation 9.6, where *Y* or *y* is output,

$$S/N = -10\log_{10}\left[\frac{\sum(1/y^2)}{n}\right] \qquad 9.6$$

- **Nominal is best**, in these cases, our objective is to come up with a nominal solution of interest, i.e. nominal shaft dimension are to be manufactured, or an operating point where efficiency goes down in either case (fluctuations from nominal settings) etc. as presented in equation 9.7 or 9.8,

$$S/N = 10\log_{10}\left[\frac{\bar{y}^2}{s^2}\right] \qquad 9.7$$

or

$$S/N = -10\log_{10}[s^2] \qquad 9.8$$

- **Smaller is better**, in these cases, gas mileage cost, cost of the project, the weight of the bridge, stress or stress concentration in a structure, waste after manufacturing, noise of the refrigerator etc. are to be minimized as presented in Equation 9.9,

$$S/N = -10\log_{10}\left[\frac{\sum(y^2)}{n}\right] \quad 9.9$$

Notice that, when we consider the *S/N* ratio, our objective is to maximize the *S/N* ratio by minimizing *noise*, this is valid for all three scenarios presented above. Now, we can go ahead and proceed to an example *Taguchi* analysis now, by evaluating *S/N* values in an example analysis.

Let's go ahead and use the same example problem utilized in *Mean Response Analysis* case in early section, i.e. where, there were 7 factors i.e. *A,B,C,D,E,F* and *G* and each at 2 levels (LOW and HIGH), considered within the experimental design space in terms of LOW and HIGH levels (bounds) as given in Table 9.14.

In the following, L-8 Taguchi design array again is presented for eight possible (input) design combinations and their corresponding outputs are calculated according to Equation 9.6 for *S/N* ratio as shown in Table 9.14,

TABLE 9.14 Taguchi L-8 array (inputs/output)

EXPT. #	A	B	C	D	E	F	G	*Y*	*S/N*
1	1	1	1	1	1	1	1	*77.3*	$S/N_1 = 37.7636$
2	1	1	1	2	2	2	2	*107.3*	$S/N_2 = 40.6120$
3	1	2	2	1	1	2	2	*103.4*	$S/N_3 = 40.2904$
4	1	2	2	2	2	1	1	*83.4*	$S/N_4 = 38.4233$
5	2	1	2	1	2	1	2	*98.4*	$S/N_5 = 39.8599$
6	2	1	2	2	1	2	1	*102.4*	$S/N_6 = 40.2060$
7	2	2	1	1	2	2	1	*114.3*	$S/N_7 = 41.1609$
8	2	2	1	2	1	1	2	*80.3*	$S/N_8 = 38.0943$

Please recall that our experiments have yielded eight output values for *Y* and associated *S/N* outputs are calculated using Equation 9.6, clearly these values depend on the design input levels as presented in Table 9.14. Now, our objective is to maximize the *Y,* in this case, using the *S/N* ratio's (the bigger is better).

In the following, the calculation of both $(S/N)_1$ and $(S/N)_8$ ratios are shown, again using Equation 9.6, where n = 1, i.e. we got only one output,

$$(S/N)_1 = -10\log_{10}\left[\frac{1/77.3^2}{n=1}\right] = 37.7636$$

and

$$(S/N)_8 = -10\log_{10}\left[\frac{1/80.3^2}{n=1}\right] = 38.0943$$

Now, using L-8 Taguchi design we found *S/N* ratios. We can assess the effects of each input variable at LOW and HIGH values as an alternative way to *mean response calculations*, by averaging the corresponding columns LOW and HIGH output results for each design variables, as shown in Table 9.15,

TABLE 9.15 Total effects of input variables on output *S/N ratios* at LOW and HIGH values

#	LOW	HIGH
A	$A_1 = (S/N_1 + S/N_2 + S/N_3 + S/N_4)/4$ $A_1 = (37.7636+40.6120+40.2904+38.4233)/4$ $\mathbf{A_1 = 39.27}$	$A_2 = (S/N_5 + S/N_6 + S/N_7 + S/N_8)/4$ $A_2 = (39.8599+40.2060+41.1609+38.0943)/4$ $\mathbf{A_2 = 39.83}$
B	$B_1 = (S/N_1 + S/N_2 + S/N_5 + S/N_6)/4$ $B_1 = (37.7636+40.6120+39.8599+40.2060)/4$ $\mathbf{B_1 = 39.61}$	$B_2 = (S/N_3 + S/N_4 + S/N_7 + S/N_8 /4$ $B_2 = (40.2904+38.4233+41.1609+41.1609)/4$ $\mathbf{B_2 = 39.49}$
C	$C_1 = (S/N_1 + S/N_2 + S/N_7 + S/N_8) / 4$ $C_1 = (37.7636+40.6120+41.1609+38.0943)/4$ $\mathbf{C_1 = 39.41}$	$C_2 = (S/N_3 + S/N_4 + S/N_5 + S/N_6)/4$ $C_2 = (40.2904+38.4233+39.8599+40.2060)/4$ $\mathbf{C_2 = 39.69}$
D	$D_1 = (S/N_1 + S/N_3 + S/N_5 + S/N_7)/4$ $D_1 = (37.7636+40.2904+39.8599+41.1609)/4$ $\mathbf{D_1 = 39.77}$	$D_2 = (S/N_2 + S/N_4 + S/N_6 + S/N_8)/4$ $D_2 = (40.6120+38.4233+40.2060+41.1609)/4$ $\mathbf{D_2 = 39.33}$
E	$E_1 = (S/N_1 + S/N_3 + S/N_6 + S/N_8)/4$ $E_1 = (37.7636+40.2904+40.2060+38.0943)/4$ $\mathbf{E_1 = 39.09}$	$E_2 = (S/N_2 + S/N_4 + S/N_5 + S/N_7)/4$ $E_2 = (40.6120+38.4233+39.8599+41.1609)/4$ $\mathbf{E_2 = 40.01}$
F	$F_1 = (S/N_1 + S/N_4 + S/N_5 + S/N_8)/4$ $F_1 = (37.7636+38.4233+39.8599+38.0943)/4$ $\mathbf{F_1 = 38.54}$	$F_2 = (S/N_2 + S/N_3 + S/N_6 + S/N_7)/4$ $F_2 = (37.7636+40.2904+40.2060+41.1609)/4$ $\mathbf{F_2 = 40.57}$
G	$G_1 = (S/N_1 + S/N_4 + S/N_6 + S/N_7)/4$ $G_1 = (37.7636+38.4233+40.2060+41.1609)/4$ $\mathbf{G_1 = 39.39}$	$G_2 = (S/N_2 + S/N_3 + S/N_5 + S/N_8)/4$ $G_2 = (40.6120+40.2904+39.8599+38.0943)/4$ $\mathbf{G_2 = 39.71}$

These results can be re-arranged into the classic form of *Response Table for S/N Ratios* similar to *Response Table for Means* format so that the relationship between inputs and output can be seen more clearly in *Rank*, as shown in Table 9.16,

TABLE 9.16 Response table for *S/N* ratio effects

	A	*B*	*C*	*D*	*E*	*F*	*G*
LOW	39.27	**39.61**	39.41	**39.77**	39.09	38.54	39.39
HIGH	**39.83**	39.49	**39.69**	39.33	**40.01**	**40.57**	**39.71**
| *Delta* |	0.56	0.12	0.29	0.43	0.93	2.03	0.33
%Delta	11.94	2.55	6.18	9.17	19.83	43.28	7.03
Rank	**3**	**7**	**6**	**4**	**2**	**1**	**5**

Again, after obtaining Response Table for *S/N* Ratios, we can proceed with selecting the optimum design variable combinations to be consistent with *S/N ratio maximization* logic, which is always what we target! Then we can go ahead and choose either LOW or HIGH input values in Table 9.16 so that each input helps to contribute the *S/N maximization.* In fact, globally we will be maximizing our objective function (i.e. output), in exactly the way we required!

Optimum Design Input Combinations for S/N_{max} → A_2, B_1, C_2, D_1, E_2, F_2, G_2

Please also note that using Equation 9.3, again we can obtain an average value of *S/N* ratios as indicated by *T*, where S/N_i, $i=1,8$, we calculate *T* as,

$$T = \frac{(S/N_1 + S/N_2 + S/N_3 + S/N_4 + S/N_5 + S/N_6 + S/N_7 + S/N_8)}{8}$$

$$T = 39{,}551$$

Now, using L-8 Taguchi design analysis, we can assess the effects of each input variable at LOW and HIGH values, by averaging the corresponding columns LOW and HIGH output results for each design variable as shown in Table 9.15. Using Equation 9.4, we can find the maximum *S/N* ratio as follows,

$$S/N_{max} = T + (A_2 - T) + (B_1 - T) + (C_2 - T) + (D_1 - T) + (E_2 - T) + (F_2 - T) + (G_2 - T)$$

$$S/N_{max} = 41.892$$

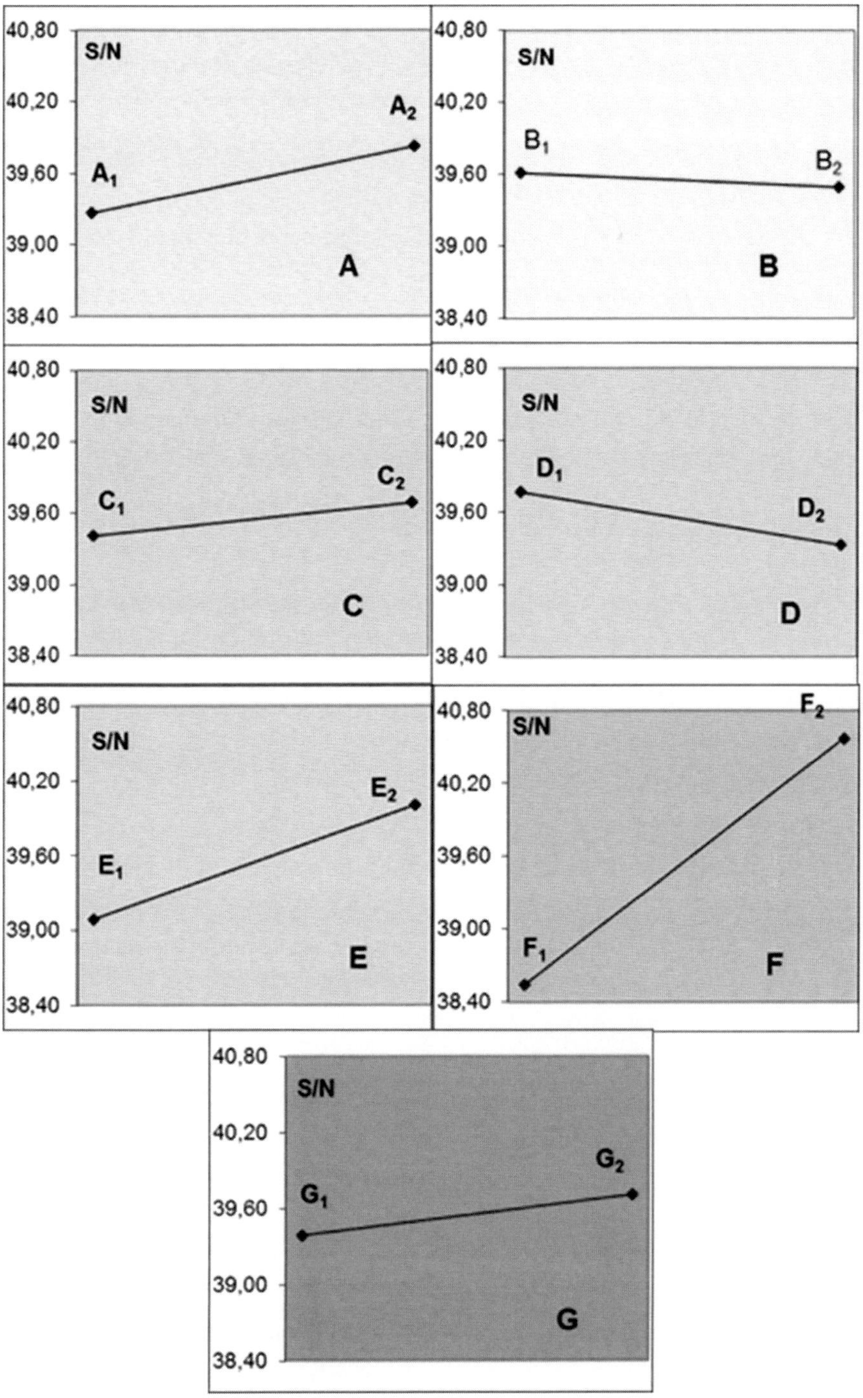

FIGURE 9.4 *S/N* Ratios effect graphs for inputs/factors

9.5 Minitab Taguchi Analysis using Mean Response Effects

As far as any statistical analysis is concerned, the main idea/logic, formulations and steps to be followed by the user are very well described and documented in the literature. Therefore, one can easily attain to his/her own solutions using spreadsheets and/or word processing tools.

An individual effort spent in solving any statistical problem has certainly a great advantage from the user's perspective in learning. This effort contributes user to accumulate invaluable experience in understanding what is going on in the whole solution steps. Once this experience is acquired, now an individual can go ahead and use any commercial packages, because he/she already knows how these calculations are performed in these packages.

Now, let's perform the same analysis using *Minitab*, for the parts that have been demonstrated in section 9.3 and 9.4. In this walkthrough solution procedure, we can emphasize four main benefits as listed below;

- For verification, increases self-confidence!
- Get a better grip in the understanding of Taguchi analysis, which provides practical improvements
- Faster/practical for usage and presentation, especially helpful in hectic work environments
- Laying out the foundation for professional use in future real-life problems

Then, using our computer,

1. Let's start up **Minitab** software that is already installed on our computer. We can go ahead and start up the Minitab software from the Windows Start menu by clicking as in the following order, and follow the GUI flow of instructions as presented below.

\> **All Programs>Minitab>Minitab 1X Statistical Software**

2. Click on **Stat > DEO > Taguchi >** T$_g$ **Create Taguchi Design…** tabs,

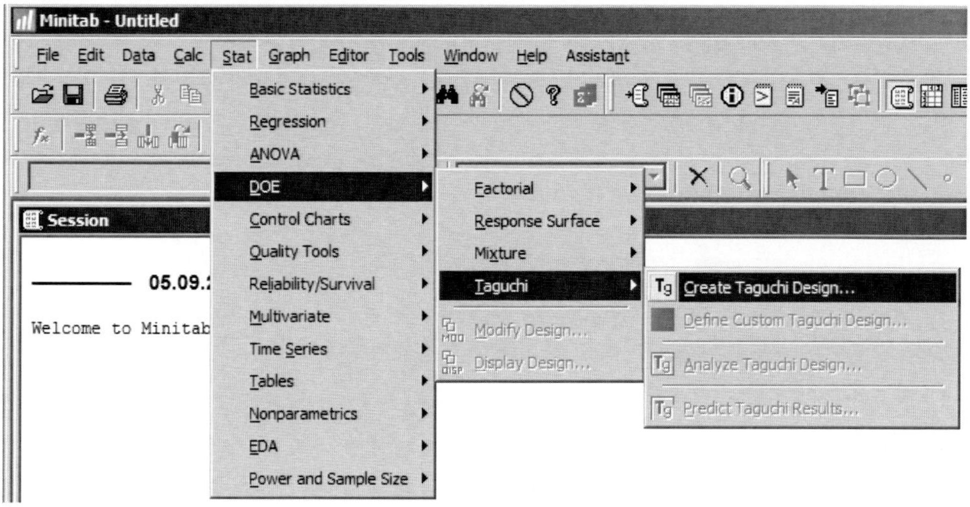

3. Now **Taguchi Design** main window will pop-up, where you are going to specify, ***x*-Level Design**, (*x=2* for Low, High levels or *x=3* for Low, Medium, High levels). From *Type of Design* options, choose ***2*-Level Design** for our problem. Please also set **Number of Factors:** 7, which corresponds to the number of our example inputs (factor, design) variables. Now, let's focus on **Taguchi Design** window options in detail.

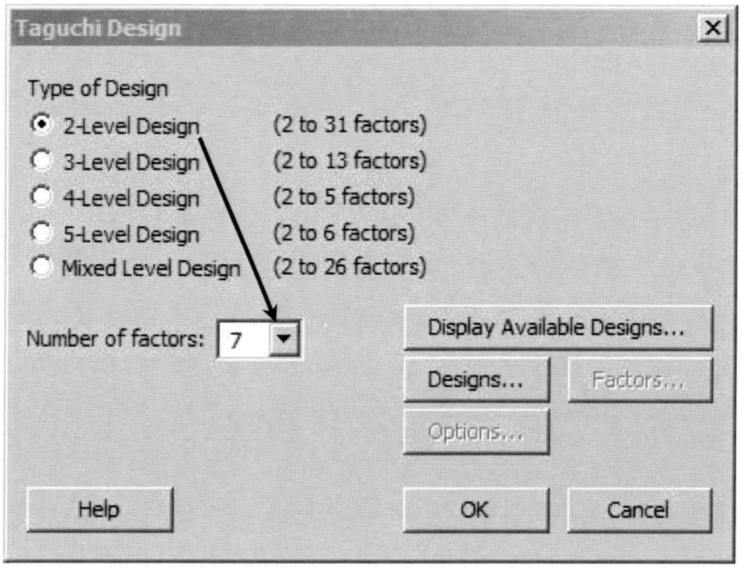

3.1. Under **Taguchi Design** main window, explore **Display Available Designs...** Tab by clicking on it,

Taguchi Design – Available Designs window pops-up with **Available Taguchi Designs (with Number of Factors)** for our chosen configuration, i.e. in **Single-level** tab notice that **2 level** column will be having our candidate Taguchi Design Arrays, i.e. L-8, L-12, L-16, and L-32 arrays, (Not L-4! since we have more than 3 inputs, program does not know this yet!). Hit OK to close the window.

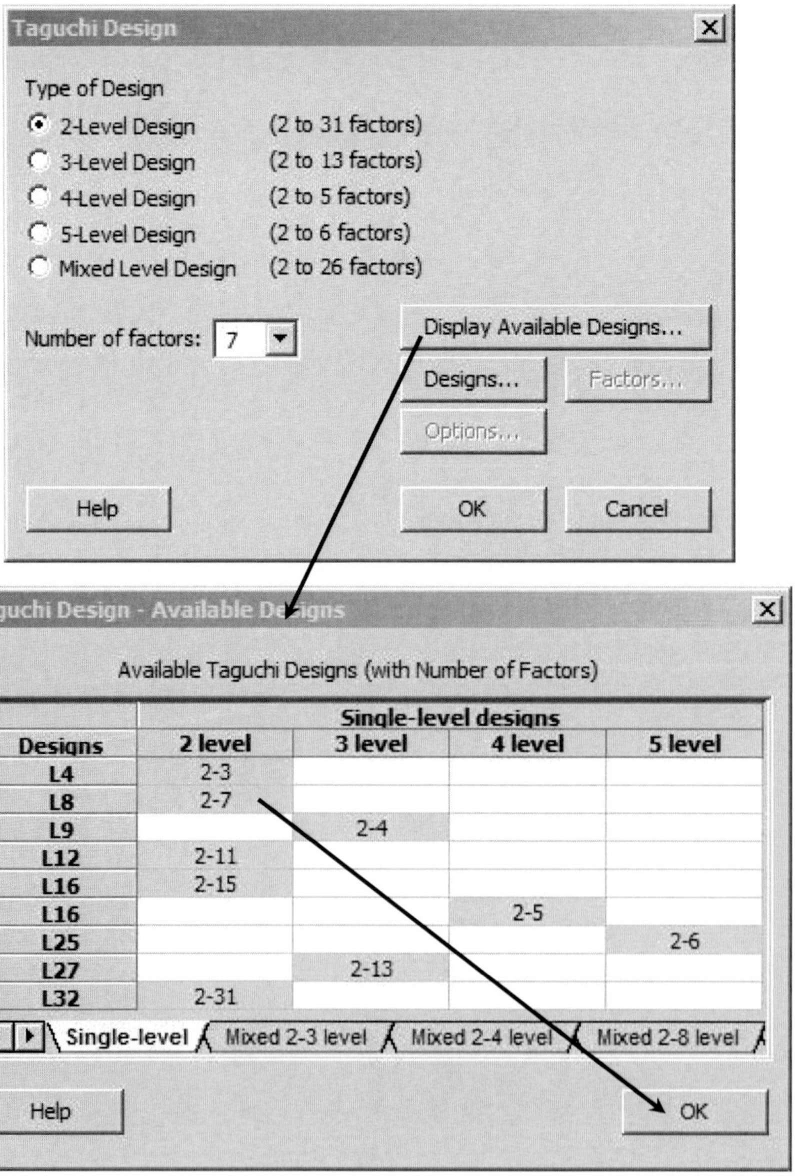

3.2. Again, Under **Taguchi Design** main window, explore **Designs…** Tab by clicking on it.

Taguchi Design – Design Window pops-up where choose **L-8** which requires the least number of experiments (corresponding to 2^7 designs). Hit OK to close the window.

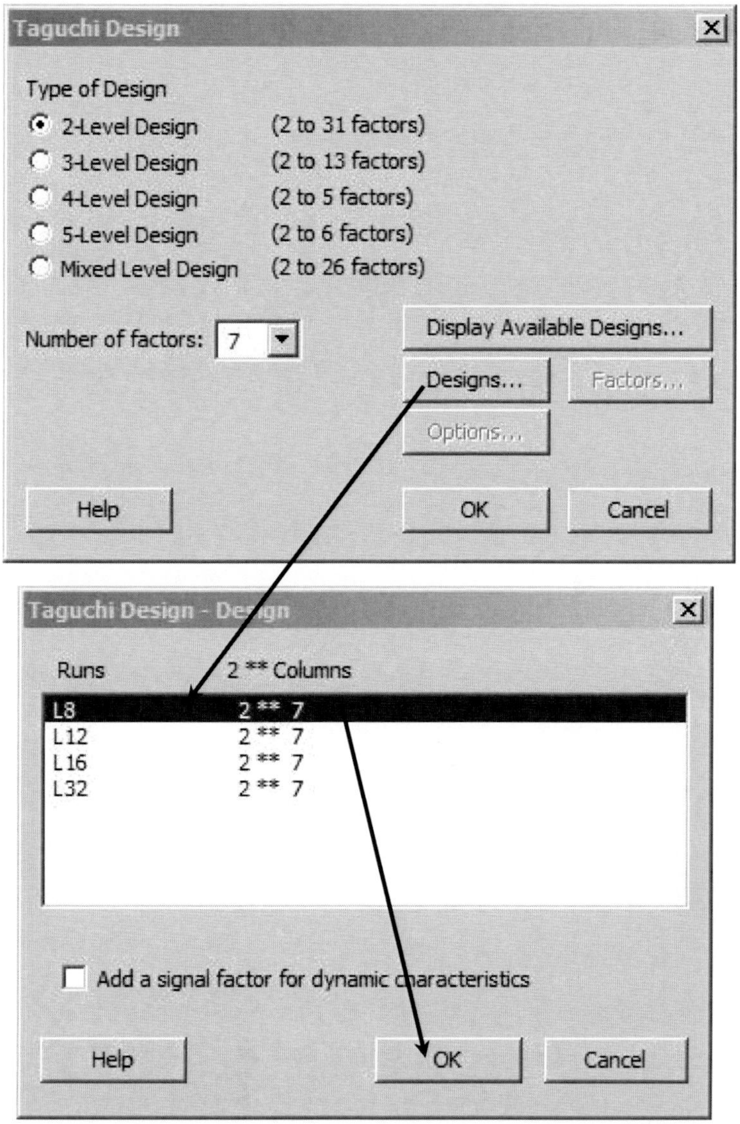

*** Notice that under **Taguchi Design** main window, NOW **Factors…** and **Options…** tabs become active (executable)!

3.3. Again, Under **Taguchi Design** main window, explore **Factors...** Tab by clicking on it.

Taguchi Design – Factors Window pops-up where **Fact** represents A,B,C,D,E,F and G input variables with Level values 1 refers to LOW and Level values 2 refers to a HIGH level. (You may scroll down to see F and G input values).

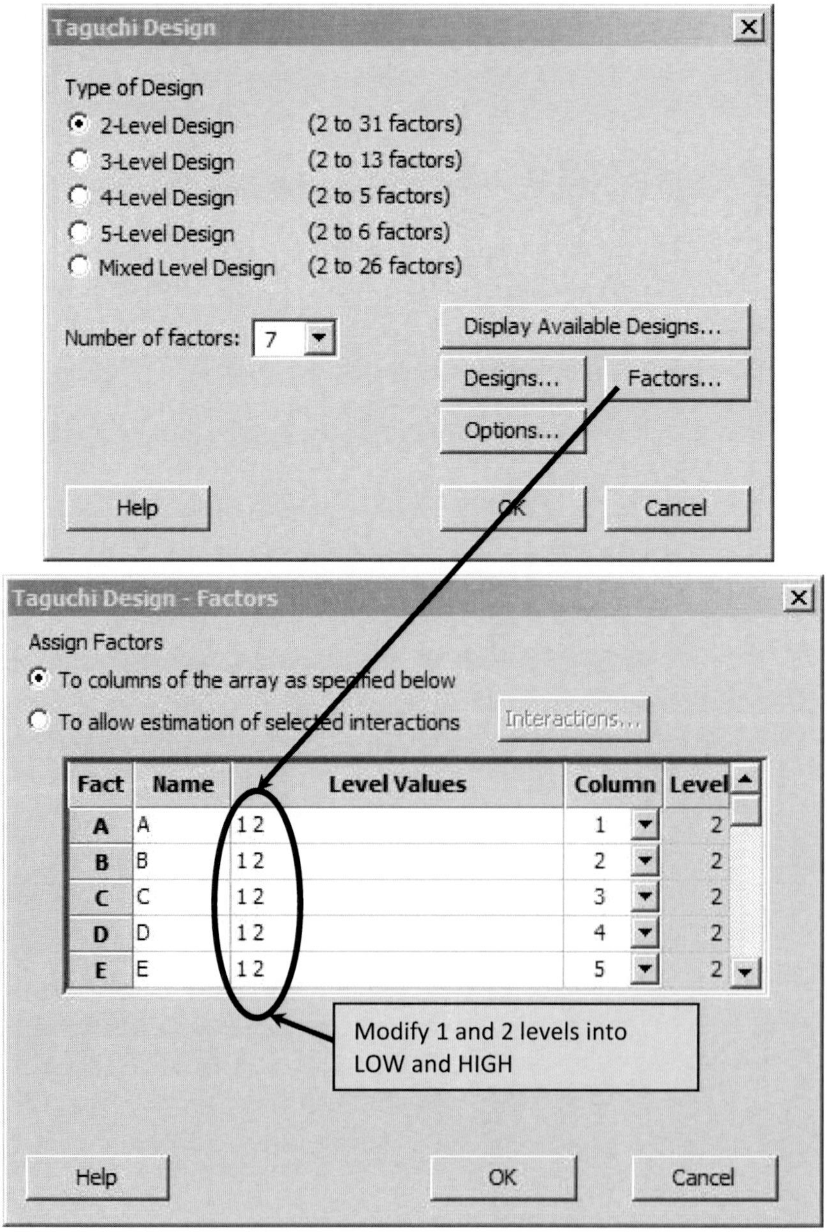

Please note that in **Taguchi Design – Factor** window, 1 and 2 level values can easily be modified into real LOW and HIGH values given in the following table. How? click on these numbers, they will be highlighted and you will see the cursor, now you can insert (edit) these values for this problem into LOW/HIGH values given in the following.

LOW	HIGH
$A_1 = 2$	$A_2 = 5$
$B_1 = 10$	$B_2 = 20$
$C_1 = 0.1$	$C_2 = 0.8$
$D_1 = 50$	$D_2 = 100$
$E_1 = 2$	$E_2 = 4$
$F_1 = 60$	$F_2 = 80$
$G_1 = 3$	$G_2 = 6$

Again, also note that you need to scroll all the way down to see remaining two factors i.e. F and G, which are not seen in the following window, due to visibility limitation in this window. Hit OK to close the window!

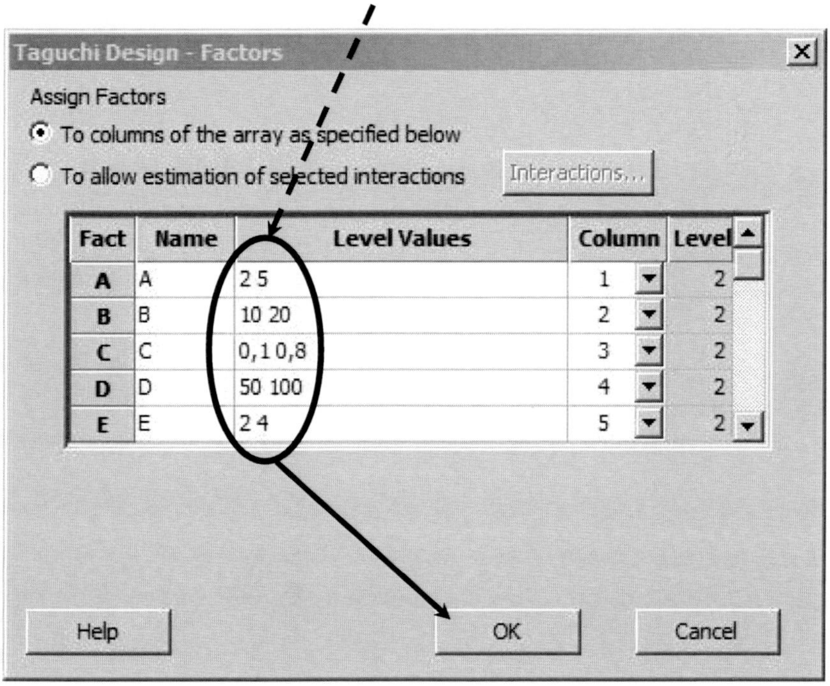

3.4. Again, Under **Taguchi Design** main window, explore **Options...** Tab by clicking on it.

Taguchi Design – Options Window pops-up where you get this window with **Store design in worksheet** option clicked on so that you can store Taguchi design array into the worksheet. Hit OK to close the window!

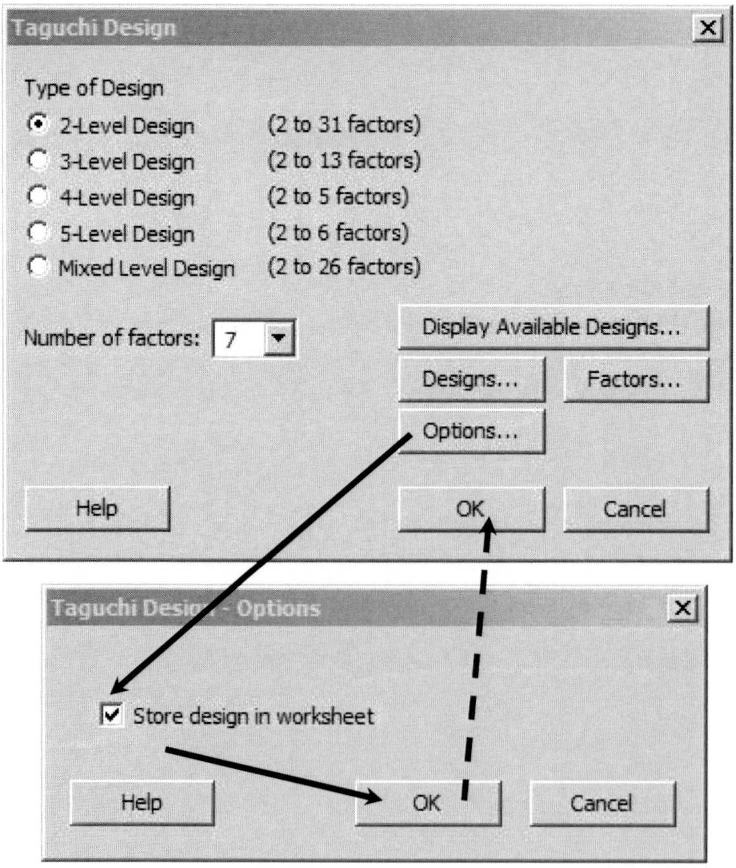

4. Close **Taguchi Design** window above by hitting OK tab. Finally, you will be getting a message from Minitab indicating that, we end up with Taguchi Design with the L-8 array, i.e. 7 factors (input variables) and 8 runs (experiments, analyses).

Note that LOW and HIGH values for L-8 array are now stored in your Minitab file, as *Worksheet1*. Also, note that all seven columns of the L-8 array are used. Here, *no interaction(s)* between the columns(inputs, factors) are assumed. This means that all the factors (input variables) are decoupled or in other words, they are mutually exclusive, i.e. they have nothing in common!

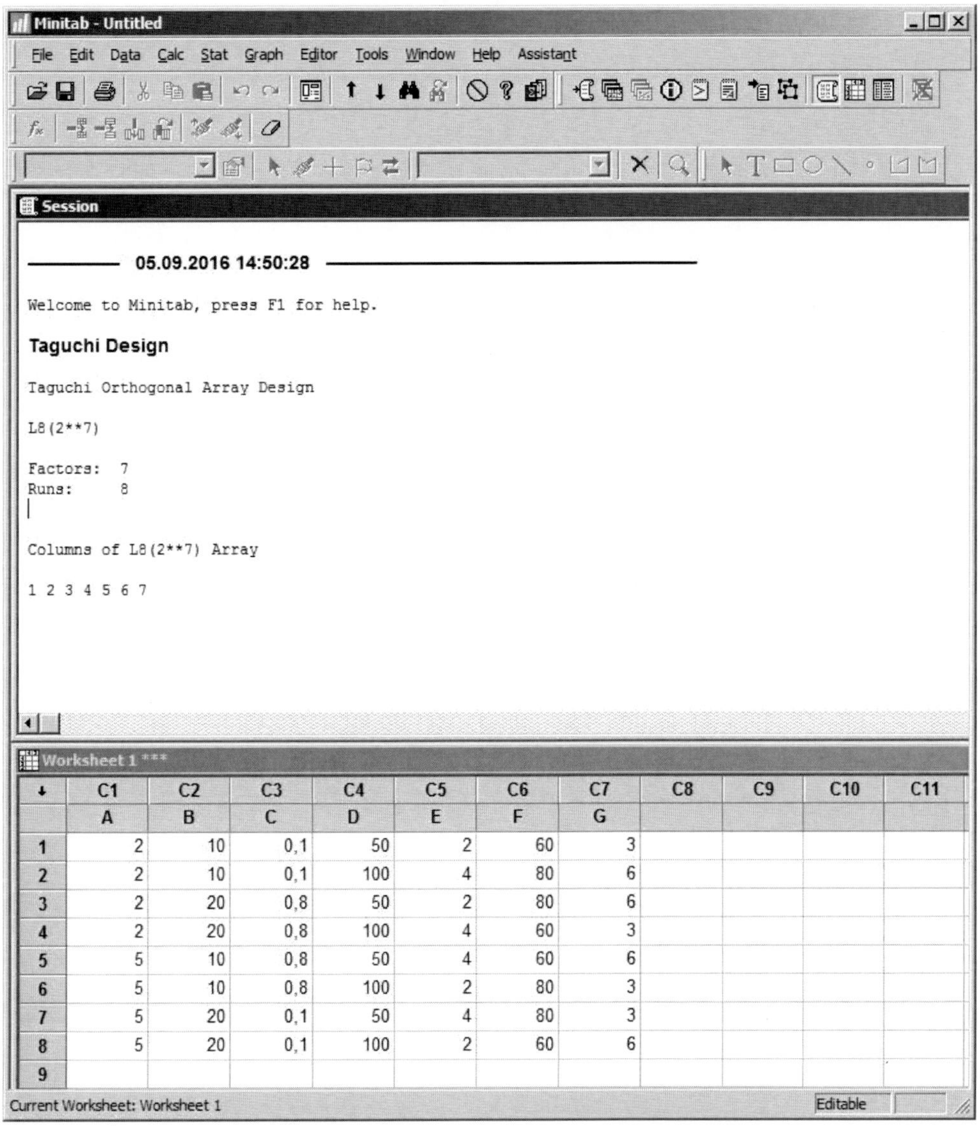

5. Now, we are done in constructing and defining L-8 Taguchi Design array, we may proceed and enter inputs/factors, next!

Click on **Stat** > **DEO** > **Taguchi** > **T_{gc} Define Custom Taguchi Design…** tabs,

5.1. **Define Custom Taguchi Design** window will pop-up with all 7 inputs/factors, while holding the shift key, select all of the factors(inputs) with mouse and hit a select button as shown. A-G inputs/factors will go in **Factors:** area. Hit OK to close the window.

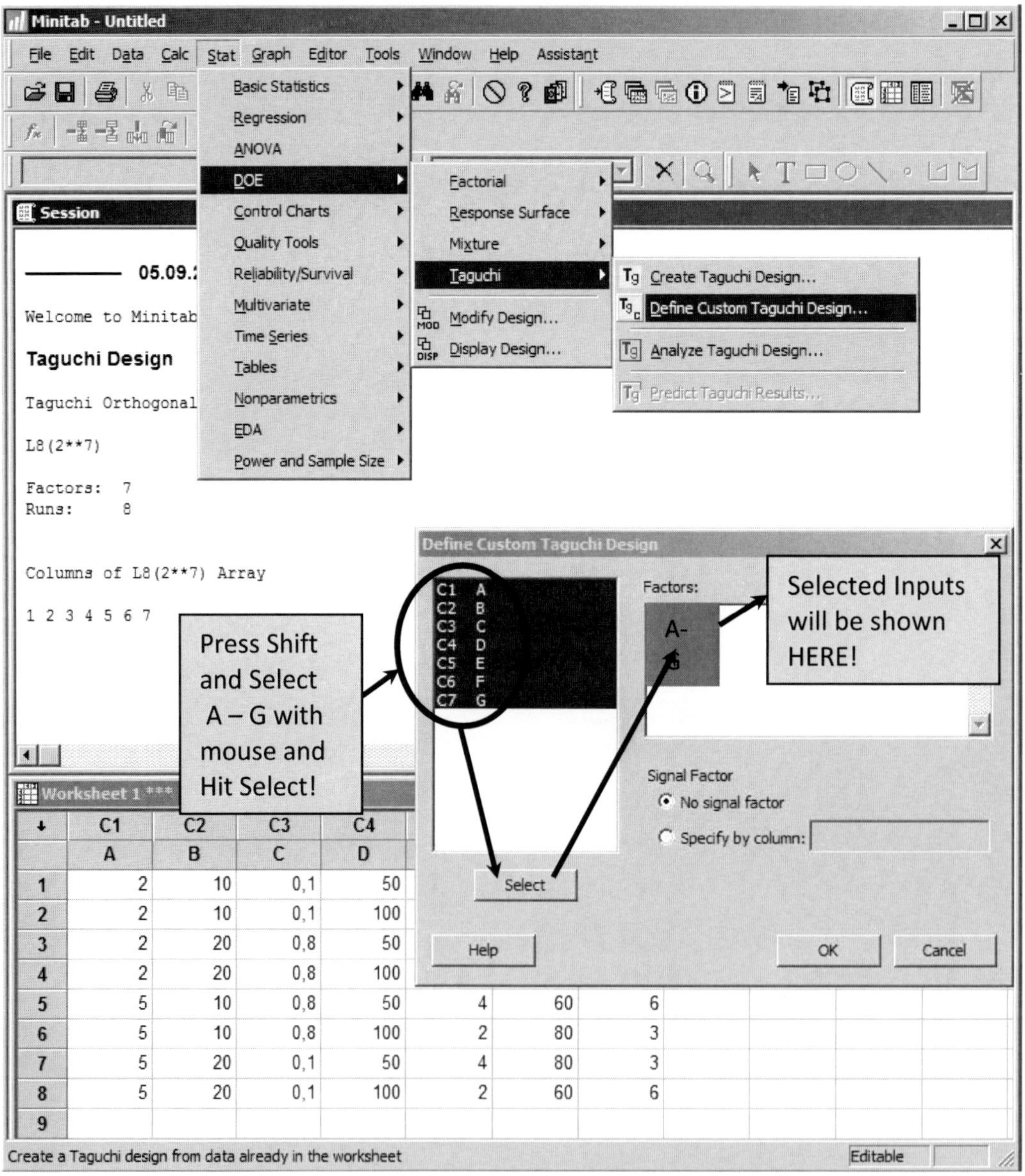

6. Now, you can copy your single ***Output*** column from another *Excel* document that you got, or you may enter them manually to the **C8** column of *Minitab* file. You are ready to let the program know what the *Output* will be!

   6.1. Click on **Stat** > **DEO** > **Taguchi** > $\mathbf{T_g}$ **Analyze Taguchi Design...** tabs, **Analyze Taguchi Design** window will pop-up and select **C8 Output** as a response by clicking with the mouse and Hit Select button! After that, **Output** will move to the **Response data are in:** area, as shown in the following.

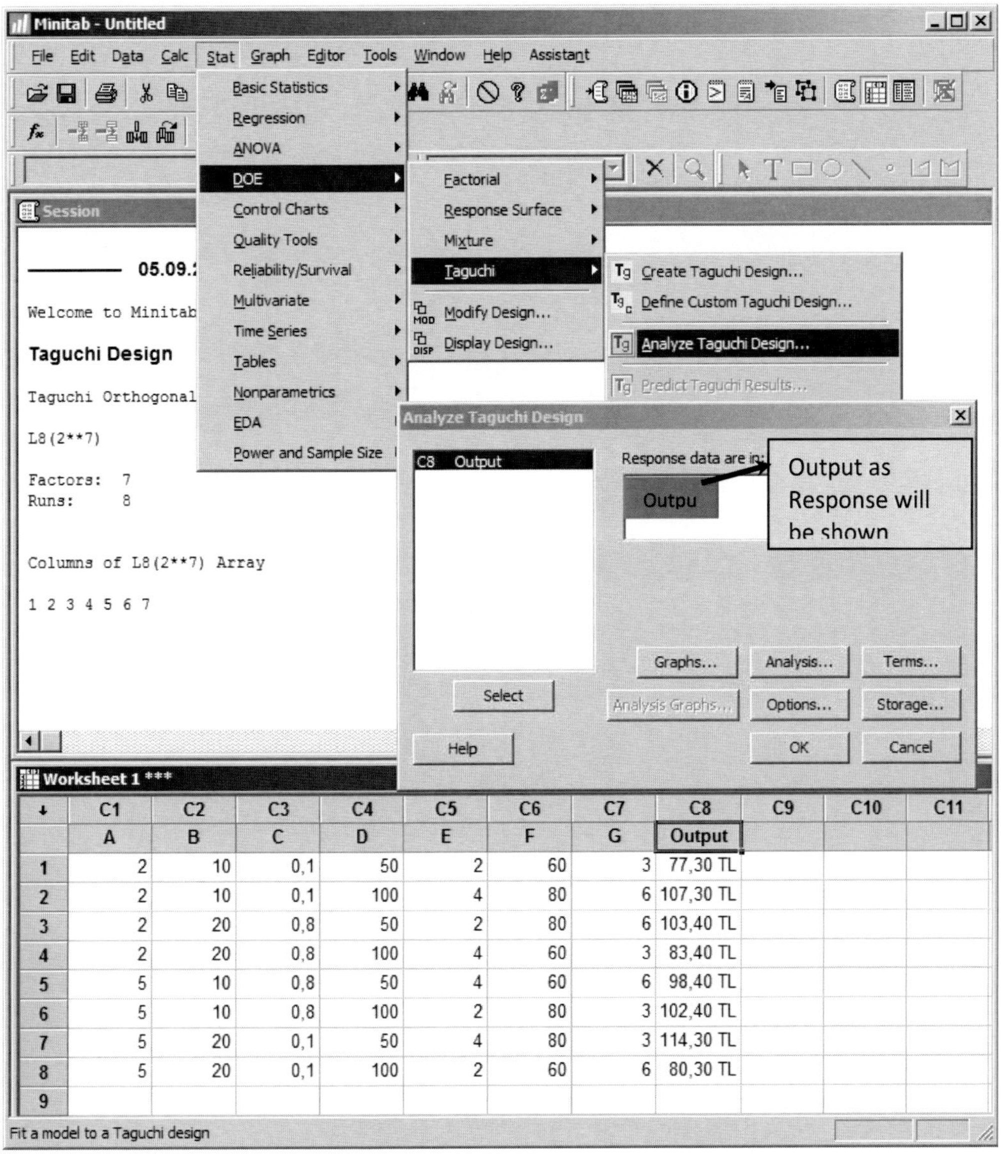

6.2. Still in **Analyze Taguchi Design** window tab, when you Click on **Graphs…** tab you shall first see the options to use in Taguchi response calculations for graphing.

- **Signal to Noise (S/N) Ratios** and **Means** options come as Ticked, in DEFAULT. Please ADOPT these two options AS THEY ARE.
- **Standard deviations,** this options comes as NOT Ticked, do not CHANGE!

Next, you are required to choose Plotting options either for group or separate graph outputs, please ADOPT AS IS. Hit OK to close the window!

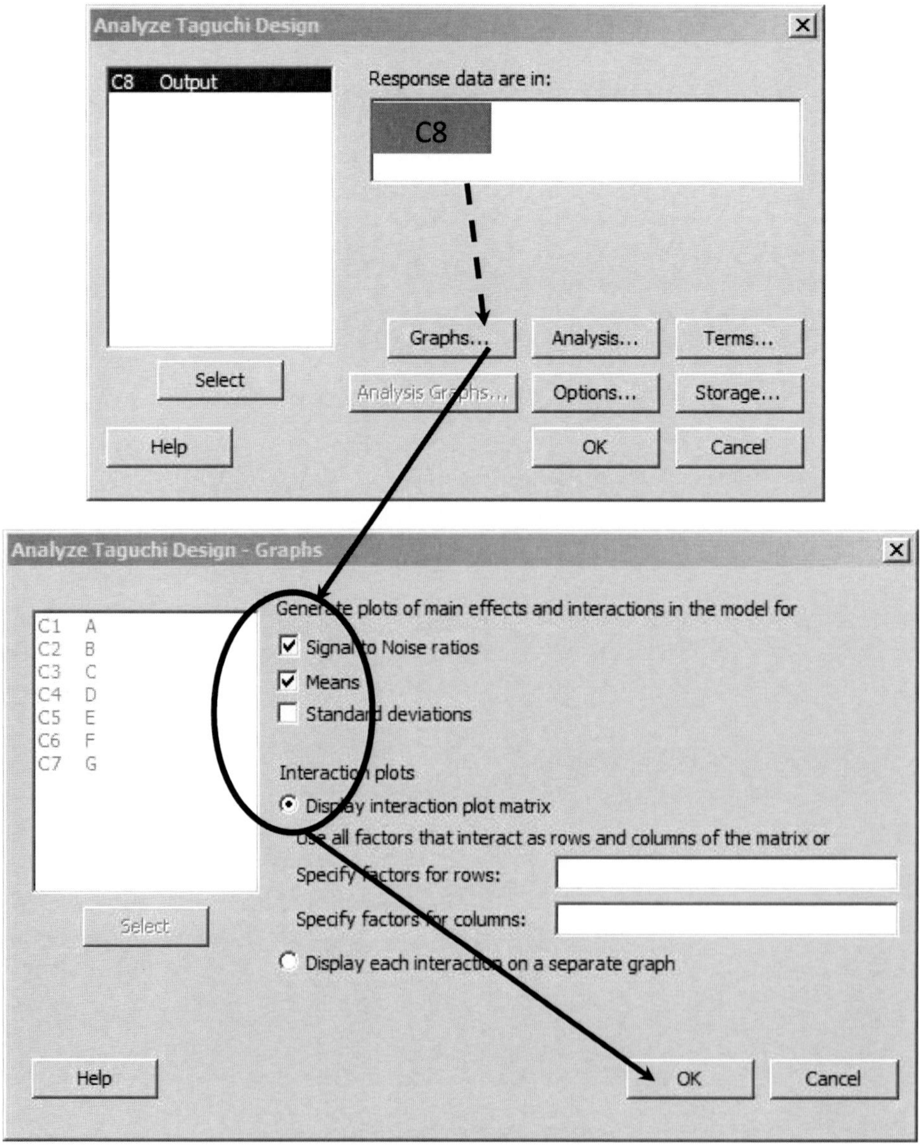

6.3. Continuing in **Analyze Taguchi Design** window, if you click on **Analysis…** tab, you will be getting **Analyze Taguchi Design – Analyze** tab. Please note that following two options are ticked as DEFAULT,

- *Signal to Noise ratios*,
- *Means*,

Hit OK to close this window to proceed to other analysis options, as this would be shown on the next page!

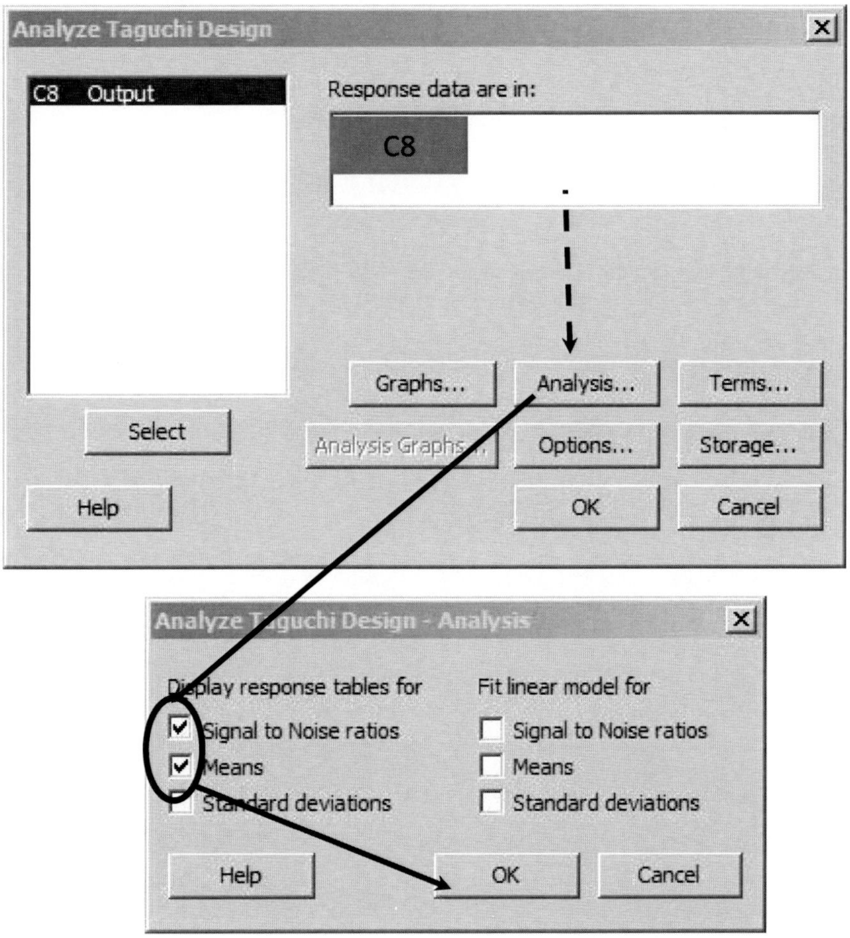

6.4. Continuing in **Analyze Taguchi Design** window, if you click on **Terms…** tab, you will be getting **Analyze Taguchi Design – Terms** window, where all terms for constructing Taguchi design are displayed with interactions, for this problem just AS DEFAULT all seven design (input, factor) variables are SELECTED! since there is no coupling term (i.e. *no interaction* is assumed between the input variables) in this example, please proceed with this DEFAULT selection, as presented in the following. Hit OK to proceed to other options.

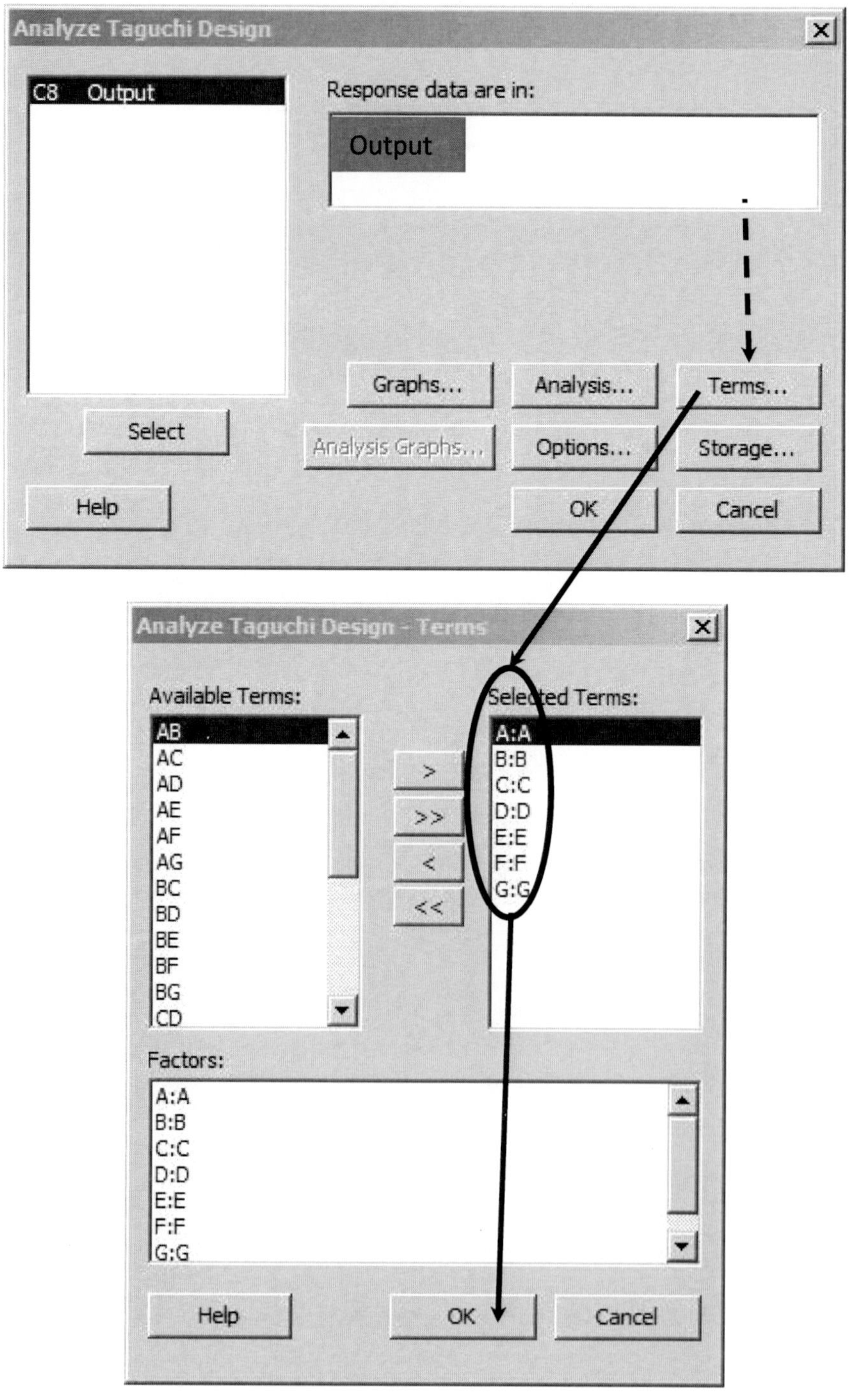

6.5. Continuing in **Analyze Taguchi Design** window, if you click on **Options…** tab, You will get **Analyze Taguchi Design - Options** window, where you will notice three possible combinations for Response (or objective function) to be, which are

- A *S/N* formulation for *Larger is better* case,
- Two *S/N* formulations for *Nominal is best* case,
- A *S/N* formulation for *Smaller is better* case,

Please choose the first option, which is ***Larger is better!*** Notice that *S/N* ratio formulations used in this software are also given here in this window. Hit OK to close the window!

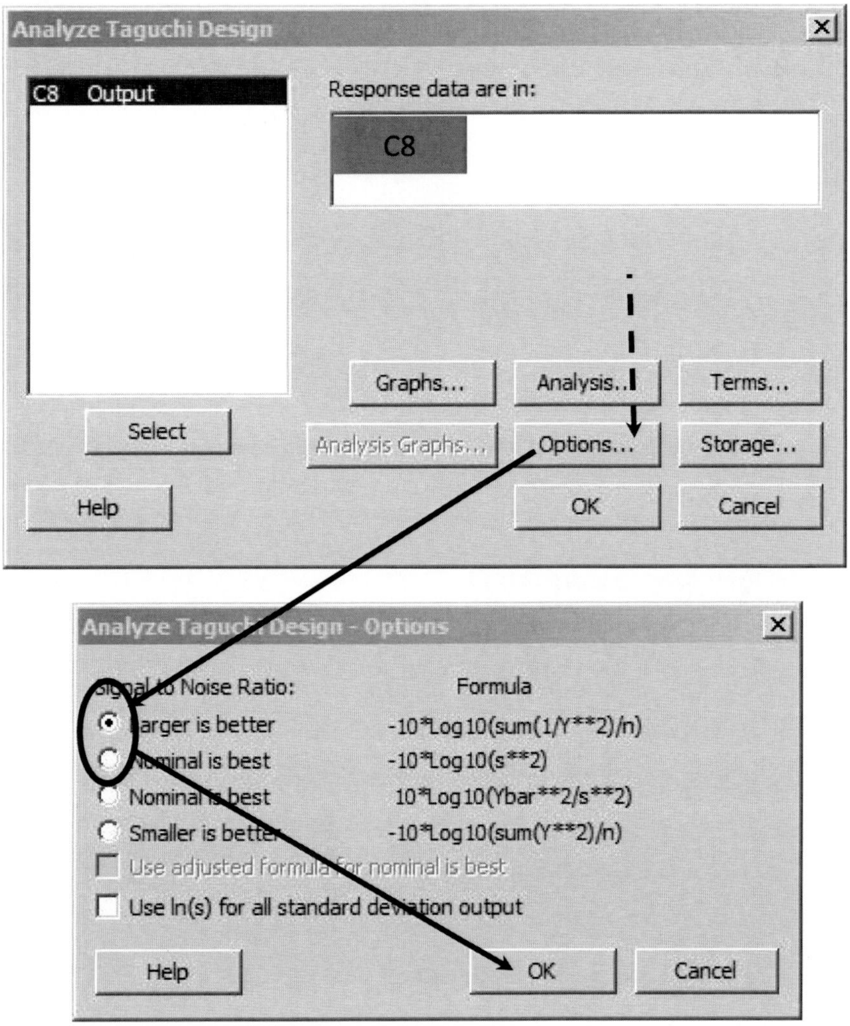

6.6. Still continuing in **Analyze Taguchi Design** window, if you click on **Storage…** tab, you will be getting **Analyze Taguchi Design – Storage** window, where *Signal to Noise ratios* and *Means* options are ticked by USER, as demonstrated in the following. Just use remaining options, AS IS and hit OK to close the window! Now you are done in setting up your Taguchi design and ready to execute the analysis.

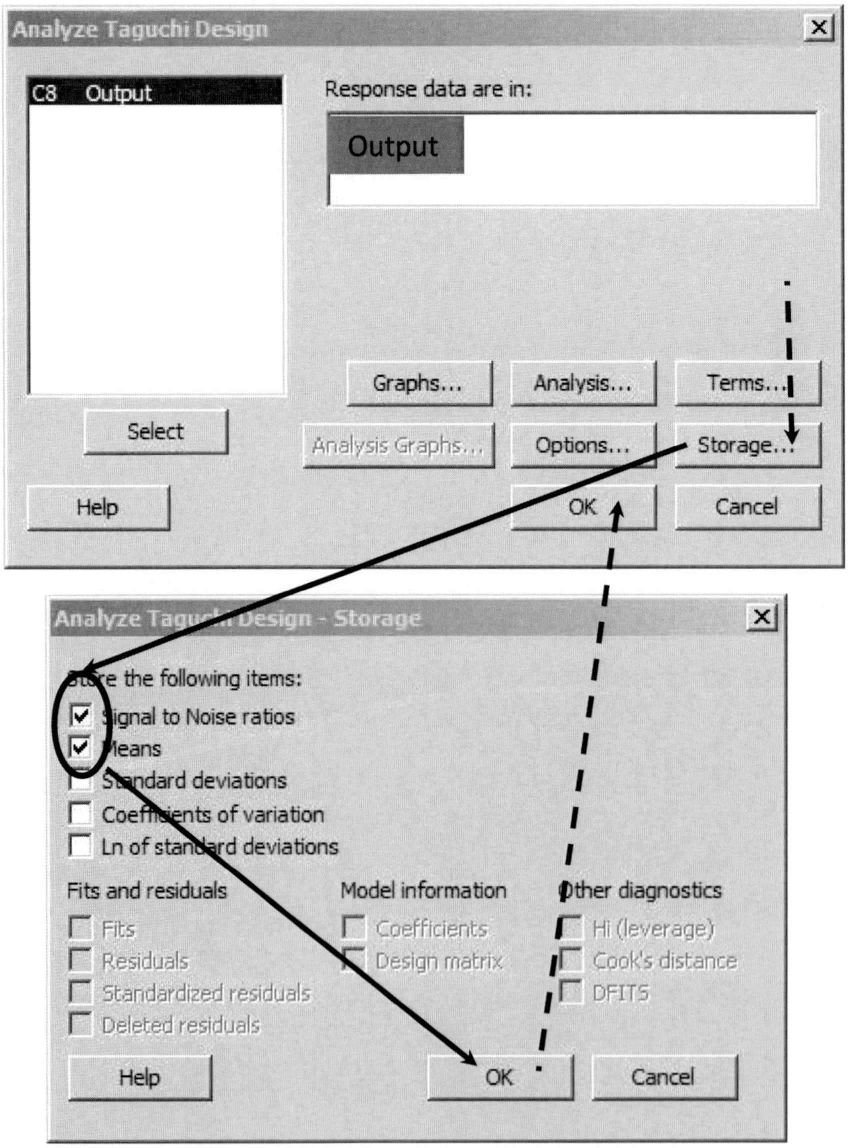

6.7. Once you close the **Analyze Taguchi Design** window, Taguchi analysis will run and you will get the overall results as presented next.

Response Table for S/N ratios and **Response Table for Mean** are calculated and stored in Worksheet1. Associated **Main Effects Plot for S/N ratios** and **Main Effects Plot for Means Plots** for all inputs/factors are displayed as given in the following page.

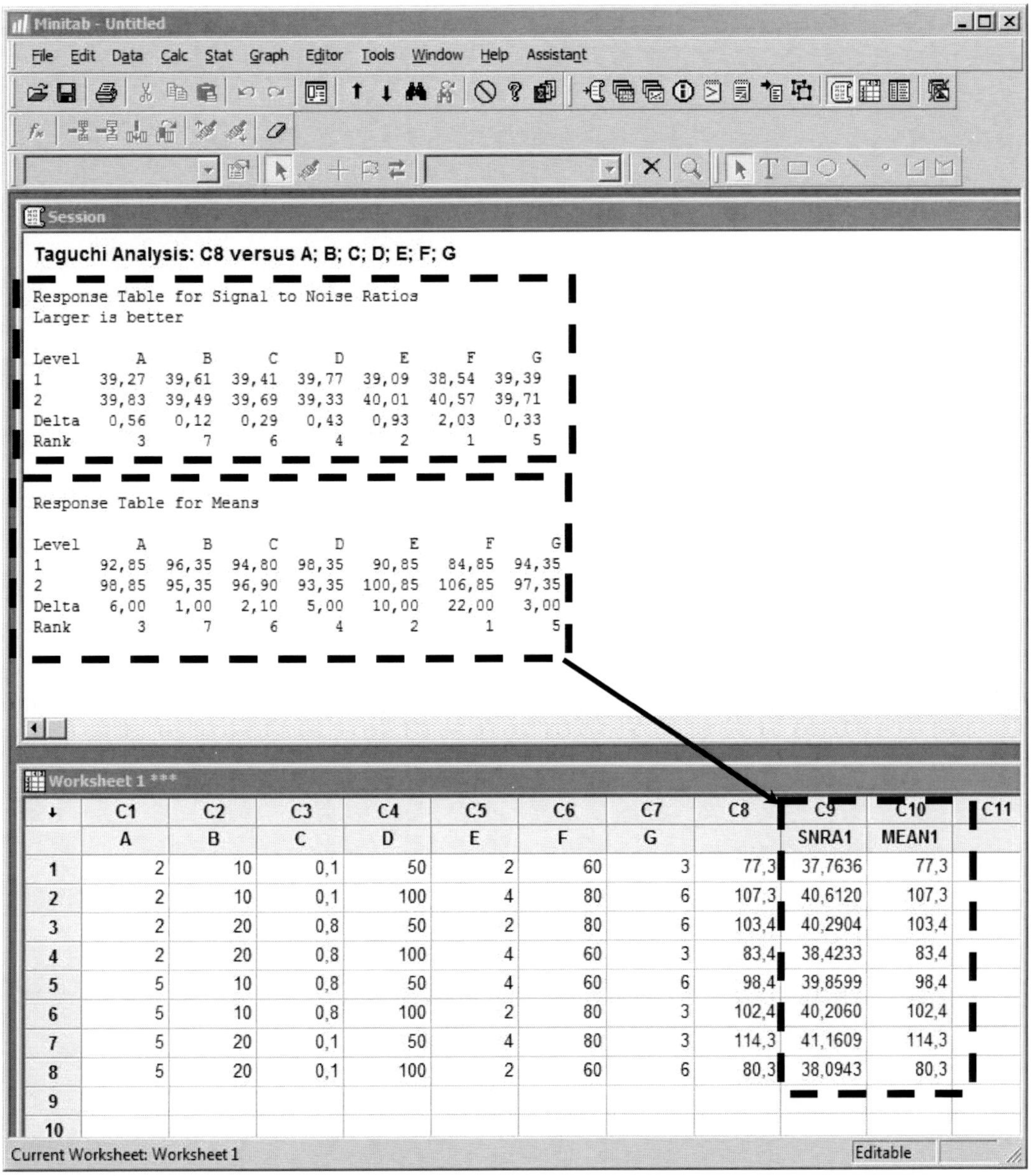

Taguchi Analysis: C8 versus A; B; C; D; E; F; G

Response Table for Signal to Noise Ratios
Larger is better

Level	A	B	C	D	E	F	G
1	39,27	39,61	39,41	39,77	39,09	38,54	39,39
2	39,83	39,49	39,69	39,33	40,01	40,57	39,71
Delta	0,56	0,12	0,29	0,43	0,93	2,03	0,33
Rank	3	7	6	4	2	1	5

Response Table for Means

Level	A	B	C	D	E	F	G
1	92,85	96,35	94,80	98,35	90,85	84,85	94,35
2	98,85	95,35	96,90	93,35	100,85	106,85	97,35
Delta	6,00	1,00	2,10	5,00	10,00	22,00	3,00
Rank	3	7	6	4	2	1	5

↓	C1	C2	C3	C4	C5	C6	C7	C8	C9	C10	C11
	A	B	C	D	E	F	G		SNRA1	MEAN1	
1	2	10	0,1	50	2	60	3	77,3	37,7636	77,3	
2	2	10	0,1	100	4	80	6	107,3	40,6120	107,3	
3	2	20	0,8	50	2	80	6	103,4	40,2904	103,4	
4	2	20	0,8	100	4	60	3	83,4	38,4233	83,4	
5	5	10	0,8	50	4	60	6	98,4	39,8599	98,4	
6	5	10	0,8	100	2	80	3	102,4	40,2060	102,4	
7	5	20	0,1	50	4	80	3	114,3	41,1609	114,3	
8	5	20	0,1	100	2	60	6	80,3	38,0943	80,3	
9											
10											

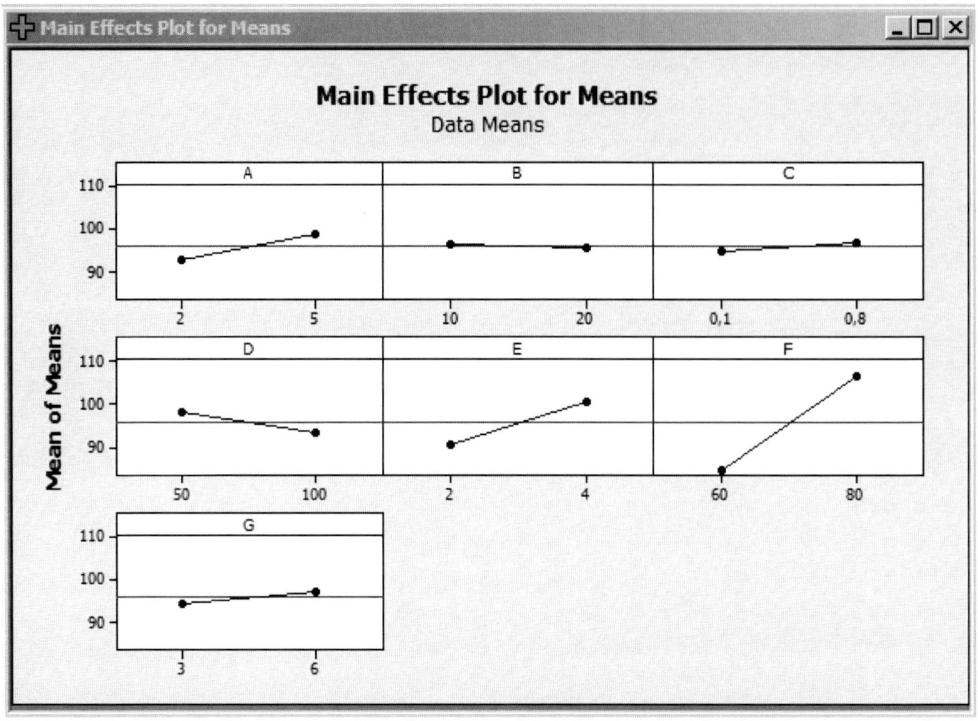
Main Effects Plot for Means
Main Effects Plot for Means
Data Means
Mean of Means
A
B
C
D
E
F
G
110
100
90
2
5
10
20
0,1
0,8
50
100
2
4
60
80
3
6

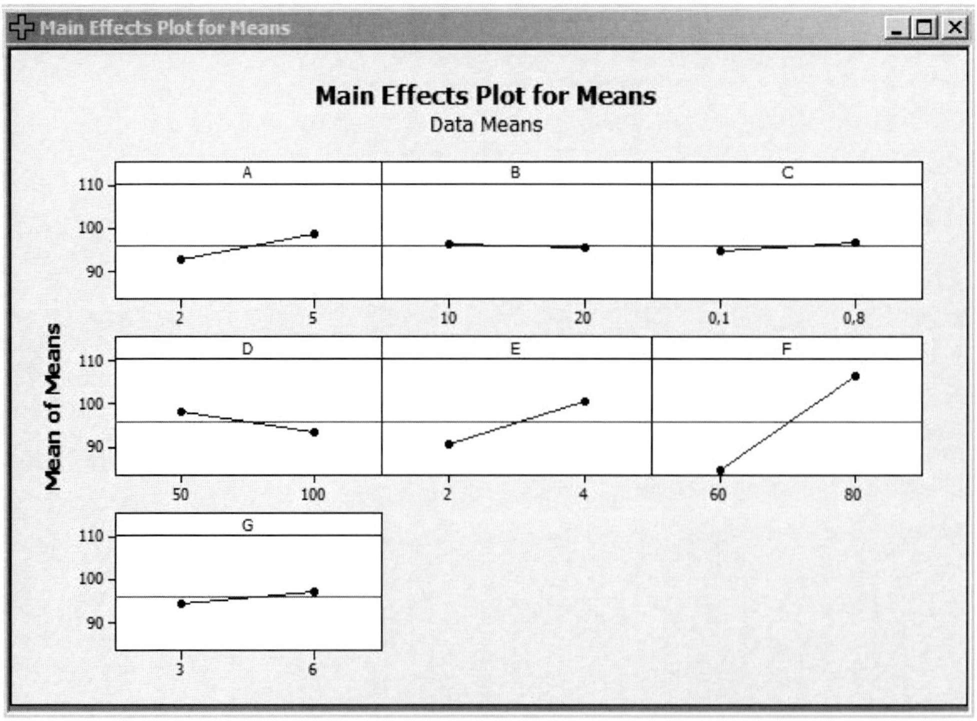
Main Effects Plot for Means
Main Effects Plot for Means
Data Means
Mean of Means
A
B
C
D
E
F
G
110
100
90
2
5
10
20
0,1
0,8
50
100
2
4
60
80
3
6

Notice that these results are fully consistent with the results presented in section 9.3 and 9.4 as expected. Now, let's go ahead and make some *Prediction* using our calculated overall results and estimate any *new input's* output level.

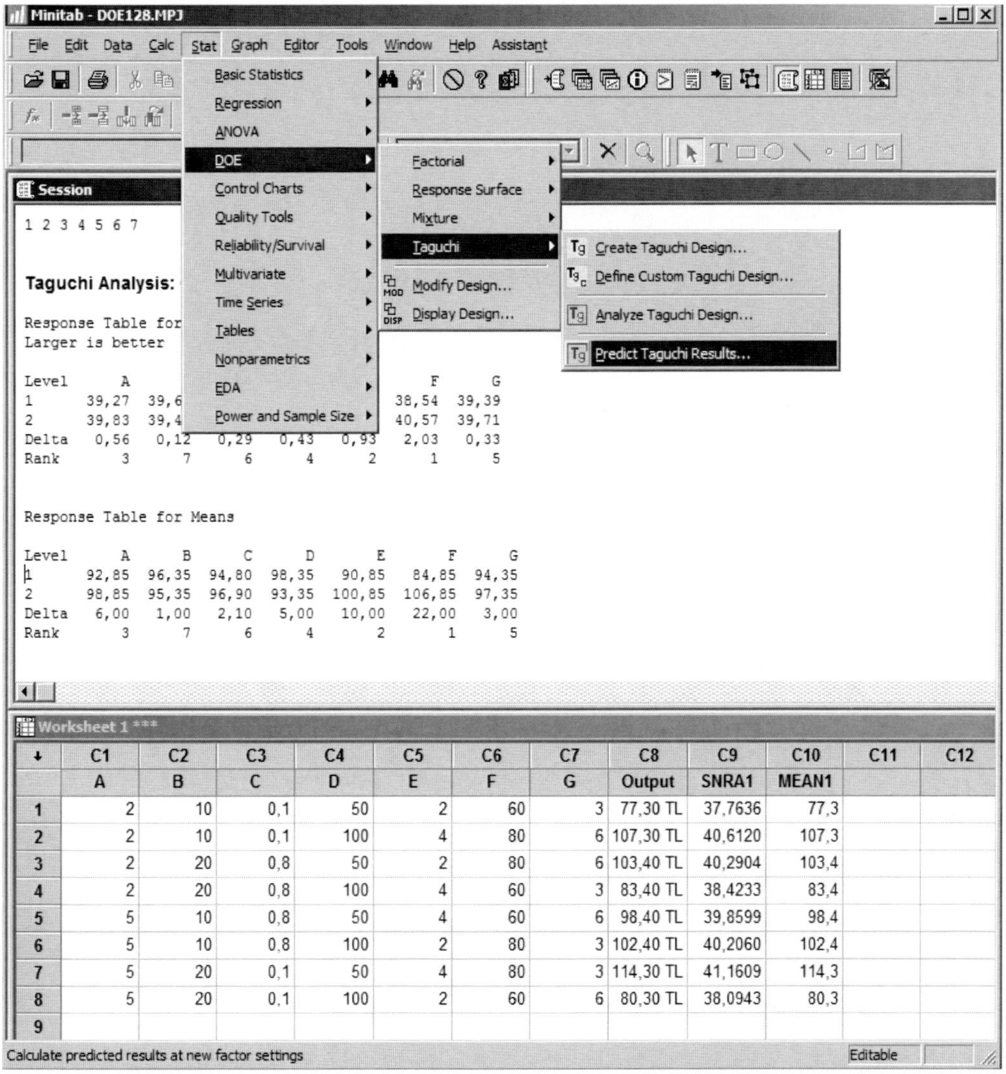

7. **Predict Taguchi Results...** option is very useful so that we can use optimized results (i.e. optimal input design) those were obtained in our previous **Analyze Taguchi Design** section and estimate optimum/near optimum outputs. Please keep in mind that this estimation still needs to be verified by conducting the real experiment/simulation in order to see the validity of our Taguchi predictions.

7.1. Click on **Stat** > **DEO** > **Taguchi** > Tg **Predict Taguchi Results...** tabs, **Predict Taguchi Results** window will pop-up and unselect *Standard deviation* and *Ln of standard deviation* options.

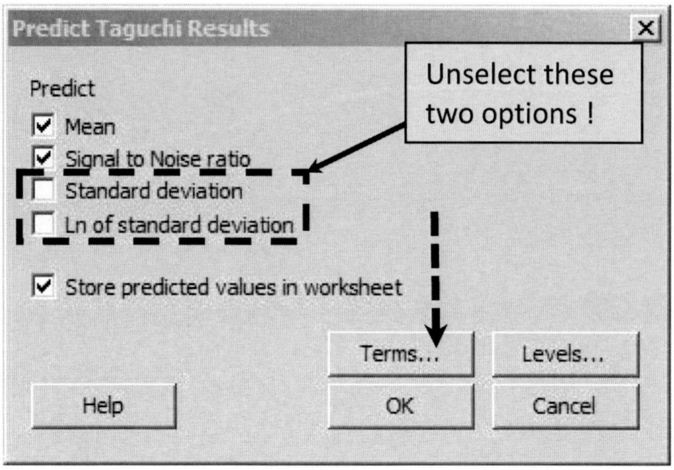

7.2. Next, click on the **Terms...** tab and select the terms (which are inputs in our case) those would be used in the predictions. Remember that *no interaction* term(s) is selected since we assumed the inputs do not interact with each other. Hit OK to close the window!

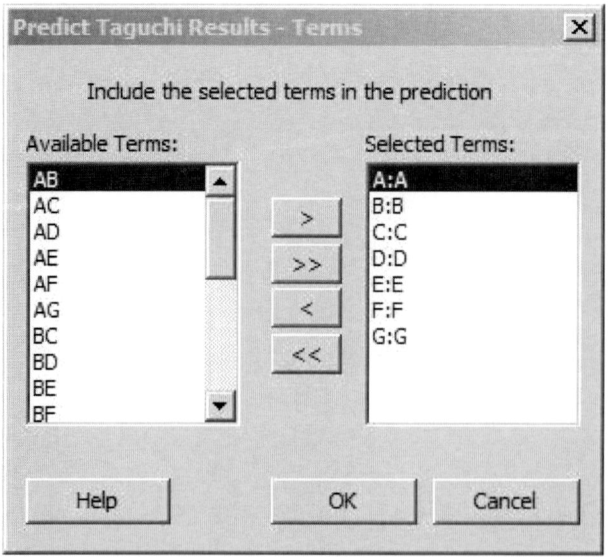

7.3. Next, click on the **Levels...** tab, you will get **Predict Taguchi Results – Levels** window. First, click on **Select levels from list** option to enter optimal input design levels to use in the predictions. Remember the Taguchi Analyses results that we obtained are as follows:

Taguchi Analysis: C8 versus A; B; C; D; E; F; G

Response Table for Signal to Noise Ratios
Larger is better

Level	A	B	C	D	E	F	G
1	39,27	39,61	39,41	39,77	39,09	38,54	39,39
2	39,83	39,49	39,69	39,33	40,01	40,57	39,71
Delta	0,56	0,12	0,29	0,43	0,93	2,03	0,33
Rank	3	7	6	4	2	1	5

Response Table for Means

Level	A	B	C	D	E	F	G
1	92,85	96,35	94,80	98,35	90,85	84,85	94,35
2	98,85	95,35	96,90	93,35	100,85	106,85	97,35
Delta	6,00	1,00	2,10	5,00	10,00	22,00	3,00
Rank	3	7	6	4	2	1	5

Based on these results, Optimal Designs for *S/N* and Mean Analyses are the same as follows, (Please note that they do not necessarily have to be the same at all times). Hit OK to close the window!

Maximizing the Output at A_2,B_1,C_2,D_1,E_1,F_1 and G_2

In **Predict Taguchi Results – Levels** window, specify the new factor units by selecting *Select levels from a list* option. Since there were two levels for each input, select appropriate levels by clicking on *downwards arrow tabs* for each *factor levels* in the window. Enter Optimum value combinations which maximize our output.

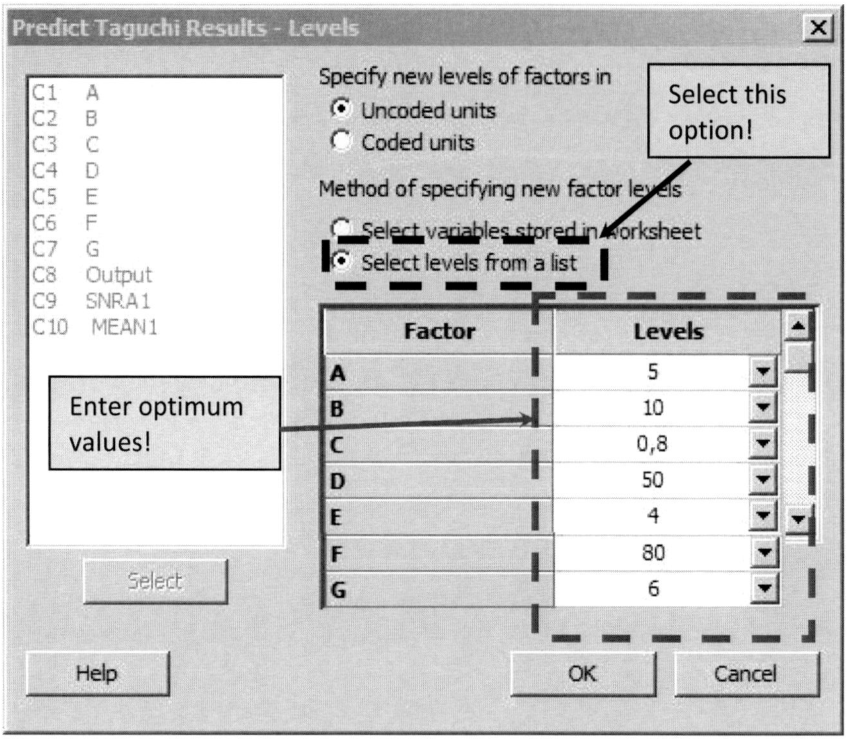

7.4. Hit OK to close **Predict Taguchi Results – Levels** window, you will get both ***S/N*** and ***Mean*** predictions as follows. Note that Maximum predicted values are the same values as those were obtained in previous sections. For example, *mean* is predicted as 120.4, which is exactly the same!

Up to this point, we have proven to ourselves that we can obtain exactly the same results by doing manual calculations and *Minitab.* Now, we obtained a firsthand experience for performing any Taguchi analyses and more confident to run more complicated analyses of our interest in the future.

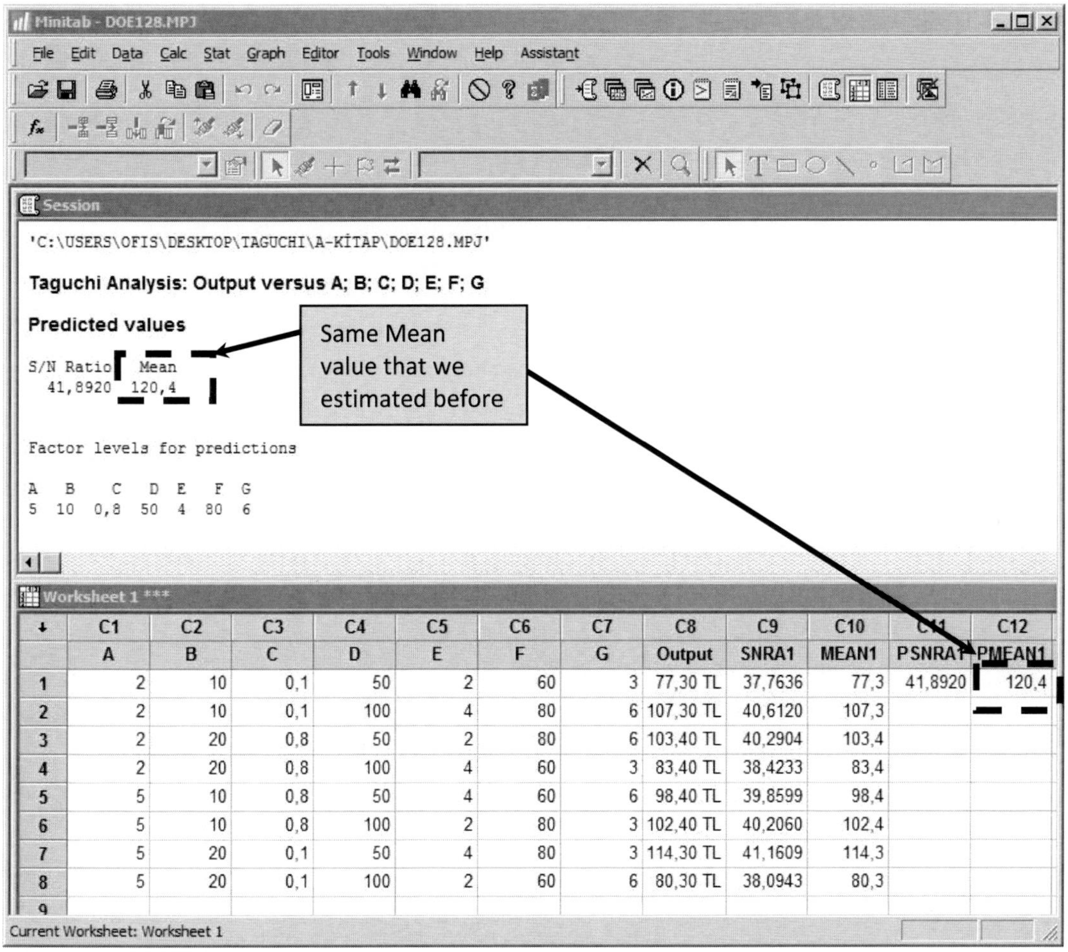

9.6 Exercises

1) When would you prefer using the full factorial design? Explain.

2) Explain why and when Taguchi design is preferred? What is the main idea behind Taguchi design?

3) Consider an experiment where you have 5 factors/inputs and this experiment yields a single Y (i.e. 77.3,107.3,103.4,83.4,98.4,102.4,114.3,80.3) output of strength values in MPa.

Your inputs and their ranges defining the experimental design space are as follows:

Input #	Input Name	Bounds
1	Material type	A and B
2	Temperature	50 , 150 ^{0}C
3	% additive	20 , 40
4	Bonding additive	0.05 , 0.08 gr
5	Number of Comp. Steps	3 , 6

You are asked to perform Taguchi experimental analysis and *maximize* the Y, i.e. the strength value!

- Describe your overall strategy for Taguchi analysis and explain the steps that you are going to follow.
- For this purpose, obtain both, the effects of each input on output and locate optimum design and its corresponding value using the following tables, (assume no interactions).
    - Response Table for Mean ratios
    - Response Table for S/N ratios
- Verify your results

Chapter 10

Poka-Yoke

Poka-Yoke is a mechanism that is designed to prevent *human errors* in such a way by preventing, correcting, stopping, signaling or drawing attention to faulty/defective parts or errors from being manufactured/assembled/done. It is mostly a *simple* configured system and can easily identify a flaw or error in each of manufacturing/assembly steps as the whole process continues (i.e. built upon one another).

Consider a plant where manufacturing and assembly are both performed where Poka-Yoke systems can be actively designed/used to eliminate product errors/defects. Poke-Yoke systems involve the design of products, working practices, setups (i.e. tools, jigs, and fixtures) by offering two-fold advantages;

- Primarily, the objective is avoidance/prevention of errors which may result in defects/flaws. Poke-yoke systems can be operated by an operator with an average/marginal level of skill
- Secondarily, it reveals any defects/flaws quickly without requiring expertise or experience

Clearly, Poka-Yoke does this by saying/showing/screaming/inhibiting/avoiding, i.e. *"Hey attention please, something is wrong here!"*. In Poke-Yoke systems, control mechanisms (Sensors, PLC etc.) are so popular and placed to stop the next step of a process/operation from occurring.

As for the supporting logic/argument behind Poka-Yoke, it is said "*an ounce of prevention is worth a pound of cure*", any early/timely investment in prevention of expected errors/illnesses can be more effective than any actions taken to fix the problems after they were identified. Worst of all, *in such a state of rush*, so many wealth/resources are spent/wasted/spared/lost. A vaccination is a good example of such.

One more supporting argument says, *"Prevention is better than detection",* because the important point is not to let "bad" things happen, in the beginning. Detecting bad/unwanted things after it happens, is not a true virtue (i.e. we just try to fix up our faults and avoid their possible consequences at elevated cost). We must have initiated/taken preventative measures very early!

But in real-life, we act upon according to our knowledge and experience, but nevertheless, we cannot foresee everything before they happen. Therefore design is always considered as *an iterative* process. We always wish if it was a single (feed-forward) iteration. Based on the complexity of the product/process, this is unfortunately not possible! But, what we can do is to minimize the iteration count. Now, let question the causes of errors.

10.1 What causes errors in Production?

Depending on the application area or product itself, which can be so vast, we can witness to so many things which can cause the errors. When looking at from manufacturing perspective, we can detect the following main reasons;

a. Weak/poor procedure and standards, which are mostly deemed lightly by people
b. Human errors – i.e. inexperience, poor supervision, rush, phycological mood, cell-phone, internet
c. Machines, missing/faulty jigs and fixtures
d. Inappropriate/faulty materials
e. Tool wears, equipment failures
f. Color/content mismatch
g. Temperature/pressure/humidity

10.2 Source of Errors

Some errors stand/act alone, some errors interact with others and create new errors those can never be anticipated. However, there may be many sources of errors as summarized below;

a. Human, to err is human
b. Skipping one operation
c. Process or processing errors

d. Setup errors
e. Missing/incomplete materials
f. Wrong parts
g. Machining wrong parts
h. Poor/insufficient machining
i. Tuning/calibration errors
j. Wrong adjustment of machines and auxiliary equipment
k. Tooling etc. errors

10.3 Human Errors

Human-based systems are so vulnerable to errors. You cannot foresee everything. Therefore, Poka-Yoke presents mostly easy! but great "error proofing/stopping/correction" messages to the users. Some human based errors include;

a. Forgetfulness
b. Wrong understanding, i.e. error in communication
c. Definition errors
d. Inexperience
e. Insisting on mistakes, and not acknowledging them
f. Unintentional errors/mistakes
g. Slow respond/movement
h. Poor standards
i. Surprise/unexpected errors
j. Intentional errors

10.4 Desired Characteristics of Poka-Yoke

Poka-Yoke systems should be cheap and easily applicable, preferably not operator dependent ones. Hence, from a practical point of view, we can outline the basic characteristics of a sample Poka-Yoke applications. They;

- should be a low cost
- should be an integral part of the specific process, and should not require special attention
- should be as simple as possible and always be used by any operator (Not operator dependent!)
- should depend on the enthusiasm/creativeness of the operator, not necessarily high cost.

10.5 Poka-Yoke tools

- Using varying types and diameters pins for required placement of machining/assembly/jig/fixture etc. parts.
- Fault/error detection and alarming/signaling systems
- Control switches, PLC etc.
- Part counters
- Check-listing equipment for monitorization
- Simple processing rules and order when required

10.6 Where do Poka-Yoke tools work well?

- At manual operations requiring operator focus
- where the probability of wrong positioning exists
- At the operations requiring adjustment
- At the operations where either the use of SPC is difficult or not effective to use
- At the places with a high cost for operator training and too much circulation in human work-force
- At the places with too many product models
- At the situations when the exterior error cost is much greater than the interior error costs

10.7 Three Poka-Yoke rules

- 60% success rate is pretty good, implement the idea!
- Do not wait for the best solution, Implement right-away!
- Do it now! Improve later by time!

Sometimes quick start is the best strategy, even the success rate is not at the desired level and we do not get the best solution. Engineers are thinkers, debuggers and tend to not let go the things unless the best(satisfactory) solution is obtained.

10.8 Poka-Yoke Examples

Although there are many example problems, we are going to give more general and easy to understand everyday life examples as presented in the following.

10.8.1 Automobile Drive-lock Security Measures

In automatic-transmission automobiles, when parking the automobile, the driver always put drive-shift gear to the P-Park position and then turns the ignition key off, as shown in Figure 10.1(A). Assume you parked your automobile in a rush and you turned the ignition key off, and you left the drive-shift gear engaged at the D-Drive position as also shown in Figure 10.1(B).

You are still in the automobile, suddenly you realized that you have forgotten to buy something that's why you need to go back to the shopping right away and you wanted to restart the engine. It won't restart! You try several times, no luck! then you panicked and gave up and you wanted to go out of the automobile! Now, you cannot get your engine key removed from the it. Your automobile warns/alerts that you did not do what you are supposed to do when parking, i.e. you should have put drive-shift gear to the P-Park position, before turning your engine off. Can you imagine, this security feature integrated into your car, can prevent you from having so many problems?

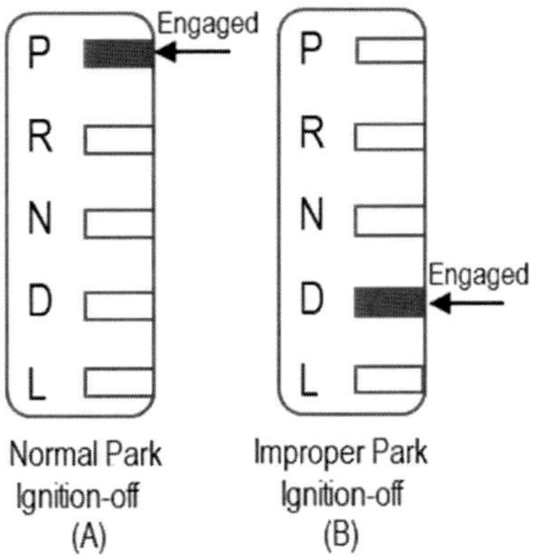

FIGURE 10.1 Automatic-transmission drive letters

10.8.2 Automobile Low/High Beam Warning

Remember again, if you forget low/high beam (lights) on, after turning your ignition key off in your automobile, you hear an alert sound (beeps), so alerting you turn low/high beams off right-away. As depicted in Figure 10.2, your car says in words;

"Hey, you cannot proceed unless you do the things properly i.e. turn off the low/high beam so that you do not eat-up/kill your battery and/or make your life uncomfortable! The longer the battery life, the cleaner environment, and savings in money and time".

FIGURE 10.2 Low/high beam warning

10.8.3 Clearance Hangers

One more simple application that we encounter frequently in the closed parking lots or tunnel entrances where leveled wooden part is hung (i.e. with caution written on it) as demonstrated in Figure 10.3. For this example, it has a maximum limiting vehicle height in clearance of 10 ft or 3.5 m for the parking lot or tunnel entrances.

If the vehicle driver is not aware of this limitation when he/she tries to enter the closed parking lot or tunnel, his/her high-vehicle hits the wooden part hanger and this hanger-hit warns the driver that he/she cannot proceed, because his/her vehicle is too high and not allowed to enter the closed parking lot or tunnel.

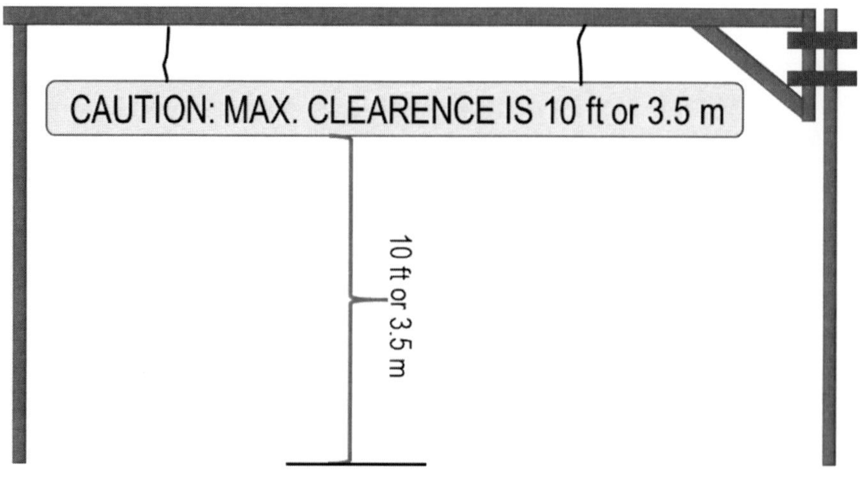

FIGURE 10.3 Vehicle height eligibility for entrance

10.8.4 Hearth By-Pass Surgery

By time/age, due to inferior health care conditions, lack of sportive activities, mal-nutrition etc., some arteries in the heart can be blocked/clogged. This puts the patient's life in danger. In such a situation, if non-invasive methods (i.e. medicine, stent) does not work, then a person may need open heart by-pass surgery. Where clogged artery(plaque build-up) is detected, hearth surgeons add a second artery (i.e. by-pass the clogged region in the artery) by using a spare artery, taken from some other part of the same person's body.

In normal health condition, pressured blood flows through the original artery even it is partially clogged as shown in Figure 10.4(A). Depending upon the magnitude of blockage percentage, there may be a partial secondary blood flow as well from the by-pass artery, which depends on optimum blood flow energy balance level in the body.

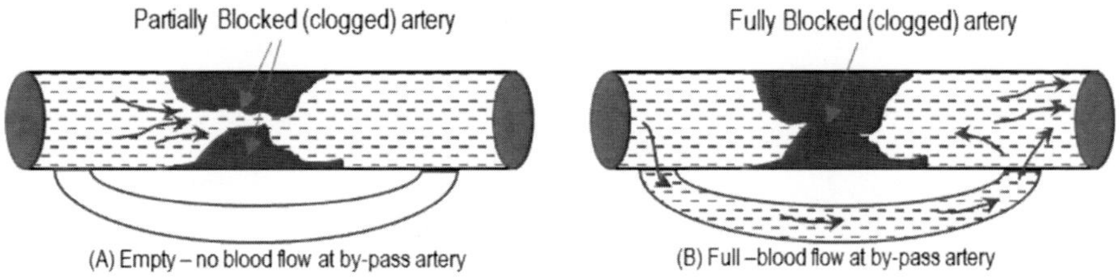

FIGURE 10.4 By-Pass Heart Surgery Representation

But, in a case when it is fully clogged, then pressured blood which carries oxygen etc. cannot proceed/flow through its natural artery, so it fully takes the detour (follows through the second artery which is placed in by-pass operation) and continues its flow so that heart diseases and stroke due to high-pressure blood stoppage/increase can be avoided/prevented by letting blood flow from the by-pass artery as shown in Figure 10.4(B). Obviously, this is a life-saving Poka-Yoke incident, not cheap but vital precaution(preventative application) for human beings.

10.9 Exercises

1) Describe Poka-Yoke by your own terms.

2) Develop your own example of Poka-Yoke system, draw/explain how it works? What advantages it presents considering its cost, complexity etc.

3) Why quality control comes after Poke-Yoke in importance? Please state your thoughts and give a brief explanation when considering the order of any manufacturing process flow for a factory.

4) In an automated glass-bottle milk filling process line, as shown in Figure 10.5, all bottles must be filled 1±0.03 liter of milk. But in rare incidents, a few bottles may be partially or not filled with milk, due to filling machine problems. One needs to evaluate the filled-milk amount if it is outside of the tolerance levels.

For such occurrences, we do not want partially filled milk bottles to reach the stores for sale, because they are returned at seller's cost. As also for the fairness, an on-site engineer, to prevent the bottle from miss filling, develop a simple Poka-Yoke system for detecting miss-filled bottles just after right after a fill-up. In case such a bottle is detected, stop the bottling line (i.e. movement) and alert the operator for correction.

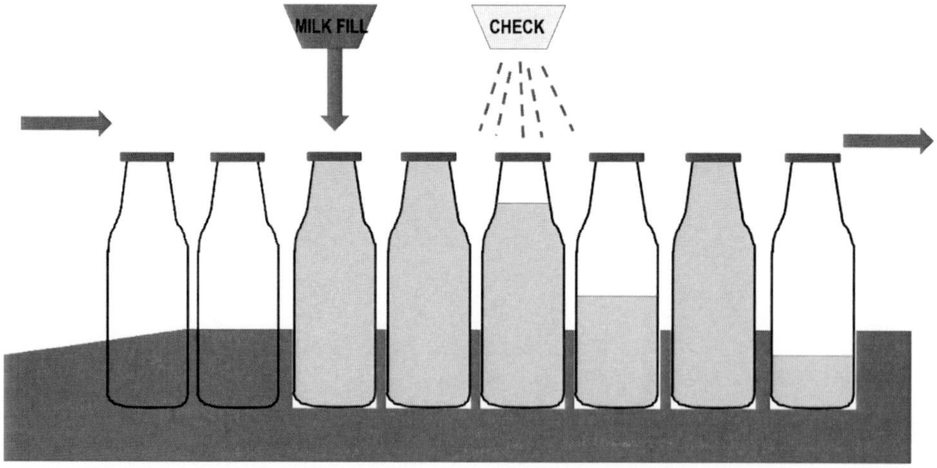

FIGURE 10.5 Glass-bottle milk filling machine

Chapter 11

Control Charts

The product or service quality has been a major concern (or *objective*) in the industry and service businesses for the last decades. Quality even can be regarded as more important than the cost, where downtimes due to interruptions, warranty returns/repairs etc. in the products have been a major concern due to a considerable amount of profit losses. Therefore, quality improvements have become a *"live or die"* issue for many companies. Why? Because of customer satisfaction! Which is strongly tied up to the *quality* of the product as well as to its *low cost* as highlighted many times in earlier chapters. As far as these two measures are concerned, global competition is never ending or slowing down. Survival is an essential part of all businesses. Therefore as a first step, for any process (i.e. manufacturing), we have to first make sure it is a *stable* (repeatable, consistent, well-established) and a capable of operating on the design target(s) with an acceptable minimum level of variation. Needless to remind, variations are not welcome! Since they directly affect/reduce the overall product or service quality.

The subject of quality is widely studied under Statistical Quality Control (SQC). Statistical quality control employs statistical techniques for measuring the process variables (thus we can also monitor them). Once these two objectives are achieved, we can improve/control the quality of processes/products. SQC includes Statistical Process Control (SPC) is being one of the most important tools.

SPC offers an active problem-solving capability. SPC provides monitoring process capability and achieving process stability by measuring/analyzing the variation and to reduce the loss (less waste). SPC should also be supported with never-ending *eagerness* and *motivation* from the institutional point of view for commitment to continuing improvements in high product/process quality and productivity.

Although there are several SPC tools (i.e. Scatter diagram, histogram, check sheet etc.) which have been introduced in Chapter 2 and 3, the *control charts* are proven to be the most powerful tool, as it would be introduced in the following sections, in detail.

11.1 Overview of Control Charts

We have already introduced the use of various well-known data display graphs in the earlier chapters. In this section, very useful *control charts* graph is also introduced to study inherent changes in a process/product (by evaluating its data) over a certain period of time. Control charts have long been used in manufacturing, total quality management etc. and have found a great deal of acceptance in the industry, especially in the *process capability analysis*.

Basically, control charts are used to monitor the stability of a process by utilizing line plots to investigate the relationships among the variables affecting the stability. Control charts can offer an *online* and *active* way of dealing with *nuisance* factors which cause variability. A statistical control chart can help both monitoring the process variation and generating valuable information on the stability (current state) of the process. Control chart alerts the user to the situations in which data fall out the control limits. Keep in mind that unstable process (means no or less control) causes defective products, poor product/process quality thus we get non-repetitive, inconsistent production in output targets. We never want to lose control! Therefore, when using control charts, traceable characteristic data points outside the control limits *alert* us to find a root cause for this unusual event. The remedy is to make effective changes/modifications in the process that is being studied to prevent such conditions from happening again. Keep in mind that our ultimate goal of using SPC is, if possible, the elimination or reduction of the variability in the process/products by using control charts very actively and effectively.

A Control Chart is a line plot of a summary statistics of sequentially taken data samples during a sampling period (in a timely manner). Each sample data can very well be in *subgroups*. On control charts, usually, a *sample mean* and a measure of the sample variability such as *standard deviation* or the *range* is plotted. The upper control limit (UCL) and the lower control limit (LCL), are statistically derived a pair of boundaries which determine the limits for assessing the inherent variability in the data obtained from a process. We also prefer a randomly distributed pattern on a control chart. Centerline (CL) is calculated using the sample mean of the data. An example of a control chart is shown in Figure 11.1 in the graphical display of inherent quality characteristics of the data vs. time, individual sample or subgroup numbers.

These control limits (which are acceptable statistical criterions based on normal variation assumption) set the intervals. If data points fall in between this control limit intervals and distributed without a systematic pattern (randomly), the process is assumed to be *in-control* or *stable*. However, a data point, falling outside of the control limit interval, is interpreted as evidence that the process is *out-of-control* or near *out-of-control*. In such a situation, we need to investigate/intervene and take immediate corrective action to eliminate the undesired cause of border violation which affects the vital process characteristics of our interest so that it does not occur again.

The control chart alarms the user when excessive fluctuations/violations occur due to unexpected breakdown or uncounted dramatic process changes. The control charts alert to put process/product data back to acceptable range/level, as shown in Figure 11.1, where CL: control line (or limit) corresponding to the process average(mean), UCL: upper control line and LCL: lower control line.

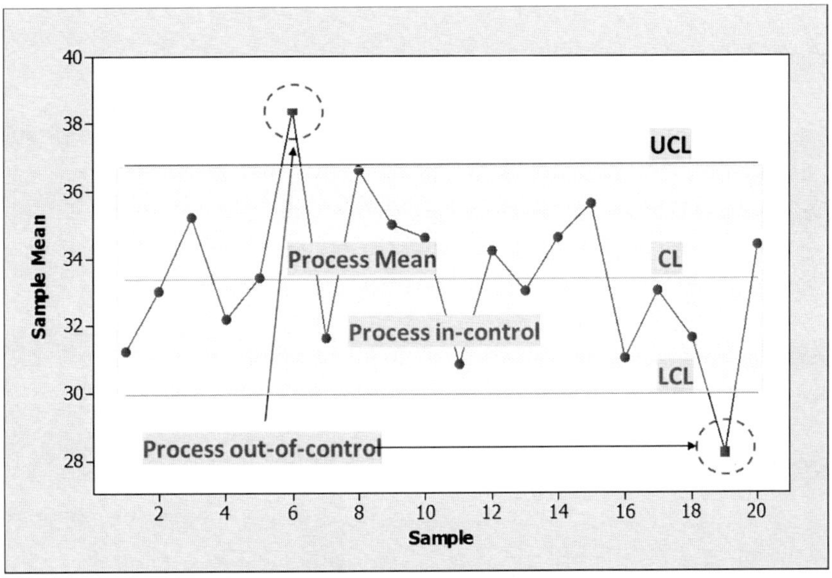

FIGURE 11.1 Sample control chart

Why control charts are extensively used in the industry, can be summarized as follows:

- Online and active monitoring of a process help reduce scrap or rework for non-conforming measures, thus productivity is improved and defects are reduced to much lower levels. Control charts also help engineers to perform purposeful/ smart changes in the process to improve process performance.

- Anything doesn't contribute to the output is considered as *waste*. If by using control charts, we reach our objectives, then we do not need to worry about the process. Thus, we can prevent ourselves from *"over doing"* in the process monitoring or control. Hence our efforts and resources can be used (with an optimum level of care) at somewhere else needed!

- Control Charts provide valuable information/management capability when performing *estimation* for process capability to product or process engineers in the industry.

Assume **x** representing process statistics that we are studying (i.e. temperature, dimension, strength, hardness, humidity, material composition etc.), then control limits can be obtained using Equation 11.1 as follows,

$$\begin{aligned} \text{UCL} &= \mu_X + 3\sigma_X \\ \text{CL} &= \mu_X \\ \text{LCL} &= \mu_X - 3\sigma_X \end{aligned} \qquad 11.1$$

Where μ_X and σ_X indicate *an estimate* of *average(mean)* and *standard deviation* of *sample* **x** data, respectively. Why estimate? Because the *true population average* and *standard deviation* (μ,σ) usually are not known! In Equation 11.1, control limit rules are established using normal probability distribution assumption, in which 99.73 percent of data will be within $\pm 3\sigma_X$ (*sample standard deviation*) from the average μ_X.

Keep also in mind that control limits are calculated from process data, and customer expectations or specification limits have nothing to do with it.

11.2 Control Charts for Individual Measurements

When the sample size for the process data is *single*, then we talk about the individual (not subgroups) measurements that we frequently encounter in real-life studies. In this situation, the individual control chart (or also called an $\bar{X}$*-Chart)* is used. In constructing the control chart we do employ Equation 11.1 where we can use either σ_S (total standard deviation is good to estimate spread and measures variation in all data) or σ_C (short term standard deviation is good to estimate the capability and measures the variation in successive data points) for estimating σ_X. σ_C (obtained from moving range, *MR*) eliminates the variation due to special causes - i.e. cycles, shifts, etc. – therefore for any existing data, σ_c is used as the better estimate of the common cause of variability in drawing control charts.

σ_c uses the *moving range* of two successive data measurements (*m* total number of measurements i.e. in practice *m=20-30* process data measurements are suggested) to estimate the process variability, hence to calculate σ_c as given in Equation 11.2 where the moving range is defined as $mr_i = |x_i - x_{i-1}|, i = 2, \dots, m$ and $\overline{mr}$ is the average of *m-1* moving ranges,

$$\sigma_c = \frac{\sum_{i=2}^{m} |x_i - x_{i-1}| / (m-1)}{d_2} = \frac{\overline{mr}}{1.128} \quad 11.2$$

Note, for two successive data measurements, control chart factors that depend on the sample size of *n=2* is given in Appendix 5 and also summarized in Table 11.1, where d_2=1.128.

TABLE 11.1 Control chart factors for *n=2*

Chart Factors	d_2	D_3	D_4
n=2	1.128	0	3.267

In obtaining $\bar{X}$-Chart. As $\bar{x}$ is the average of all *x* measurements, the UCL and LCL for an individual control chart, or I-Chart is given as;

$$UCL = \bar{x} + 3\sigma_c = \bar{x} + 3\frac{\overline{mr}}{1.128} \quad 11.3$$

$$CL = \bar{x}$$

$$LCL = \bar{x} - 3\sigma_c = \bar{x} - 3\frac{\overline{mr}}{1.128}$$

We can also plot the control chart for moving ranges or MR-Chart where control chart factors are presented in Appendix 5 and Table 11.1 as D_3=0 and D_4=3.267 and CL,UCL and LCL lines are given as;

$$UCL = D_4\overline{mr} = 3.267\overline{mr} \quad 11.4$$

$$CL = \overline{mr}$$

$$LCL = D_3\overline{mr} = 0$$

From practical standpoint, when plotted together, both charts are called *I-MR Control Chart* in the applications. Now, let's also demonstrate all of these control chart calculations and drawings on an example in the next section!

TABLE 11.2 Coating thickness test data

Time	x		mr
1	3.78		-
2	4.55		0.77
3	3.79		0.76
4	5.07		1.28
5	2.92		2.15
6	2.78		0.14
7	4.01		1.23
8	3.41		0.6
9	3.65		0.24
10	3.89		0.24
11	4.77		0.88
12	3.11		1.66
13	3.22		0.11
14	4.02		0.8
15	3.66		0.36
16	3.44		0.22
17	4.76		1.32
18	3.55		1.21
19	2.88		0.67
20	3.89		1.01
21	3.01		0.88
22	3.33		0.32
23	4.22		0.89
24	2.32		1.9
$\bar{x}$ =	**3.668**	$\overline{mr}$ =	**0.8539**
		σ_c =	**0.7570**

Example 11.1: Coating Thickness Data Control Chart

Cutting tool coating thickness is very important in the *TiN* surface coating process to increase the cutting tool life and cutting conditions as well. For this purpose, only one measurement is taken and recorded on a one-hour intervals for 24 hours. Measured coating thicknesses (i.e. **x**'s and also **mr** values*)* are given in *μm,* as presented in Table 11.2.

We can proceed and calculate the control limits for **x** data and obtain I-Chart, using $\bar{x}$ and $\overline{mr}$ from Table 11.2 as follows;

$$UCL = 3.668 + 3 * \frac{0.8539}{1.128} = 5.939$$

$$UCL = 3.668$$

$$UCL = 3.668 - 3 * \frac{0.8539}{1.128} = 1.397$$

$$\sigma_c = \frac{0.8539}{1.128} = 0.757$$

We can also calculate the limits for *MR* Chart as follows;

$$UCL = 3.267 * 0.8539 = 2.790$$

$$m\bar{r} = 0.8539$$

$$LCL = 0$$

The *I-MR charts* for individual coating thickness process are presented in Figure 11.2, where it is seen that the process is *stable* and *TiN* coating process is perfectly *under control* at the end of the sample day. Of course, what tomorrow brings is totally unknown! Be alert!

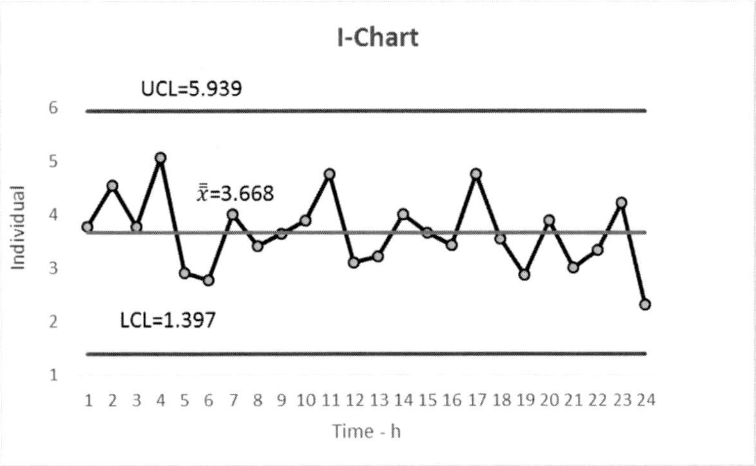

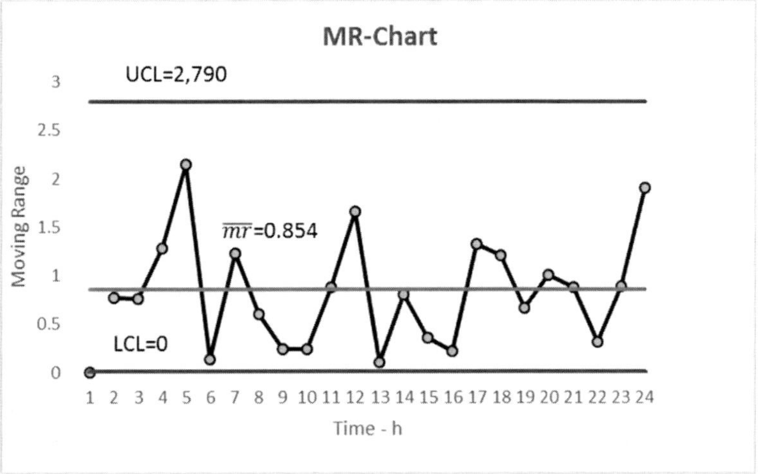

FIGURE 11.2 I-MR Control Charts for *TiN* coating thickness example

While, we can use *Excel* or *Minitab* in our calculations extensively, it is good idea to introduce one simple "*sample form of control chart*", where all of the results obtained above can be performed by using hand calculations as well. In the next page, Figure 11.3 briefly introduces usage of such a *sample form* to be used in the industry.

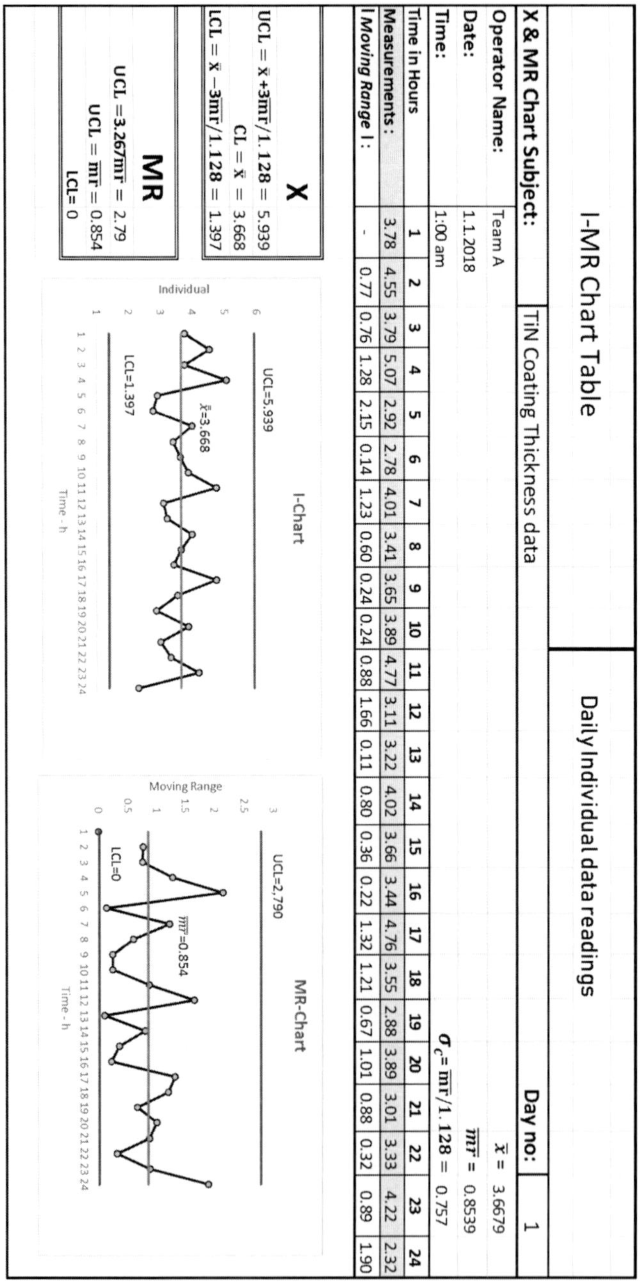

I-MR Chart Table		Daily Individual data readings		
X & MR Chart Subject:	TiN Coating Thickness data		**Day no:**	1
Operator Name:	Team A		$\bar{x}$ =	3.6679
Date:	1.1.2018		$\overline{mr}$ =	0.8539
Time:	1:00 am		$\sigma_c = \overline{mr}/1.128$ =	0.757

Time in Hours	**1**	**2**	**3**	**4**	**5**	**6**	**7**	**8**	**9**	**10**	**11**	**12**	**13**	**14**	**15**	**16**	**17**	**18**	**19**	**20**	**21**	**22**	**23**	**24**
Measurements :	3.78	4.55	3.79	5.07	2.92	2.78	4.01	3.41	3.65	3.89	4.77	3.11	3.22	4.02	3.66	3.44	4.76	3.55	2.88	3.89	3.01	3.33	4.22	2.32
I *Moving Range* I :	-	0.77	0.76	1.28	2.15	0.14	1.23	0.60	0.24	0.24	0.88	1.66	0.11	0.80	0.36	0.22	1.32	1.21	0.67	1.01	0.88	0.32	0.89	1.90

X

$UCL = \bar{x} + 3\overline{mr}/1.128 =$ 5.939

$CL = \bar{x} =$ 3.668

$LCL = \bar{x} - 3\overline{mr}/1.128 =$ 1.397

MR

$UCL = 3.267\overline{mr} =$ 2.79

$UCL = \overline{mr} =$ 0.854

LCL= 0

FIGURE 11.3 Filled I-MR control charts form for *TiN* coating thickness example

11.3 Control Charts for $\bar{X}$ (Mean), R (Range) and S (Standard Deviation)

When we concentrate on quality measures of our interest (i.e. weight, temperature, contamination, dimension, concentration, hardness etc.), then an $\bar{X}$ control chart provides valuable information using the average value of quality of the output and its variability. Process variability control can be performed using either *R-Chart* (Range) or *S -Chart* (standard deviation) depending on the approach for *estimating population standard deviation.*

Suppose we have *m* samples each with size of *n* (subgroups where data is taken, i.e. *n=2,3,4,5,6,7,8,9...* times for each *m* sample). $\bar{\bar{\mathbf{x}}}$, grand mean as shown with *double bar* where $\bar{x}_j$ is subgroup average for j^{th} sample measurement where $j=1,...,m$

$$\bar{\bar{\mathbf{x}}} = \frac{\bar{x}_1 + \bar{x}_2 + \cdots + \bar{x}_m}{m} \quad 11.5$$

Similarly if the range of each sample is denoted as $r_j=max(x_j)-min(x_j)$ where $j=1,...,m$. Then, the average range $\bar{r}$ is computed as,

$$\bar{r} = \frac{r_1 + r_2 + \cdots + r_m}{m} \quad 11.6$$

Similarly, if the standard deviation is denoted as s_j where $j=1,...,m,$. Then, the average standard deviation $\bar{s}$ is computed as,

$$\bar{s} = \frac{s_1 + s_2 + \cdots + s_m}{m} \quad 11.7$$

$\bar{X}$-R Control Chart

Then $\bar{X}$ control chart using $\bar{r}$ (range), the center line, upper and lower control limits are given as,

$$\begin{aligned} UCL &= \bar{\bar{x}} + A_2\bar{r} \\ CL &= \bar{\bar{x}} \\ LCL &= \bar{\bar{x}} - A_2\bar{r} \\ \hat{\sigma} &= \frac{\bar{r}}{d_2} \end{aligned} \quad 11.8$$

Where A_2 and d_2 are the control chart factors given for *n* sample sizes (subgroups) as presented in Appendix 5.

Similarly, *R* control chart depending on average sample ranges, the center line, upper and lower control limits are given as,

$$\begin{aligned} UCL &= D_4\bar{r} \\ CL &= \bar{r} \\ LCL &= D_3\bar{r} \end{aligned} \qquad 11.9$$

Where D_4 and D_3 are the control chart factors given for *n* sample sizes (subgroups) as presented in Appendix 5.

$\overline{X}$-S Control Chart

Alternatively, we can calculate $\overline{X}$ control chart using $\bar{s}$ (standard deviation), in this case, the center line, upper and lower control limits are given as,

$$\begin{aligned} UCL &= \bar{\bar{x}} + \frac{3\bar{s}}{C_4\sqrt{n}} \\ CL &= \bar{\bar{x}} \\ LCL &= \bar{\bar{x}} - \frac{3\bar{s}}{C_4\sqrt{n}} \\ \hat{\sigma} &= \frac{\bar{s}}{C_4} \end{aligned} \qquad 11.10$$

Where C_4 is the control chart constant given for *n* sample sizes (subgroups) as presented in Appendix 5.

Finally, *S* control chart, the center line, upper and lower control limits are given again using C_4. Now, let's also demonstrate all of these control chart calculations and drawings on an example next!

$$\begin{aligned} UCL &= \bar{s} + \frac{3\bar{s}\sqrt{1-C_4^2}}{C_4} \\ CL &= \bar{s} \\ LCL &= \bar{s} - \frac{3\bar{s}\sqrt{1-C_4^2}}{C_4} \end{aligned} \qquad 11.11$$

Example 11.2: Compressive Strength Control Chart

The compressive strength of any concrete is an important quality/measure (in *MPa*) when designing concrete structures. Consider such data, where $m=20$ test sample measurements are recorded for twenty days. Each day four samples are taken with a sample size of $n=4$ (subgroups) and all observations/measurements are presented in Table 11.3.

Now, we are asked to make required calculations and draw $\overline{X}$-R and $\overline{X}$-S control charts. For ***Day 1***,

$$\bar{x}_1 = \frac{x_1 + x_2 + x_3 + x_4}{n}$$

$$\bar{x}_1 = \frac{26 + 35 + 32 + 37}{4}$$

$$\bar{x}_1 = 32.5$$

$r_1 = \max(x_1, x_2, x_3, x_4) - \min(x_1, x_2, x_3, x_4) =$

$\max(26, 35, 32, 37) - \min(26, 35, 32, 37) =$

$r_1 = 37 - 26 = 11$

$s_1 = \text{std}(x_1, x_2, x_3, x_4) =$

$\text{std}(26, 35, 32, 37) =$

$s_1 = 4.80$

TABLE 11.3 Compressive strength test data

Days	x_1	x_2	x_3	x_4	$\bar{x}$	r	s
1	26	35	32	37	32.5	11	4.80
2	23	27	29	31	27.5	8	3.42
3	15	16	18	20	17.25	5	2.22
4	37	22	33	34	31.5	15	6.56
5	32	29	25	28	28.5	7	2.89
6	33	29	28	33	30.75	5	2.63
7	20	31	33	23	26.75	13	6.24
8	22	34	36	26	29.5	14	6.61
9	23	44	29	34	32.5	21	8.89
10	23	23	29	40	28.75	17	8.02
11	24	41	23	39	31.75	18	9.57
12	26	33	26	24	27.25	9	3.95
13	55	41	44	27	41.75	28	11.53
14	36	25	39	33	33.25	14	6.02
15	43	37	23	19	30.5	24	11.36
16	24	32	41	20	29.25	21	9.29
17	36	33	44	19	33	25	10.42
18	33	38	25	24	30	14	6.68
19	37	27	24	36	31	13	6.48
20	17	21	23	33	23.5	16	6.81
				$\bar{\bar{x}}$ =	**29.84**		
					$\bar{r}$ =	**14.9**	
						$\bar{s}$ =	**6.72**

In Table 11.3, we calculated $\bar{\bar{x}}$=29.84 *MPa* (i.e. x double bar is equivalent to the average of $\bar{x}_j, j = 1, \dots, m$. $\bar{r}$ is the average of ranges, $r_j, j = 1, \dots, m$. $\bar{s}$ is the average of standard deviations, $s_j, j = 1, \dots, m$).

Table 11.4 is given below obtained from Appendix 5 displays the control chart factors for this example.

TABLE 11.4 Control chart factors for *n=4*

Chart Factors	A_2	d_2	D_3	D_4	C_4
n=4	0.729	2.059	0	2.282	0.9213

$\overline{X}$-R Control Chart

Let's start with drawing $\overline{\text{X}}$-control chart using $\bar{\text{r}}$ statistics. The center line, upper and lower control limits can be calculated using the control chart factors as given in Table 11.4 and using Equation 11.8.

$$\text{UCL} = 29.84 + 0.729 * 14.9 = 40.7$$
$$\text{CL} = 29.84$$
$$\text{LCL} = 29.84 - 0.729 * 14.9 = 18.98$$
$$\hat{\sigma} = \frac{\mathbf{14.9}}{\mathbf{2.059}} = \mathbf{7.2365}$$

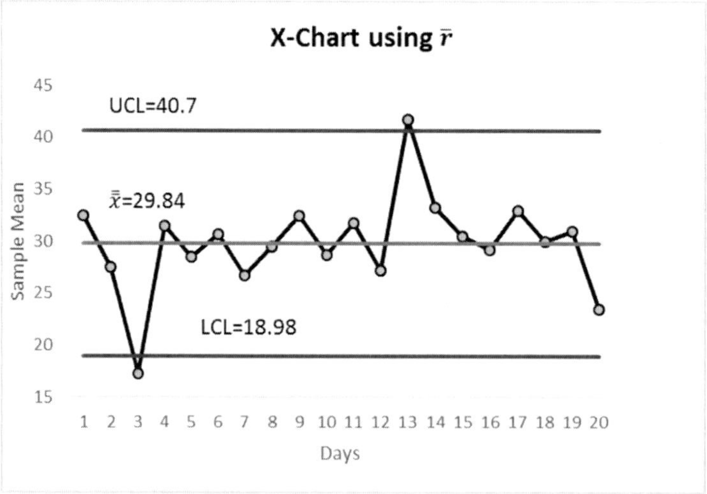

FIGURE 11.4 $\overline{\text{X}}$-Control chart using $\bar{r}$

By examining $\overline{\text{X}}$-Control Chart given in Figure 11.4, we notice that there are violations at both control levels at days 3 and 13 for which corrective measures needs to be taken to put the process back on the track!

Similarly, R-Control chart depending on average sample range, the center line, upper and lower control limits are calculated using Equation 11.9 and plotted as shown in Figure 11.5.

$$\text{UCL} = 2.282 * 14.9 = 34$$
$$\text{CL} = 14.9$$
$$\textbf{LCL} = \mathbf{0} * \mathbf{14.9} = \mathbf{0}$$

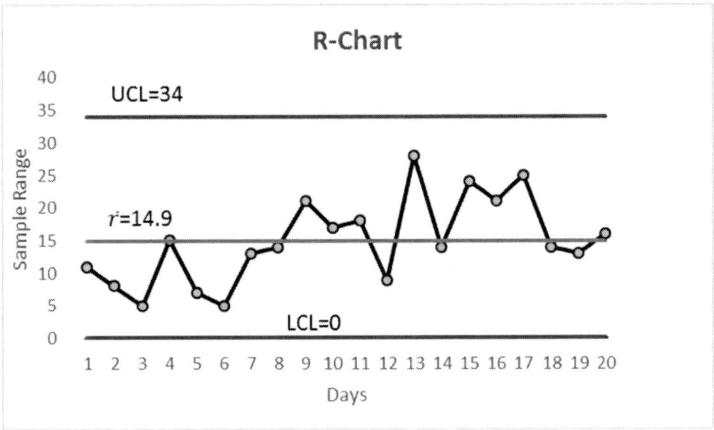

FIGURE 11.5 R-Control chart

$\overline{X}$-S Control Chart

Let's again start with drawing $\overline{\text{X}}$-control chart now using $\overline{\text{s}}$ (standard deviation) statistics. The center line, upper and lower control limits are calculated using the control chart factors as given in Table 11.4 and also using Equation 11.10,

$$\text{UCL} = 29.84 + \frac{3 * 6.72}{0.9213\sqrt{4}} = 40.78$$

$$\text{CL} = 29.84$$

$$\text{LCL} = 29.84 - \frac{3 * 6.72}{0.9213\sqrt{4}} = 18.9$$

$$\boldsymbol{\hat{\sigma} = \frac{6.72}{0.9213} = 7.294}$$

By examining $\overline{\text{X}}$-Control chart given in Figure 11.6, we notice the same violations at both control levels at days 3 and 13. Also note that UCL and LCL line values look very close for both $\overline{\text{X}}$-control charts using range and standard deviation.

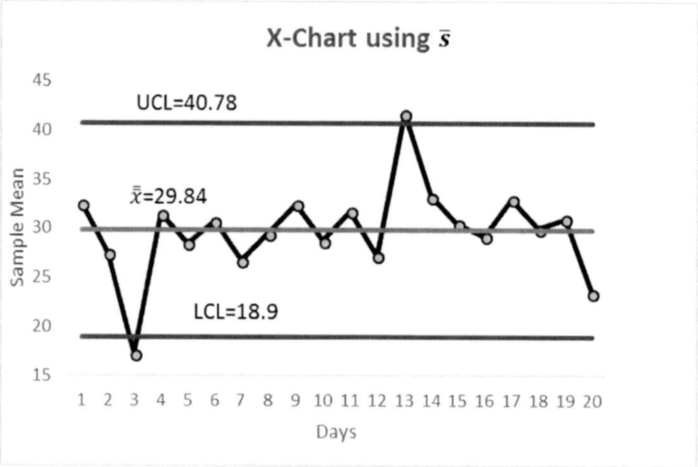

FIGURE11.6 X-Control chart using $\bar{s}$

Then, S control chart, the center line, upper and lower control limits again using C_4 are given as,

$$\text{UCL} = 6.72 + \frac{3 * 6.72\sqrt{1 - 0.9213^2}}{0.9213} = 15.22$$

$$\text{CL} = 6.72$$

$$\text{LCL} = \bar{s} - \frac{3 * 6.72\sqrt{1 - 0.9213^2}}{0.9213} = -1.79$$

$$\mathbf{LCL = max(-1.79, 0) = 0}$$

Please keep in mind that when the LCL for R or S-charts are a negative number, in such a situation, LCLs are set to zero as shown in Figure 11.7.

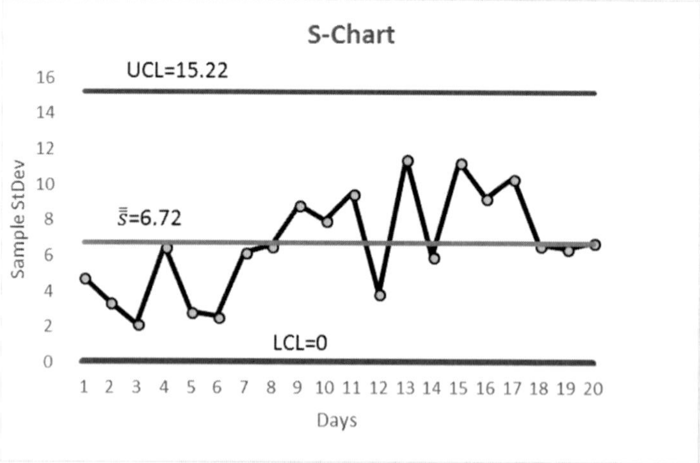

FIGURE 11.7 S-Control chart

11.4 P-Control Charts

So far we have considered the cases where continuous output data is used. What if we classify the product as either defective or non-defective on the basis our accepted standard measures. In this situation, as you notice our data will exhibit a discrete nature. Why do we need this classification/comparison? Because, we do this for simplicity as well as the economy in the quality inspection etc.

For example, when we measure the diameter of serially machined holes, we can use "*plain plug gauges*". Go and not-go *plain plug gauges* are mechanical measurement tools for accurately checking diameters of holes. These gauges are also very useful in the production lines to guarantee if the final diameter of the holes is as desired and within specific tolerances. Here we do not measure the diameter or inspect the hole surface, we rather use *plain plug gauges* and ease the final decision for dimension conformity with simple plug-in/out movements. In such situations, we use the *P*-control chart and we make our decision based on this chart either to continue or adjust the production process. This is considerable an easy task!

P-Control chart *(i.e. fraction-defective control chart)* uses $p_i = d_i / n$ ratio where d_i is the total defects in the ith sample and each sample consists of n (batch) total sample size. Also, d_i is assumed to be a *binomial random parameter* with an unknown parameter of p as presented in Equation 11.12.

$$\sigma_p = \sqrt{\frac{\bar{p}(1-\bar{p})}{n}} \qquad 11.12$$

$$\bar{p} = \frac{p_1 = \frac{d_1}{n} + p_2 = \frac{d_2}{n} + \cdots + p_m = \frac{d_m}{n}}{m} = \frac{d_{total}}{mn}$$

The sampling distribution of p can be approximated by a normal probability distribution when the sample size n is large! Then, the sampling distribution with mean $\bar{p}$ *(average defect ratio)* as center line and control limits with standard deviation of $\pm 3\sigma_p$ are given in Equation 11.13. (Note d_{total} : Total defective samples, m:sample count, n : per sample (batch) size, $m*n$: total number of samples).

$$\text{UCL} = \bar{p} + 3\sqrt{\frac{\bar{p}(1-\bar{p})}{n}}$$

$$\text{CL} = \bar{p} \qquad 11.13$$

$$LCL = \bar{p} - 3\sqrt{\frac{\bar{p}(1-\bar{p})}{n}}$$

In the section, an example of *P*-control chart is demonstrated on solar panel manufacturing example data, next.

TABLE 11.5 Solar panel production data

Sample	d_i	d_i/n	$\bar{p}$	UCL	LCL
1	15	0.125	0.15	0.24	0.05
2	19	0.158	0.15	0.24	0.05
3	18	0.150	0.15	0.24	0.05
4	22	0.183	0.15	0.24	0.05
5	33	0.275	0.15	0.24	0.05
6	12	0.100	0.15	0.24	0.05
7	11	0.092	0.15	0.24	0.05
8	15	0.125	0.15	0.24	0.05
9	23	0.192	0.15	0.24	0.05
10	10	0.083	0.15	0.24	0.05
11	19	0.158	0.15	0.24	0.05
12	22	0.183	0.15	0.24	0.05
13	11	0.092	0.15	0.24	0.05
14	14	0.117	0.15	0.24	0.05
15	25	0.208	0.15	0.24	0.05
16	17	0.142	0.15	0.24	0.05
17	23	0.192	0.15	0.24	0.05
18	15	0.125	0.15	0.24	0.05
19	16	0.133	0.15	0.24	0.05
20	13	0.108	0.15	0.24	0.05
d_{total} =	353	*Total defective samples*			
m =	20	*Sample no.*			
n =	120	*Batch size per sample*			
m*n =	2400	*Total # of samples*			
$\bar{p}$ =	0.1471	*Defect ratio*			
σ_p =	0.0323	*Standard deviation*			

Example 11.3: Solar Panel Production Data

In the production of solar panels, one of the important metrics states that all solar modules should possess an *electrical capacity* that can be measured in *Amps* and *Volts*.

In such a continuous solar panel production line, $m=20$ samples are tested each with sample(batch of solar panel) size of $n=120$. In Table 11.5 the number of defective solar panels, i.e. d_i $i=1,2,..n$ in each sample (batch) size are presented and using Equation 11.13, we get;

$$UCL = 0.147 + 3\sqrt{\frac{0.147(1-0.147)}{120}} = 0.244$$

$$CL = p = \frac{d_{total}}{mn} = \frac{353}{20 * 120} = 0.147$$

$$LCL = 0.147 - 3\sqrt{\frac{0.147(1-0.147)}{120}} = 0.05$$

$$\sigma_p = \sqrt{\frac{0.147(1-0.147)}{120}} = 0.0323$$

By examining Figure 11.8, we notice that we have out-of-control situation for sample no.5 in Table 11.5, where apparently there are excessive number of defective parts in that batch of n=120 samples. We need to go ahead and make some adjustments in the production and find out root cause for this violation. On the other hand, it is very clear that $\bar{p}$ is also high. There need to be more systematic interventions in the manufacturing strategies in order to reduce $\bar{p}$, i.e. more qualified workers, increased automation with successful management strategy etc.

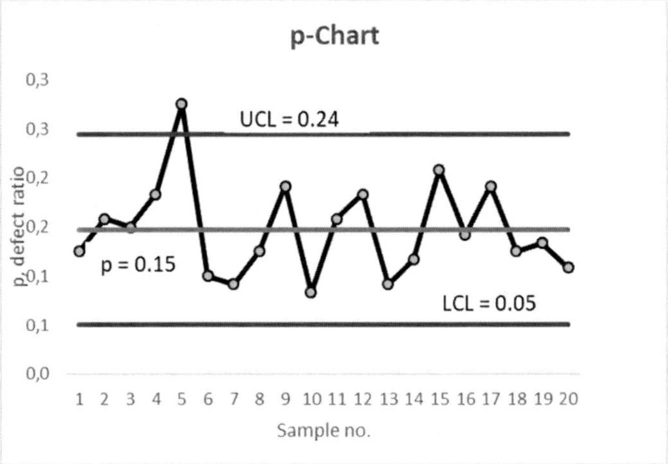

FIGURE 11.8 *P*-Control charts for solar panel production

11.5 U-Control Charts

We are continuing our discussion for another case of data with a discrete nature. In real-life, we are sometimes more interested in number of defects per unit than the defect-ratio. For example, we are interested in the control of weld defect counts in a certain length of ship body weld. In such situations, we can use *U*-Control charts for displaying the defects per-unit-measure.

In *U*-Control chart (u_i: defects-per-unit, for sample $i=1,...,n$), $u_i = c_i / n$ where c_i is the total defects in the ith sample and each sample consists of n (batch) sample units. Poisson probability distribution is well suited for such situations with parameter $\lambda = \bar{u}$, as the mean and σ_u, as the standard deviation, again $\bar{u}$: the average number of defects per unit are given by,

$$\sigma_u = \sqrt{\frac{\lambda = \bar{u}}{n}} \qquad 11.14$$

$$\bar{u} = \frac{u_1 = \frac{c_1}{n} + u_2 = \frac{c_2}{n} + \cdots + u_m = \frac{c_m}{n}}{m} = \frac{c_{total}}{mn}$$

where *m*:sample count, *n*: per sample (batch) size, *m**n*: total number of samples, c_{total} is total defects in *m**n* unit length.

$$\text{UCL} = \bar{u} + 3\sqrt{\frac{\bar{u}}{n}}$$
$$\text{CL} = \bar{u} \qquad 11.15$$
$$\text{LCL} = \bar{u} - 3\sqrt{\frac{\bar{u}}{n}}$$

TABLE 11.6 Velvet fabric rolls production

Sample	u_i	u_i/n	$\bar{u}$	UCL	LCL
1	3	0.3	0.15	0.52	0
2	0	0.0	0.15	0.52	0
3	2	0.2	0.15	0.52	0
4	2	0.2	0.15	0.52	0
5	0	0.0	0.15	0.52	0
6	0	0.0	0.15	0.52	0
7	1	0.1	0.15	0.52	0
8	2	0.2	0.15	0.52	0
9	1	0.1	0.15	0.52	0
10	3	0.3	0.15	0.52	0
11	1	0.1	0.15	0.52	0
12	2	0.2	0.15	0.52	0
13	3	0.3	0.15	0.52	0
14	1	0.1	0.15	0.52	0
15	0	0.0	0.15	0.52	0
16	2	0.2	0.15	0.52	0
17	1	0.1	0.15	0.52	0
18	2	0.2	0.15	0.52	0
19	1	0.1	0.15	0.52	0
20	3	0.3	0.15	0.52	0
c_{total} =	30	*Total defects in samples*			
m =	20	*Sample no.*			
n =	10	*Batch size per sample unit*			
m*n =	200	*Total # of samples*			
$\bar{u}$ =	0.15	*Defects per unit*			
σ_p =	0.1225	*Standard deviation*			

Example 11.4: Velvet Fabric Rolls Production

In the production of velvet fabrics, one of the metrics states that defects per unit length (fabric roll) should be controlled. Every hour ten unit velvet fabric rolls are selected and after the inspection, the number of defects in each sample of ten fabric roll units are recorded. In such a continuous production line, *m=20* velvet fabric samples are tested each with sample(batch) size of *n=10* fabric rolls. In Table 11.6 all inspection numbers are presented and using Equation 11.15, we get,

$$\text{UCL} = 0.15 + 3\sqrt{\frac{0.15}{10}} = 0.517$$

$$\text{CL} = \bar{u} = \frac{c_{total}}{mn} = \frac{30}{20 * 10} = 0.15$$

$$\text{LCL} = \max\left(0.15 - 3\sqrt{\frac{0.15}{10}}, 0\right) = 0$$

$$\sigma_p = \sqrt{\frac{0.15}{10}} = 0.1225$$

By examining Figure 11.9, we notice that we have things under control(stable). On the other hand ū still looks high! We need to concentrate on lowering it when planning quality improvement actions.

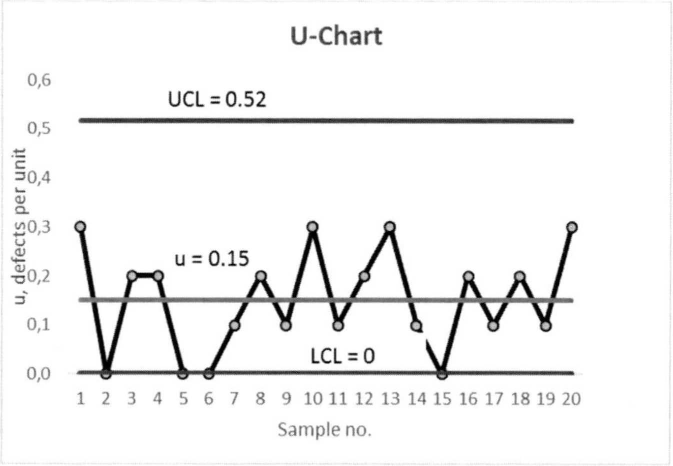

FIGURE 11.9 *U*-Control charts for velvet fabric production

11.6 Supplementary Material for *Excel* Commands

Example 11.1: Coating Thickness Data Control Chart

Note, in the following *Excel* sheet, **Colum-J** lists all the variables used and in **Column-K** their coded *Excel commands* with required formulations are given.

	A	B	C	D	E	F	G	H	I	J	K
1	Time	x	mr	$\bar{x}$	UCL	LCL	$\overline{mr}$	UCL	LCL	*i=1*	i_{th} EXCEL FORMULA
2	1	3.78	-	3.668	5.939	1.397	0.854	2.790	0	$\bar{x}_1$	= B$26
3	2	4.55	0.77	3.668	5.939	1.397	0.854	2.790	0	mr_{i+1}	= ABS(B3-B2)
4	3	3.79	0.76	3.668	5.939	1.397	0.854	2.790	0	$\bar{\bar{x}}_1$	= B$26
5	4	5.07	1.28	3.668	5.939	1.397	0.854	2.790	0	UCL_1	= B$26+3*C$27/1.128
6	5	2.92	2.15	3.668	5.939	1.397	0.854	2.790	0	LCL_1	= B$26-3*C$27/1.128
7	6	2.78	0.14	3.668	5.939	1.397	0.854	2.790	0	$m\bar{r}_1$	= C$27
8	7	4.01	1.23	3.668	5.939	1.397	0.854	2.790	0	UCL_1	= 3.267*C$27
9	8	3.41	0.6	3.668	5.939	1.397	0.854	2.790	0	LCL_1	= 0
10	9	3.65	0.24	3.668	5.939	1.397	0.854	2.790	0		
11	10	3.89	0.24	3.668	5.939	1.397	0.854	2.790	0		
12	11	4.77	0.88	3.668	5.939	1.397	0.854	2.790	0		
13	12	3.11	1.66	3.668	5.939	1.397	0.854	2.790	0		
14	13	3.22	0.11	3.668	5.939	1.397	0.854	2.790	0		
15	14	4.02	0.8	3.668	5.939	1.397	0.854	2.790	0		
16	15	3.66	0.36	3.668	5.939	1.397	0.854	2.790	0		
17	16	3.44	0.22	3.668	5.939	1.397	0.854	2.790	0		
18	17	4.76	1.32	3.668	5.939	1.397	0.854	2.790	0		
19	18	3.55	1.21	3.668	5.939	1.397	0.854	2.790	0		
20	19	2.88	0.67	3.668	5.939	1.397	0.854	2.790	0		
21	20	3.89	1.01	3.668	5.939	1.397	0.854	2.790	0		
22	21	3.01	0.88	3.668	5.939	1.397	0.854	2.790	0		
23	22	3.33	0.32	3.668	5.939	1.397	0.854	2.790	0		
24	23	4.22	0.89	3.668	5.939	1.397	0.854	2.790	0		
25	24	2.32	1.9	3.668	5.939	1.397	0.854	2.790	0		
26	$\bar{x}$ =	3.668								$\bar{x}$	= AVERAGE(B2:B25)
27		$\overline{mr}$ =	0.8539	Factors	A_2	d_2	D_3	D_4	C_4	$\overline{mr}$	= AVERAGE(C3:C25)
28		σ_c =	0.7570	*n=2*	1.88	1.128	0	3.267	0.9213	σ_c	= C27/1.128

Example 11.2: Compressive Strength Control Chart
$\bar{X}$-R Control Chart

Note, in the following *Excel* sheet, **Colum-N** lists all the variables used and in **Column-O** their coded *Excel commands* with required formulations are given.

	A	B	C	D	E	F	G	H	I	J	K	L	M	N	O
1	Days	X_1	X_2	X_3	X_4	$\bar{x}$	r	$\bar{\bar{x}}$	UCL	LCL	$\bar{r}$	UCL	LCL	*i*=1	i_{th} EXCEL FORMULA
2	1	26	35	32	37	32.5	11	29.84	40.7	18.98	14.9	34	0	$\bar{x}_1$	= AVERAGE(B2:E2)
3	2	23	27	29	31	27.5	8	29.84	40.7	18.98	14.9	34	0	r_1	=MAX(B2:E2)-MIN(B2:E2)
4	3	15	16	18	20	17.25	5	29.84	40.7	18.98	14.9	34	0	$\bar{\bar{x}}_1$	= F$22
5	4	37	22	33	34	31.5	15	29.84	40.7	18.98	14.9	34	0	UCL_1	= F$22+0.729*G$23
6	5	32	29	25	28	28.5	7	29.84	40.7	18.98	14.9	34	0	LCL_1	= F$22-0.729*G$23
7	6	33	29	28	33	30.75	5	29.84	40.7	18.98	14.9	34	0	$\bar{r}_1$	= G$23
8	7	20	31	33	23	26.75	13	29.84	40.7	18.98	14.9	34	0	UCL_1	= 2.282*G$23
9	8	22	34	36	26	29.5	14	29.84	40.7	18.98	14.9	34	0	LCL_1	= 0
10	9	23	44	29	34	32.5	21	29.84	40.7	18.98	14.9	34	0		
11	10	23	23	29	40	28.75	17	29.84	40.7	18.98	14.9	34	0		
12	11	24	41	23	39	31.75	18	29.84	40.7	18.98	14.9	34	0		
13	12	26	33	26	24	27.25	9	29.84	40.7	18.98	14.9	34	0		
14	13	55	41	44	27	41.75	28	29.84	40.7	18.98	14.9	34	0		
15	14	36	25	39	33	33.25	14	29.84	40.7	18.98	14.9	34	0		
16	15	43	37	23	19	30.5	24	29.84	40.7	18.98	14.9	34	0		
17	16	24	32	41	20	29.25	21	29.84	40.7	18.98	14.9	34	0		
18	17	36	33	44	19	33	25	29.84	40.7	18.98	14.9	34	0		
19	18	33	38	25	24	30	14	29.84	40.7	18.98	14.9	34	0		
20	19	37	27	24	36	31	13	29.84	40.7	18.98	14.9	34	0		
21	20	17	21	23	33	23.5	16	29.84	40.7	18.98	14.9	34	0		
22					$\bar{\bar{x}}$ =	29.84								$\bar{\bar{x}}$	= AVERAGE(F2:F21)
23						$\bar{r}$ =	14.9	Factors	A_2	d_2	D_3	D_4	C_4	$\bar{r}$	= AVERAGE(G2:G21)
24						$\hat{\sigma}$ =	7.2365	*n*=4	0.729	2.059	0	2.282	0.9213	$\hat{\sigma}$	= G23/2.059

$\overline{X}$-S Control Chart

Note, in the following *Excel* sheet, **Colum-N** lists all the variables used and in **Column-O** their coded *Excel commands* with required formulations are given.

	A	B	C	D	E	F	G	H	I	J	K	L	M	N	O
1	Days	x_1	x_2	x_3	x_4	$\bar{x}$	s	$\bar{\bar{x}}$	UCL	LCL	$\bar{s}$	UCL	LCL	i=1	i_{th} EXCEL FORMULA
2	1	26	35	32	37	32.5	4.80	29.84	40.78	18.9	6.72	15.22	0	$\bar{x}_1$	= AVERAGE(B2:E2)
3	2	23	27	29	31	27.5	3.42	29.84	40.78	18.9	6.72	15.22	0	s_1	= STDEV(B2:E2)
4	3	15	16	18	20	17.25	2.22	29.84	40.78	18.9	6.72	15.22	0	$\bar{\bar{x}}_1$	= F$22
5	4	37	22	33	34	31.5	6.56	29.84	40.78	18.9	6.72	15.22	0	UCL_1	= F$22+3*K2/SQRT(4)/**0.9213**
6	5	32	29	25	28	28.5	2.89	29.84	40.78	18.9	6.72	15.22	0	LCL_1	= F$22-3*K2/SQRT(4)/**0.9213**
7	6	33	29	28	33	30.75	2.63	29.84	40.78	18.9	6.72	15.22	0	$\bar{s}_1$	= G$23
8	7	20	31	33	23	26.75	6.24	29.84	40.78	18.9	6.72	15.22	0	UCL_1	= K2+3*K2*SQRT(1-**0.9213**^2)/**0.9213**
9	8	22	34	36	26	29.5	6.61	29.84	40.78	18.9	6.72	15.22	0	LCL_1	= MAX(K2-3*K2*SQRT(1-**0.9213**^2)/**0.9213**,0)
10	9	23	44	29	34	32.5	8.89	29.84	40.78	18.9	6.72	15.22	0		
11	10	23	23	29	40	28.75	8.02	29.84	40.78	18.9	6.72	15.22	0		
12	11	24	41	23	39	31.75	9.57	29.84	40.78	18.9	6.72	15.22	0		
13	12	26	33	26	24	27.25	3.95	29.84	40.78	18.9	6.72	15.22	0		
14	13	55	41	44	27	41.75	11.53	29.84	40.78	18.9	6.72	15.22	0		
15	14	36	25	39	33	33.25	6.02	29.84	40.78	18.9	6.72	15.22	0		
16	15	43	37	23	19	30.5	11.36	29.84	40.78	18.9	6.72	15.22	0		
17	16	24	32	41	20	29.25	9.29	29.84	40.78	18.9	6.72	15.22	0		
18	17	36	33	44	19	33	10.42	29.84	40.78	18.9	6.72	15.22	0		
19	18	33	38	25	24	30	6.68	29.84	40.78	18.9	6.72	15.22	0		
20	19	37	27	24	36	31	6.48	29.84	40.78	18.9	6.72	15.22	0		
21	20	17	21	23	33	23.5	6.81	29.84	40.78	18.9	6.72	15.22	0		
22					$\bar{\bar{x}}$ =	**29.84**								$\bar{\bar{x}}$	= AVERAGE(F2:F21)
23						$\bar{s}$ =	**6.72**	Factors	A_2	d_2	D_3	D_4	C_4	$\bar{s}$	= AVERAGE(G2:G21)
24						$\hat{\sigma}$ =	**7.292**	n=4	0.729	2.059	0	2.282	**0.9213**	$\hat{\sigma}$	= G23/**2.059**

Example 11.3: Solar Panel Production Data

Note, in the following *Excel* sheet, **Colum-G** lists all the variables used and in **Column-H** their coded *Excel commands* with required formulations are given.

	A	B	C	D	E	F	G	H
1	Sample	d_i	d_i/n	$\bar{p}$	UCL	LCL		
2	1	15	0.125	0.15	0.24	0.05	d1/n	= B2/B$24
3	2	19	0.158	0.15	0.24	0.05	$\bar{p}_i$	= B$26
4	3	18	0.150	0.15	0.24	0.05	UCL_i	= B$26+3*SQRT(B$26*(1-B$26)/B$24)
5	4	22	0.183	0.15	0.24	0.05	LCL_i	= B$26-3*SQRT(B$26*(1-B$26)/B$24)
6	5	33	0.275	0.15	0.24	0.05		
7	6	12	0.100	0.15	0.24	0.05		
8	7	11	0.092	0.15	0.24	0.05		
9	8	15	0.125	0.15	0.24	0.05		
10	9	23	0.192	0.15	0.24	0.05		
11	10	10	0.083	0.15	0.24	0.05		
12	11	19	0.158	0.15	0.24	0.05		
13	12	22	0.183	0.15	0.24	0.05		
14	13	11	0.092	0.15	0.24	0.05		
15	14	14	0.117	0.15	0.24	0.05		
16	15	25	0.208	0.15	0.24	0.05		
17	16	17	0.142	0.15	0.24	0.05		
18	17	23	0.192	0.15	0.24	0.05		
19	18	15	0.125	0.15	0.24	0.05		
20	19	16	0.133	0.15	0.24	0.05		
21	20	13	0.108	0.15	0.24	0.05		
22	d_{total} =	353	*Total defective samples*				d_{total}	= SUM(B2:B21)
23	m =	20	*Sample no.*					
24	n =	120	*Batch size per sample*					
25	m*n =	2400	*Total # of samples*				mn	= B23*B24
26	$\bar{p}$ =	0.147	*Defect ratio*				$\bar{p}$	= B22/B25
27	σ_p =	0.0323	*Standard deviation*				σ_p	= SQRT(B26*(1-B26)/B24)

Example 11.4: Velvet Fabric Rolls Production

Note, in the following *Excel* sheet, **Colum-G** lists all the variables used and in **Column-H** their coded *Excel commands* with required formulations are given.

	A	B	C	D	E	F	G	H
1	**Sample**	**u_i**	**u_i/n**	**$\bar{u}$**	**UCL**	**LCL**		
2	1	3	0.3	0.15	0.52	0	u_1/n	= B2/B$24
3	2	0	0.0	0.15	0.52	0	$\bar{u}_i$	= B$26
4	3	2	0.2	0.15	0.52	0	UCL_i	= B$26+3*SQRT(B$26*(1-B$26)/B$24)
5	4	2	0.2	0.15	0.52	0	LCL_i	= 0
6	5	0	0.0	0.15	0.52	0		
7	6	0	0.0	0.15	0.52	0		
8	7	1	0.1	0.15	0.52	0		
9	8	2	0.2	0.15	0.52	0		
10	9	1	0.1	0.15	0.52	0		
11	10	3	0.3	0.15	0.52	0		
12	11	1	0.1	0.15	0.52	0		
13	12	2	0.2	0.15	0.52	0		
14	13	3	0.3	0.15	0.52	0		
15	14	1	0.1	0.15	0.52	0		
16	15	0	0.0	0.15	0.52	0		
17	16	2	0.2	0.15	0.52	0		
18	17	1	0.1	0.15	0.52	0		
19	18	2	0.2	0.15	0.52	0		
20	19	1	0.1	0.15	0.52	0		
21	20	3	0.3	0.15	0.52	0		
22	C_{total} =	30	*Total defects in samples*				C_{total}	= SUM(B2:B21)
23	m =	20	*Sample no.*					
24	n =	10	*Batch size per sample unit*					
25	m*n =	200	*Total # of samples*				mn	= B23*B24
26	$\bar{u}$ =	0.15	*Defects per unit*				$\bar{u}$	= B22/B25
27	σ_p =	0.1225	*Standard deviation*				σ_p	= SQRT(B$26/B$24)

11.7 Supplementary Material for Minitab Results

Example 11.1: Coating Thickness I-MR Control Chart

As for to demonstrate how to draw **I-MR… charts** in *Minitab*, please follow all the steps below.

- ✓ Start-up *Minitab* and enter all data in C1 as **x**. Then using **Stat >Control Charts>Variables Charts for Individuals> I-MR…** option, click here as shown in the following,

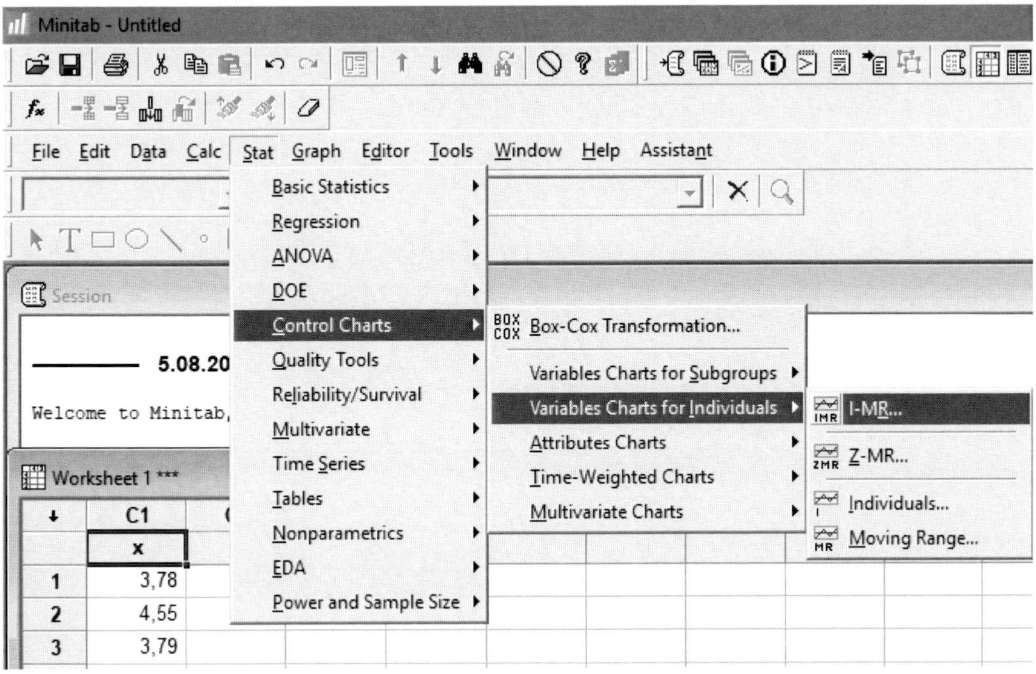

- ✓ In the pop-up **Individuals-Moving Range Chart** window, click in the window for **Variables** (following, the variable **x** will appear), and select x_1 and hit **Select** then, hit OK next,

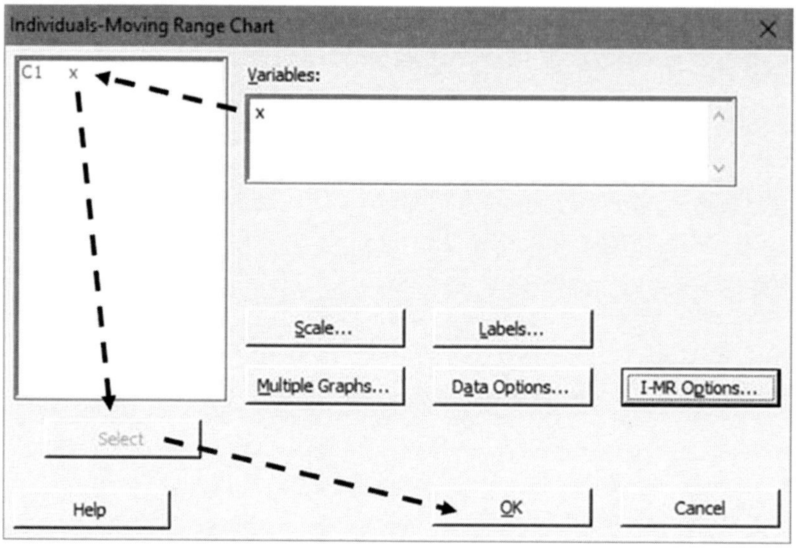

✓ Finally, you will get the following output results where **I-MR Chart** as shown in the following.

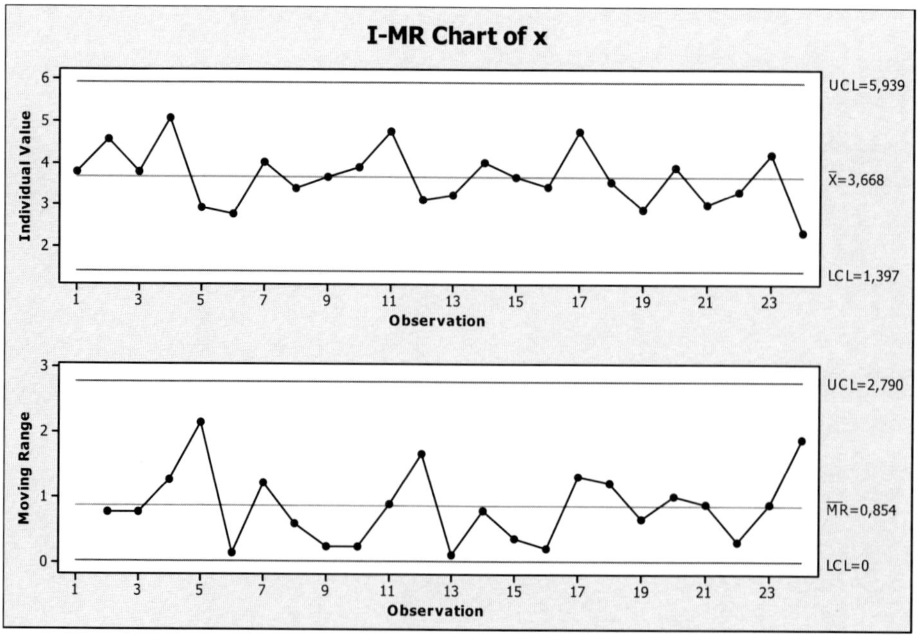

Example 11.2: Compressive Strength Example $\overline{X}$-R and $\overline{X}$-S Control Charts

As for to demonstrate how to draw **Xbar-R... charts** in *Minitab*, we are going to show all the steps below, the **Xbar-S... charts** can be formed similarly.

- ✓ Start-up *Minitab* and enter all data in C1 to C4 as x_1, x_2, x_3 and x_4 i.e. for all $n=4$ samples. Then using **Stat >Control Charts>Variables Charts for Subgroups> Xbar-R...** option, click here as shown in the following,

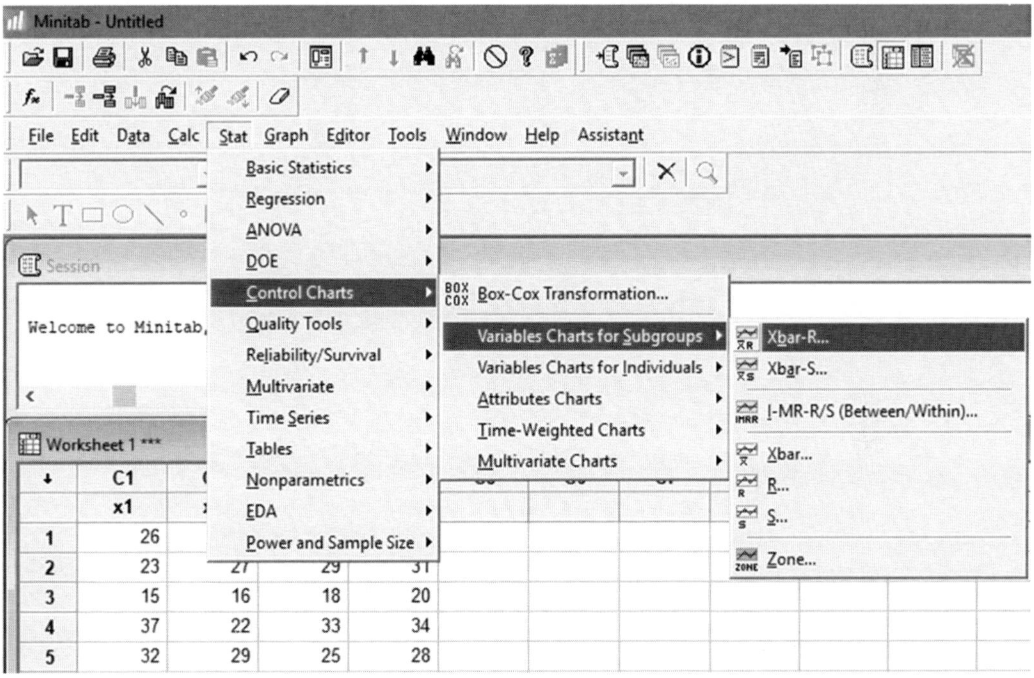

- ✓ In the pop-up **Xbar-R Chart** window, select **Observations for a subgroup are in one row columns:** option, Click in the window underneath (following, the variable **x** will appear), and select x_1 to x_4 and hit **Select** first, then click on **Xbar-R options...** tab next,

- ✓ **Xbar-R Chart** window will pop-up and click on **Estimate** tab and select **Rbar** option and hit **OK**

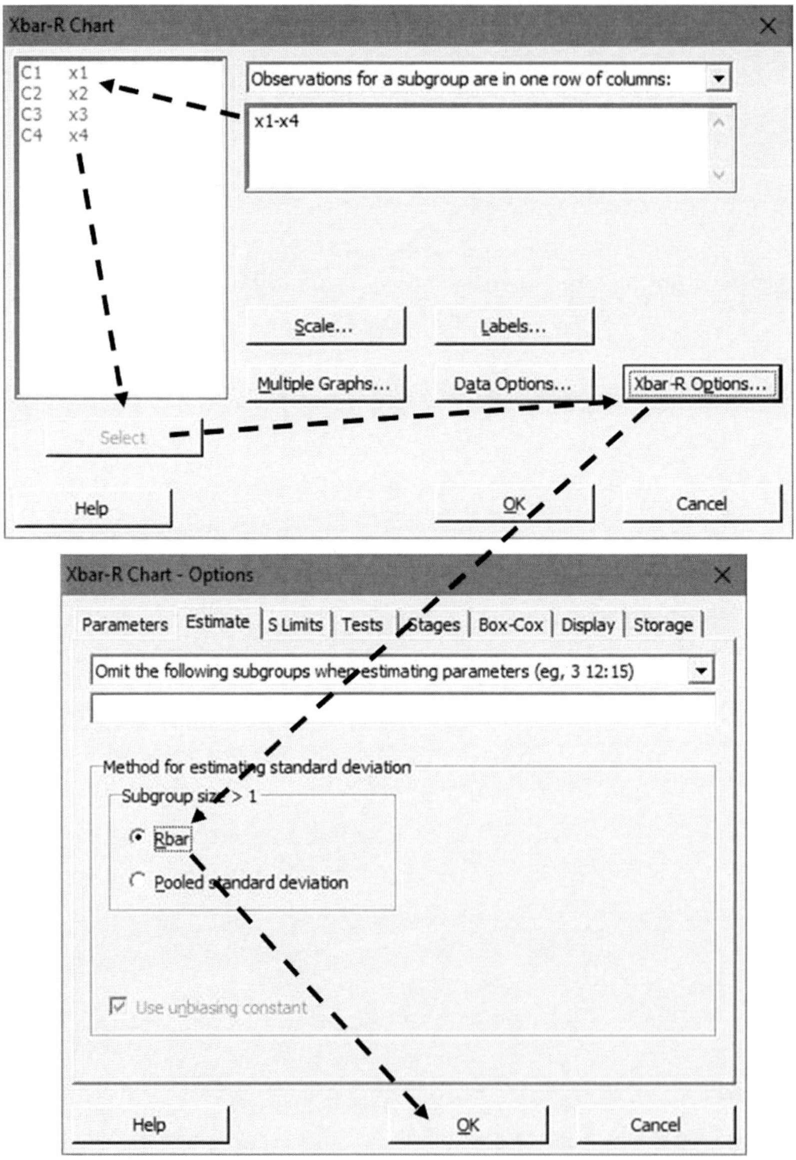

✓ Finally, you will get the following output results where **Xbar-R Chart** displayed next!

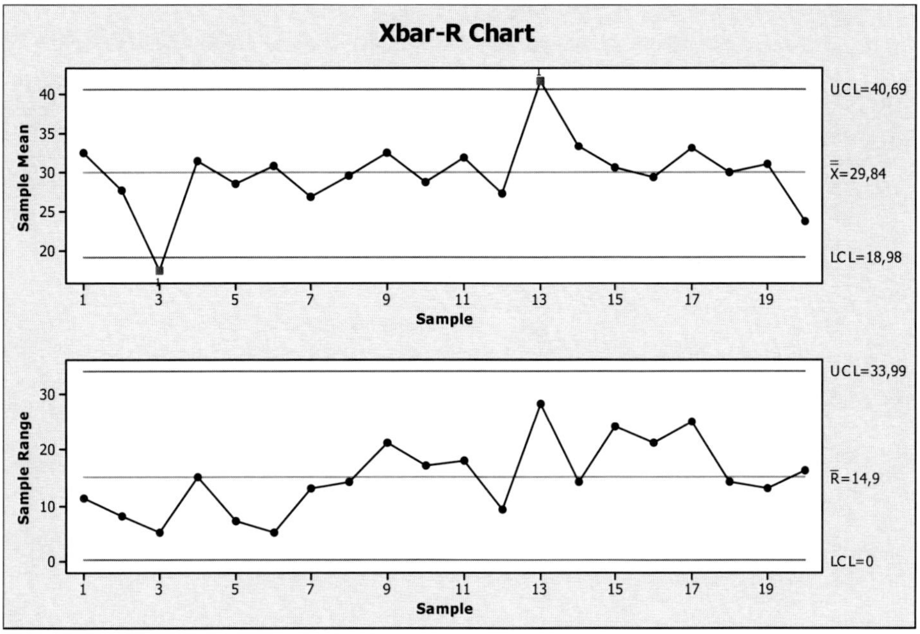

Similarly, we can draw **Xbar-S... charts** in *Minitab* as also demonstrated in the following!

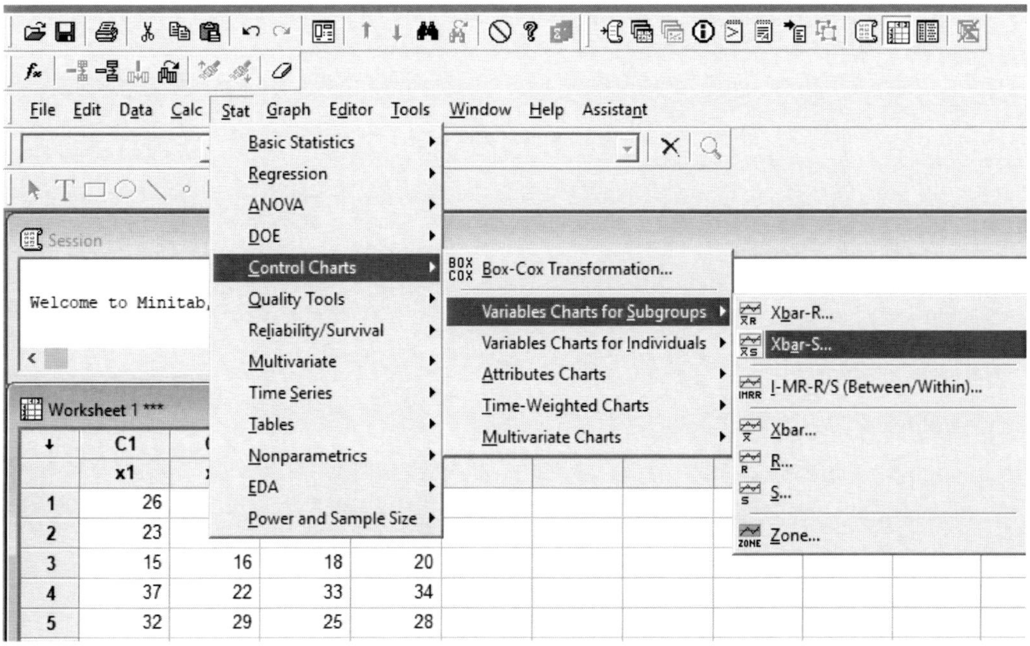

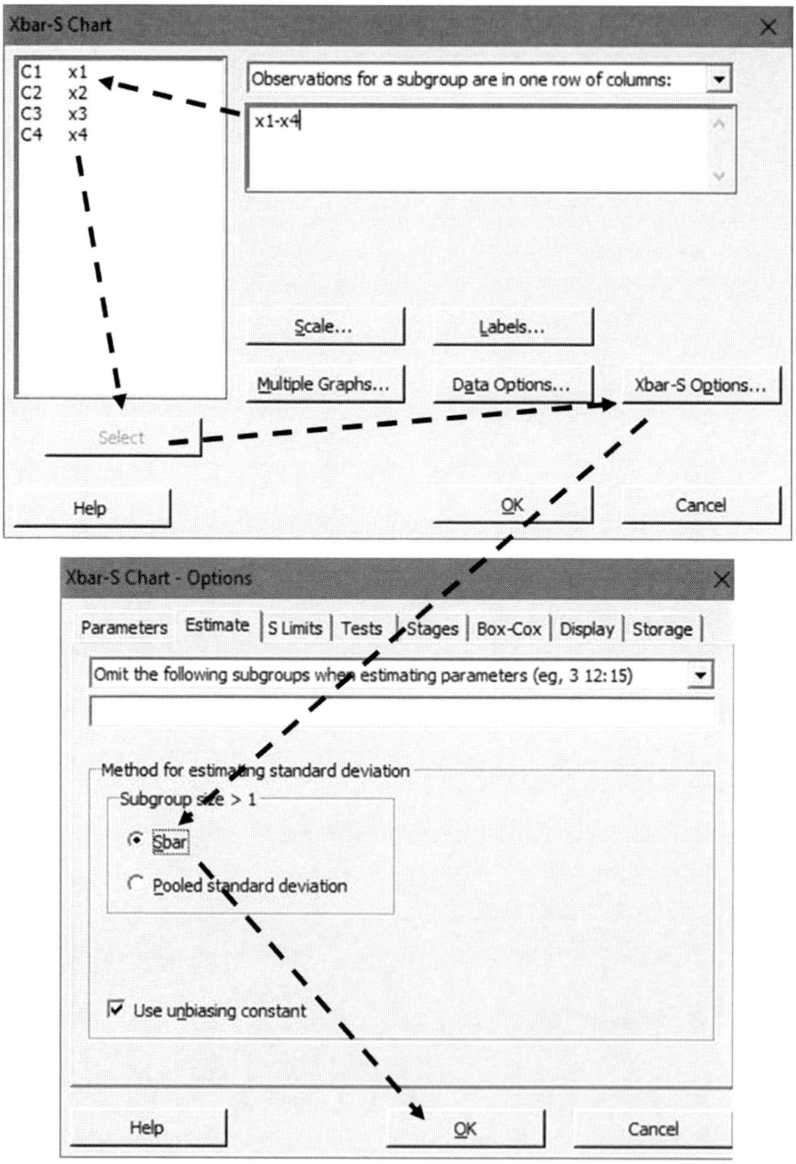

✓ Following the similar steps, you will get the following output results where **Xbar-S Chart** displayed next!

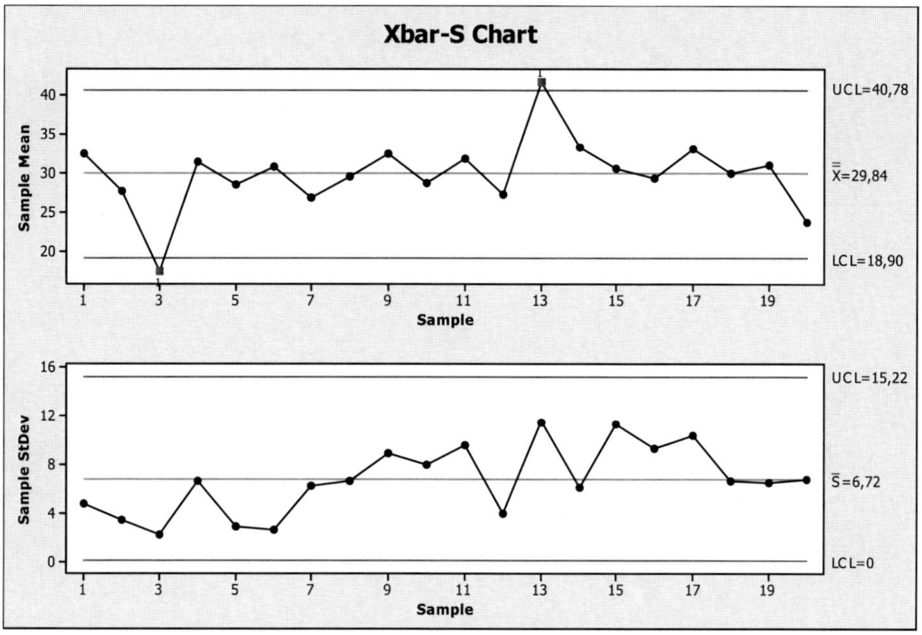

Example 11.3: Solar Panel Production Data P-Chart

As for to demonstrate how to draw **P... charts** in *Minitab*, please follow all the steps outlined below.

- ✓ Start-up *Minitab* and enter all data in C1 as **d**. Then using **Stat >Control Charts>Attributes Charts> P...** option, click here as shown in the following,

- ✓ In the pop-up **P-Chart** window, Click in **Variables:** window and select **d** and hit **Select** first. Then, enter **Subgroup sizes: 120** next and click OK!

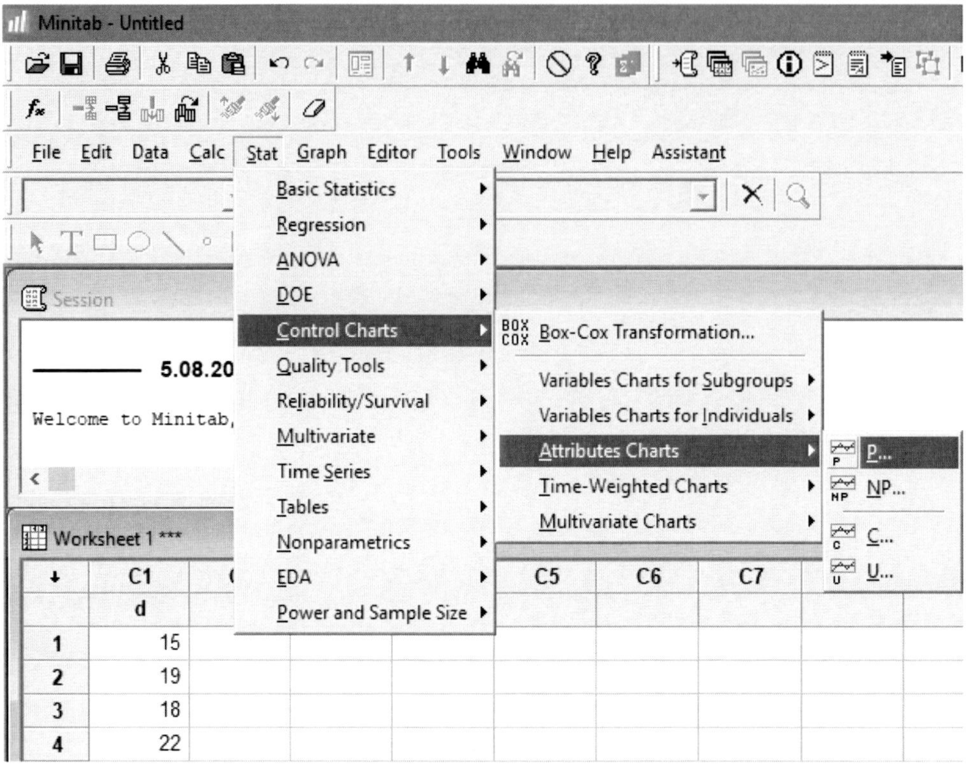

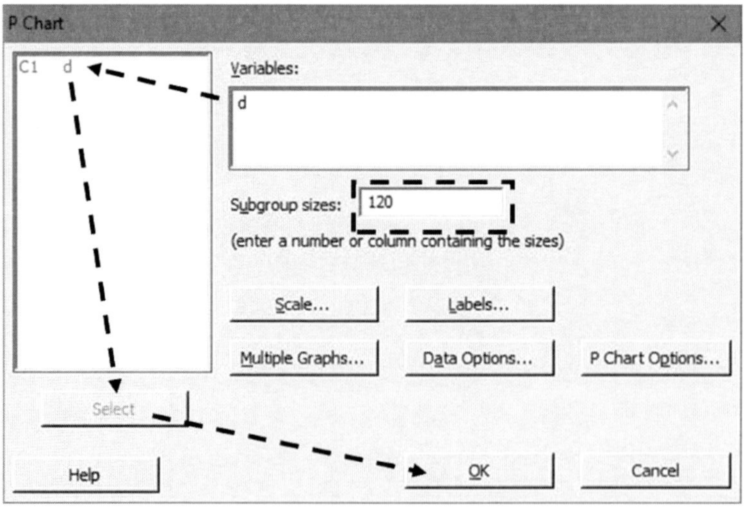

✓ Finally, you will get the following output results where **P-Chart** pops-out next!

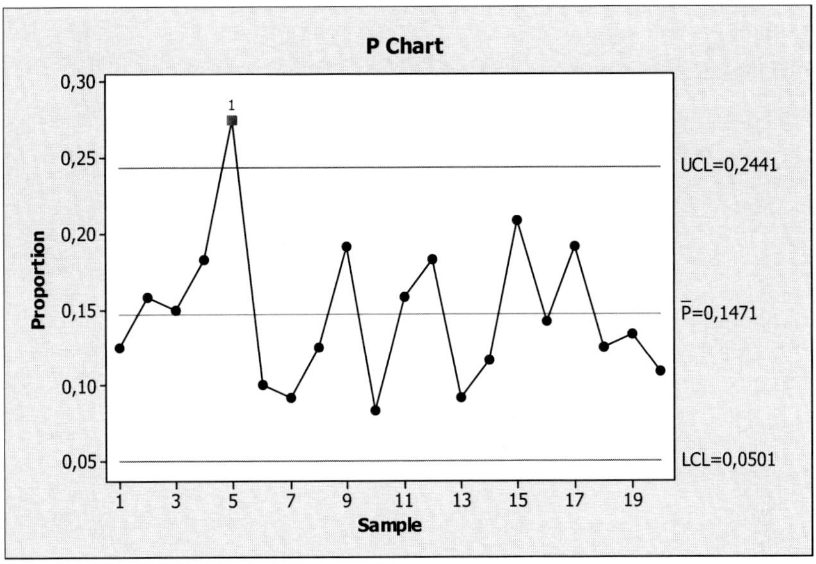

Example 11.4: Velvet Fabric Rolls Production U-Chart

As for to demonstrate how to draw **U… charts** in *Minitab*, please follow all the steps outlined below.

✓ Start-up *Minitab* and enter all data in C1 as **u**. Then using **Stat >Control Charts>Attributes Charts> U…** option, click here as shown in the following,

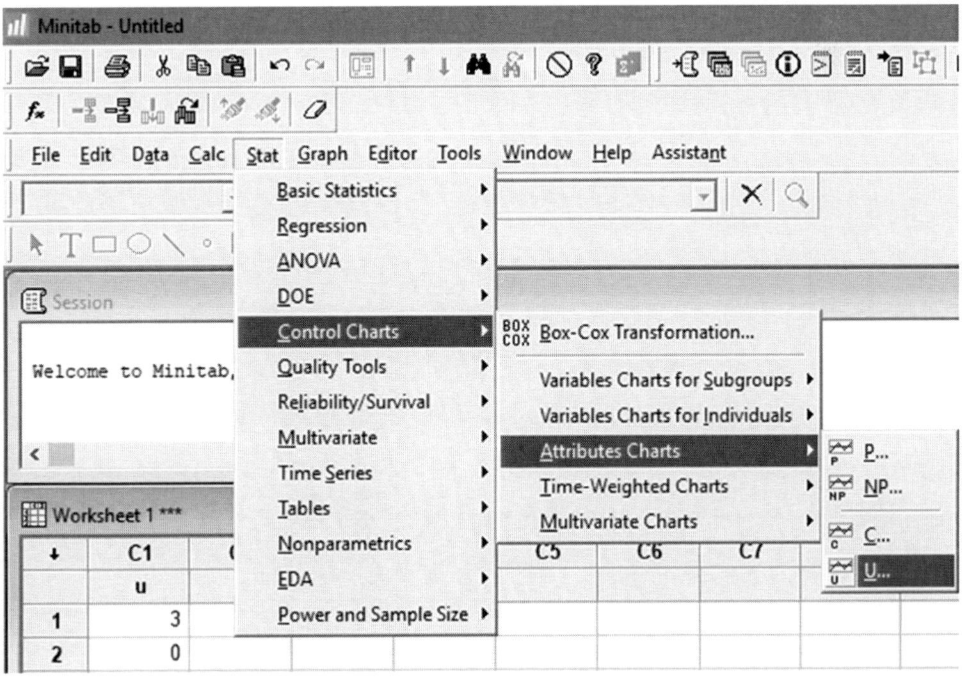

✓ In the pop-up **U Chart** window, Click in **Variables:** window and select **u** and hit **Select** first. Then, enter **Subgroup sizes: 10** next and click OK!

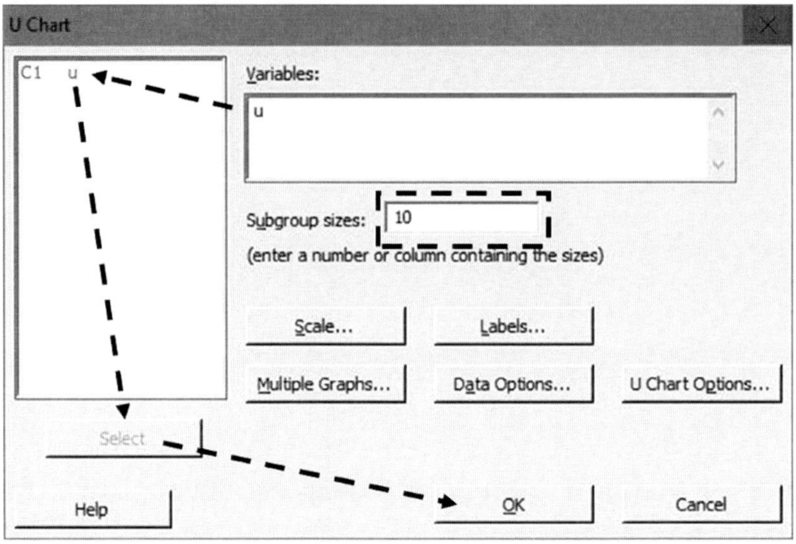

✓ Finally, you will get the following output results where **U-Chart** displayed next!

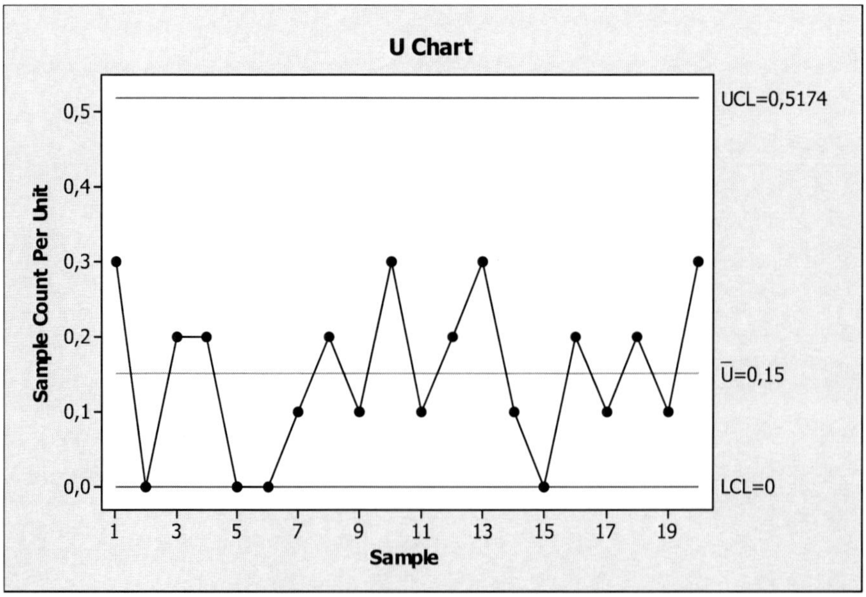

11.8 Exercises

1) Please give brief information about statistical quality control (SQC) and statistical process control (SPC) with some examples.

2) An automobile manufacturing company is testing its 110 HP gas engines in their plant to see horse power variations. For this purpose twenty consecutive days, three sample engines are tested and their data is recorded for each day, in a random manner.

The automobile engineers are interested in output HP and its distribution. These data are recorded as presented in Table 11.6, Assuming that when the manufacturing process is stable. You are expected to do the following;

- Plot $\bar{X}$-R control charts using *Excel* and Minitab
- Plot $\bar{X}$-S control charts using *Excel* and Minitab
- State your conclusions about the process using your control charts

TABLE 11.6 Engine HP manufacturing data

Days	x_1	x_2	x_3
1	110	111	105
2	118	105	100
3	111	109	120
4	106	110	110
5	107	111	110
6	118	113	109
7	119	119	101
8	108	110	114
9	118	110	116
10	104	109	117
11	119	111	116
12	111	111	115
13	110	107	111
14	113	108	112
15	103	109	113
16	107	110	122
17	109	112	100
18	119	113	109
19	117	115	103
20	110	105	105

3) In a plastic injection machine, variations occur and injected products may come out of the machine with different weights due to errors in the setting of process parameters and machine characteristics. For this purpose, only one measurement is taken and recorded on one-hour intervals for 24 hours and weight of the products are measured. Weights (i.e. *x's)* are given in *gram,* as presented in Table 11.7.

TABLE 11.7 Plastic injection part weight data

Time	1	2	3	4	5	6	7	8	9	10	11	12
X	1060	1060	1099	1110	1099	1077	1068	1092	1100	1102	1101	1044
Time	**13**	**14**	**15**	**16**	**17**	**18**	**19**	**20**	**21**	**22**	**23**	**24**
X	1009	1100	1112	1101	1135	1102	1109	1088	1089	1101	1102	1055

- Plot X-IR control charts using *Excel* and Minitab
- State your conclusions about the process using your control charts

4) In the production line of semi-conductor wafers, due to process and material variations we may get below standard quality chips. For testing this, $m=20$ samples are tested each with sample(batch size of $n=150$). In Table 11.8 the number of defective wafers in each sample (batch) size of 150 are presented,

TABLE 11.8 Semi-conductor wafers data

Sample	**1**	**2**	**3**	**4**	**5**	**6**	**7**	**8**	**9**	**10**
#*of* defects	15	5	4	5	9	21	1	6	8	11
Sample	**11**	**12**	**13**	**14**	**15**	**16**	**17**	**18**	**19**	**20**
#*of* defects	13	16	7	20	6	8	9	10	9	12

- Plot *P*-Control charts using *Excel* and Minitab
- Comment on the results and make up some rational idea regarding the outcomes.

Chapter 12

Process Capability

A process is a wisely planned and executed part of (say) a product manufacturing cycle is comprised of pre-determined sequences of interrelated activities/operations. During each of these processes, some resources are used and *each process* contributes to the ultimate objective/output/product by progressively *adding some value* to the next *process*. This goes on until the final product realization. From the manufacturing perspective, eventually, raw materials are converted into finished products to meet the needs of the society. In a process as depicted in Figure 12.1, raw data consisting of various line shapes are processed into meaningful words, next, words are cast into poems/compositions/ structures with *wisdom*!

Unfortunately, the real-life problems are not without up and downs! We usually suffer from the variations during the execution of the processes. As emphasized on several occasions in the previous chapters, in each process a certain level of variation does exist and although they are not welcome! Nevertheless, they are unavoidable. Then, as an engineer, our primary job is to identify these variations and find out the (important) parameters affecting these variations so that we can have control over them.

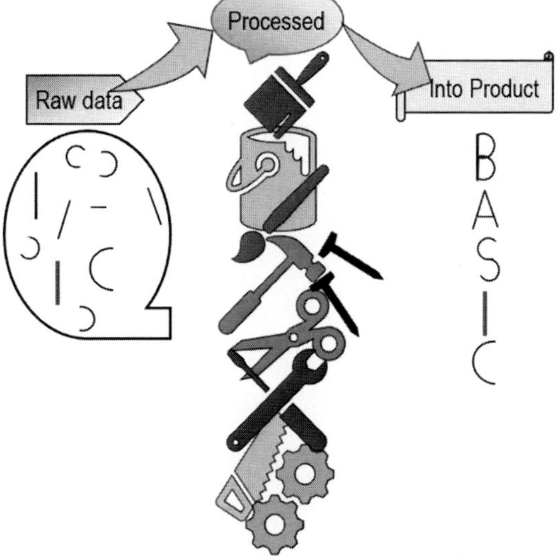

FIGURE 12.1 Raw data processed into product

Further, we also need to have the capability to tackle these variations. How? assume variations are an elephant and we are an ant. Not a good match for wrestling! At the same time, variations need to be measured accurately and all necessary precautions must be taken for keeping/handling them at reduced/acceptable levels for functionality and further for quality concerns/issues. Generally, a great deal of effort is spent toward controlling the variability in each process. Yet, there will still be an effect of some variations in the final product anyway. We should keep in mind that customer satisfaction will be directly linked to product/process quality as well as the cost. Customer satisfaction and preference is the best way of *good* reputation and advertisement for any company.

Let's consider a cheeseburger example which needs to be cooked and served at all franchise stores at the same freshness, taste and low price with reasonable short waiting time for customer satisfaction. This example is rather complicated where as you may notice that many factors directly affect the taste, price and customer service. Now, let's consider the important processes briefly, for a better understanding of the variations. Assume, the main processes for cheeseburger making (i.e. totaling five) are;

- the quality (ingredients, freshness, type etc.) of the burger bread
- the taste of the raw ground beef mixture(fat content, meat quality, and freshness, spices, cheese, mixture etc.)
- cooking process (grill, fry, temperature, cooking duration etc.)
- toppings preparation i.e. ketchup, mayonnaise, tomato slice, onion, lettuce etc.
- low price and customer satisfaction.

Now consider the variations that may occur during the cheeseburger making a process where we can talk about five different processes, each contributing to the output, i.e. cheeseburger (and its taste). If you serve this cheeseburger in many stores or even in a single store, you may have problems due to variations in each five-process and its contents as highlighted. How? changes in the suppliers, franchise personnel skills, regional effects and many more! Therefore, not every cheeseburger would have been exactly the same in taste even served in a single store. We need to make sure, all metrics (in five processes affecting the end taste) must be measured just about right so that we could control the outcoming cheeseburgers taste in all stores, and make sure all cheeseburgers taste very good and very similar.

Here, we clearly realize the importance of process control, by utilizing necessary equipment and means to take reliable/consistent measurements in controlling/evaluating each process so that we can secure the least (or no) variation on the taste of the cheeseburgers. Here *process capability analysis* can offer crucial improvements and control over the taste of the cheeseburgers by monitoring/evaluating each process to achieve great cheeseburger taste at all times. The higher quality (in taste) means the higher customer satisfaction and preference.

Further note that, process capability level is assessed, maintained and improved by using well-established statistical techniques in statistical process control. In summary, the process capability analysis can be used to control the expected consistency/ repeatability of any product features by meeting/keeping its specifications at a high level of customer satisfaction. After this analysis, we can assert some percentage level (degree) that is showing that our process is meeting our product requirements or not. All these calculations are performed to make sure the product capability.

In real-life of industrial applications, the outcome of process capability analysis is shown to be very valuable process output assessment and in many ways as it would be discussed in the next section.

12.1 Process Capability Analysis

Process Capability analysis is a set of statistical calculations performed to estimate/see if the capability of a process is able to meet a set of specifications/requirements. For these calculations, some form of crucial data has to be generated or collected. Note that product specifications/requirements are predetermined values specifying the acceptable *upper* and *lower* limits between the system is designed to operate. Customers, design engineers or necessities usually set these specifications. Specifications are numerical requirements, for reaching out our goals. These specifications should not be confused with the control limits! Why? Control limits are based on the inherent data characteristics and obtained from control charts, remember LCL and UCL. Specifications are the numerical requirements of the system set by customer or process, i.e. LSL and USL.

Capability analysis needs a *statistically stable system* first from which process/product data should be collected and checked/monitored using a control chart(s). Control charts are introduced in detail in Chapter 11. Keep in mind that capability analysis reveals heads-up what happened in the past by providing only a snapshot(instantaneous measure) of the current process. Capability analysis is a very useful SPC tool to visualize the improvement that we made in a process by smart intervention(s). It can also report;

i. the capability of a process, (what to expect)
ii. the extent of more improvement needed, (room for improvement)
iii. (and thus), the level of the improvement achieved with respect to the previous one (further achievements)

Process capability analysis a useful tool helps us reduce the variation relative to the specifications/requirements for process/product. This directly affects product cost saving due to scrap or rework as well.

Process Capability should also be accompanied with *process stability*, which is exactly what we would like to achieve for making sure our process is under control. As an example, let's consider two different Processes A and B where we would like to make sure outer dimension (i.e. diameter, $^{\phi}$d) of the power transmission shaft is manufactured based upon the specs.

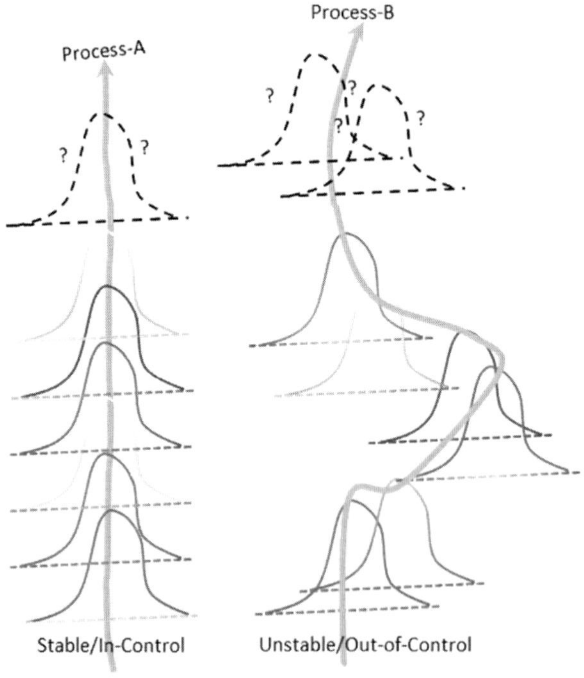

FIGURE 12.2 Stability in processes

If you examine Figure 12.2, Process-A measurements are taken at six different time intervals at certain sample days. Distributions all look like each other and exhibit a normal distribution. Measurements for shaft dimension distributions are very consistent with each other, exhibiting an almost straight line of order for *mean* locations. We can assert that Process-A is stable or in-control, based upon that, we can estimate the seventh measurement's behavior in consisted with the history of previous measurements. Apparently, this is what everyone wants to see for their process.

Whereas, in Figure 12.2, in Process-B, as for the six measurements and their distributions are alike, shaped good and with normal characteristics but they are exhibiting a fluctuating line of order for *mean* locations. We experience too much instability. Obviously, shaft dimensions do not seem under control. Excessive variations make Process-B unstable or out-of-control. Further, we have gotten six bad normal curve distribution characteristics for Process-B based on them, we have no way of estimating the behavior of the seventh measurement. This would further make our fixing up Process-B for stability rather very difficult!

What about the relationship between being capable vs. stable for a process? Process capability clearly bounds the shaft dimension distributions within the specs. But stability controls the distribution *means* location as well. By examining Figure 12.3, we can talk about 2*2= 4 options and we can read each alternative as follows;

1. No -Yes means the not-capable but stable process
2. No - No, means the not-capable but un-stable process
3. Yes - Yes means the capable and stable process
4. Yes - No means the capable and un-stable process

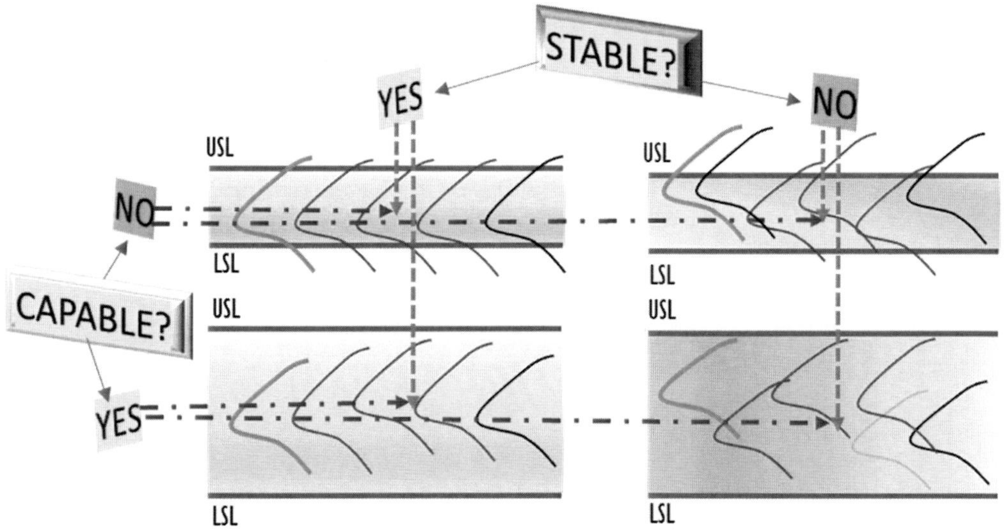

FIGURE 12.3 Relationship between stability vs. capability

Now, let's go ahead and look at the statistical indices and formulations used in a typical process capability analysis, using which we could identify our process characteristics as being capable/stable as introduced next.

12.1.1 C_p and P_p Indices

Process capability is the *ability* to produce products or providing services to meet the specifications set forth by the customer or common needs. For example, special care on the pharmaceutical ampoules for injecting drugs must be spent during manufacturing i.e. certain cut of depth in *mm/in.* for *'one point cut'* on the ampoule is done so that it is easy/healthy to break open from *'one point cut'* by hand force before the injection. In this case, capability analysis can reveal how well the one point cutting process during manufacturing of pharmaceutical ampules meets the needs/specifications. Capability analysis also provides a valuable insight into how to keep/improve the process and sustain the improvements, thus minimizing the defective units.

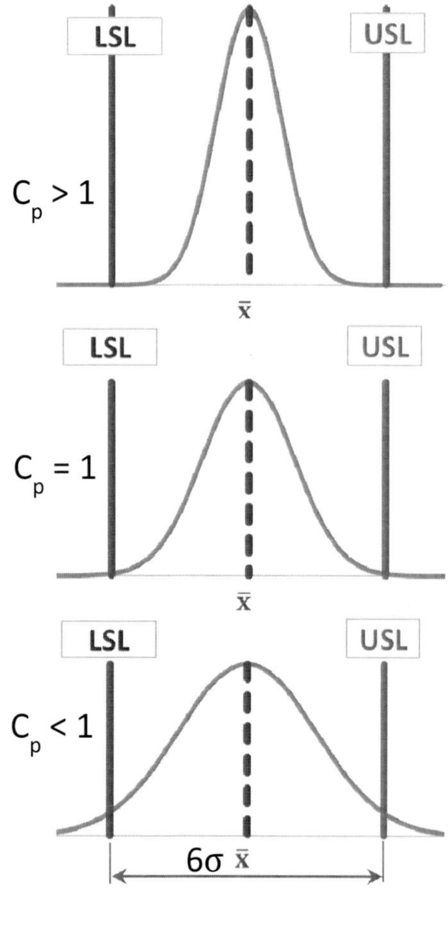

FIGURE 12.4 C_p's for spec. range = 6σ

Process capability is expressed with an *index*. C_p is called process capability and P_p is called process performance indices and they are used to verify/see if the process of our interest can meet the customer critical to quality requirements.

Assuming normal distribution with standard deviation σ and process *mean* $\bar{x}$ is **centered on the target value**, then the process capability index, C_p> = < 1. C_p index relates *the allowable spread of the specification limits* (i.e. *design tolerance*: difference in **U**pper **S**pecification **L**imit, USL – **L**ower **S**pecification **L**imit, LSL) to the variation of the process, i.e. *process tolerance*: stated by 6σ (±3σ and σ is mostly an estimate!) as shown in Figure 12.4.

C_p is used when a process is under statistical control/stable. This often happens with a process that has been going on for a while. P_p is used when a process is at its early state (not stable!) to determine if it is under statistical control. When there is not a lot of historical data or evidence we need a much larger number of samples from the process to account for variation.

If you examine Figure 12.4, you shall quickly notice that as C_p goes above 1, the distribution curve goes up and becomes more "*pointy*" and fits well in between the USL and LSL lines. This is always what we want so that the area under the curve between the USL and LSL is much closer to 1. This implies us that our process is in a state of control yet it yields less waste or rework i.e. less defective products.

On the contrary, when C_p is less than 1, it means that under the curve areas outside the LSL and USL lines are quite large. This translates into more scrap or reworks in the production. It also means quality issues in the process/production, Is not it? Yes, C_p can help reveal our process capability and helps us for improvement in the process!

As far as the formulations are concerned, C_p and P_p look similar, the only difference, in the denominator for both statistics as presented in Equation 12.1 where C_p is calculated using the **WITHIN standard deviation**, while P_p is calculated using the **OVERALL standard deviation**. The within standard deviation as the average of the subgroup data standard deviations, while the overall standard deviation represents the variation of all the data.

In these situations both C_p and P_p indexes can be calculated as;

$$C_p = \frac{USL - LSL}{6\sigma_{within}} \text{ and } P_p = \frac{USL - LSL}{6\sigma_{overall}} \qquad 12.1$$

As long as the process is centered and in-control and remember, σ_{within} is as being short term estimate for output data, we can assert that,

i. If $C_p < 1$ (poor) the process variation exceeds specification,

ii. If $C_p = 1$ (good) the process meets specifications.

iii. If $C_p > 1$ (very good) the process variation is less than the specification,

iv. Further, *Six Sigma* people suggest that $C_p \geq 2$, (i.e. z-value $= 3C_p$, design tolerance should be twice as wide as the process tolerance).

12.1.2 C_{pk} and P_{pk} Indices

What if the process mean does not *coincide* with the process *target* as it was not the case before?

In this situation, C_p does not tell us how well the process *mean*, $\bar{\mathbf{x}}$ is centered relative to the *target*. C_{pk} index exactly does this (C_{pk} takes either C_{pu} or C_{pl} values) by considering the location of the process *mean* at lower and upper specification sides as presented in Figure 12.5. Similarly, we can calculate the process performance index, P_{pk}, as well.

If you examine Figure 12.5, when the process mean starts moving away the target (assume the target is in the middle of USL and LSL, but not necessarily!) then C_{pk} goes below 0. If process mean coincides with either USL or LSL, then C_{pk} is 0. In both cases, we talk about too much scraps. Note, scrapped area corresponds to under the curve areas outside the USL and LSL lines.

If $0 < C_{pk} < 1$, then our process *mean* is in between LSL and USL. $C_{pk} = C_p = 1$ then process *mean* is centered.

But we always prefer $C_{pk} > 1$ as we are well in between the LSL and USL, this translates into very little scrap and a well-established stable process. (of course, all these assessments are made at the snapshot of the process, at a later time, things may change!)

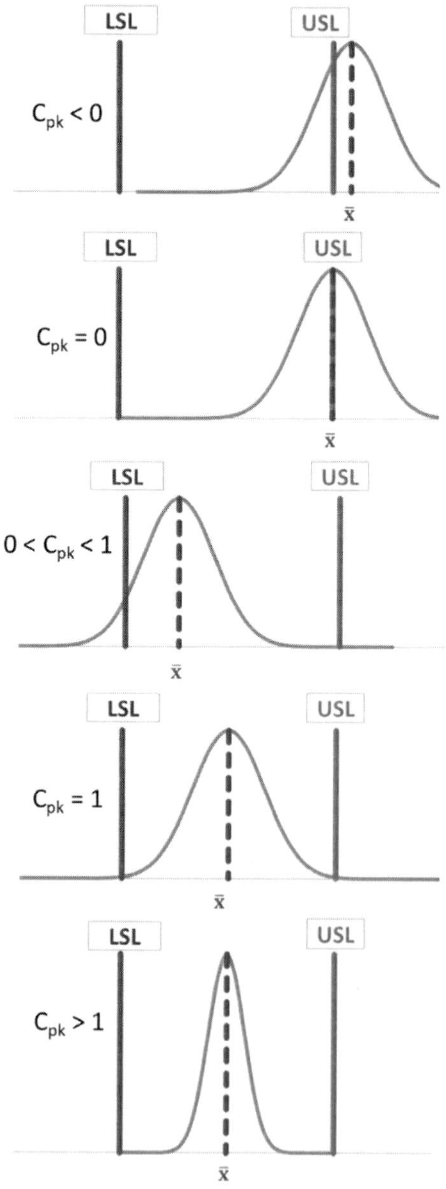

FIGURE 12.5 C_{pk}'s process *mean* is not centered

C_{pk} also accounts for the variation *within* the subgroups. On the contrary, P_{pk} accounts for the *overall* variation of all data. Process Performance Index basically tries to verify if the sample data are capable of meeting customer **C**ritical **t**o **Q**uality (i.e. CTQ) requirements. Relevant formulations are given in Equations 12.2 and 12.3, respectively.

$$C_{pk} = \min(C_{pu}, C_{pl})$$

$$C_{pu} = \frac{USL - \bar{x}}{3\sigma_{within}} \quad \text{Upper Process Capability Index} \qquad 12.2$$

$$C_{pl} = \frac{\bar{x} - LSL}{3\sigma_{within}} \quad \text{Upper Process Capability Index}$$

Where $\bar{\mathbf{x}}$ is the process *mean* and C_{pk} is equal to smaller of C_{pu} or C_{pl}. Keep in mind that C_{pk} will always be less than or equal to C_p.

$$P_{pk} = \min(P_{pu}, P_{pl})$$

$$P_{pu} = \frac{USL - \bar{x}}{3\sigma_{overall}} \quad \text{Upper Process Capability Index} \qquad 12.3$$

$$P_{pl} = \frac{\bar{x} - LSL}{3\sigma_{overall}} \quad \text{Upper Process Capability Index}$$

Where $\bar{\mathbf{x}}$ is the process *mean* and P_{pk} is equal to smaller of P_{pu} or P_{pl}. Keep in mind that P_{pk} will always be less than or equal to P_p.

Summary of Dimensionless Indices:

C_p = Process Capability is used when process *mean* is centered
C_{pk} = Process Capability Index, i.e. adjusted C_p when process *mean* is non-centered
P_p = Process Performance is used when process *mean* is centered
P_{pk} = Process Performance Index, i.e. adjusted P_p when process *mean* is non-centered

If a process is in *statistical control* then $C_{pk} = P_{pk}$ or at least very close! This implies that sample *inner* and *overall* standard deviations are identical or at least very close! In such situation, histograms and normal distribution curves will be overlapping! So far we have not really considered the ***target*** in our formulations. We only considered the cases and their corresponding C_{pk}, P_{pk} indices which indicate how much our process was in between the USL and LSL.

There is one more index called C_{pm}, Capability index which includes also ***targ***et value in the calculations as introduced in the next section!

12.1.3 C_{pm} Capability Index

There is further C_{pm}, "*capability*" index, also known as Taguchi capability index. The process capability index calculation is similar to the C_{pk} index but C_{pm} also takes into account variation between the process *mean* and a *target* value, i.e. $(\mu - T)^2$, based on Taguchi loss function. As for obtaining a *good* C_{pm}, a process needs to be right on the target, whereas remember as for *satisfactory* C_{pk} value, a process just needs to *stay* well-within the lower and upper specification lines.

Similarly, if you are on the *target*, C_{pm} index value is *the same* as both C_p and C_{pk} . But as the process *mean* moves away from the *target*, you start incurring increasing (greater) level of penalty in C_{pm} which is formulated as in Equation 12.4

$$C_{pm} = \frac{\text{USL} - \text{LSL}}{6\sqrt{\sigma^2 + (\mu - T)^2}} \qquad 12.4$$

In the calculation of Equation 12.4, σ value corresponds to $\sigma_{overall}$. Another alternate form of C_{pm} formulation is also given in Equation 12.5 where notice that the second formulation (an open form) is also given since it is easier to code in *Excel.*

$$C_{pm} = \frac{\text{USL} - \text{LSL}}{6\hat{\sigma}_{C_{pm}}} = \frac{\text{USL} - \text{LSL}}{6\sqrt{\frac{\sum_{i=1}^{mn}(x_i - T)^2}{(mn - 1)}}} = \frac{\text{USL} - \text{LSL}}{6\sqrt{\frac{\sum_{i=1}^{mn}(x_i)^2 - 2T\sum_{i=1}^{mn}x_i + mnT^2}{(mn - 1)}}} \qquad 12.5$$

Where USL: Upper Specification limit, LSL: Lower specification limit, T: target, and m: number of sample data, n:sample (batch) subgroup size, **x**: statistics of interest, μ or $\bar{x}$: population/process average/mean.

12.1.4 Summary of Indices in Table Format

When the process ***mean* is *centered*** with *target* value as depicted in Figure 12.6, we can summarize the calculation of all indices and related PPM values in Table 12.1, where if you use σ_{within} for Sigma, you get C_p, C_{pk}. Note P_p, P_{pk} and C_{pm} are only calculated using $\sigma_{overall}$ for Sigma as presented in Table 1.2. Remember, for this case, since our process *mean* is on the *target*, C_{pm} index value is *the same* as both C_p and C_{pk}.

Also, in Table 12.1 and 12.2, P(PL) and P(PU) values correspond to Z-Table readings of (already normalized) LSL and USL values. If you subtract the cumulative probabilities, i.e. P(PU) for USL - P(PL) for LSL, you can get the under the curve area in between the LSL and USL. As for ±3σ case, %99.73 area is covered and at least PPM=2700 scrapped units estimated.

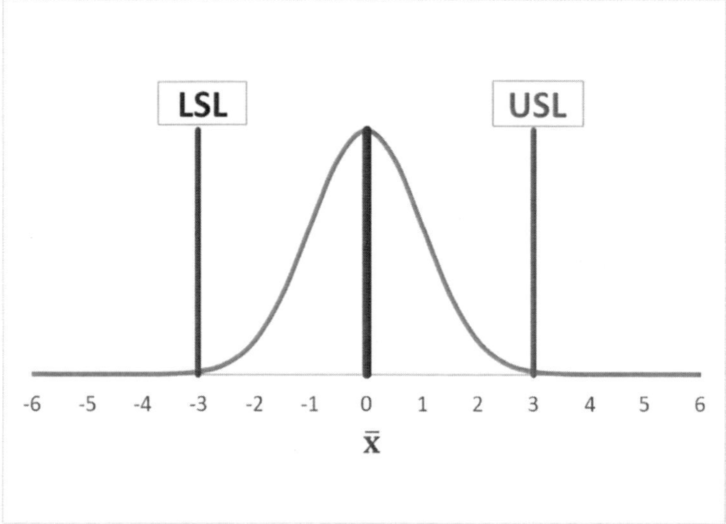

FIGURE 12.6 C_p's for spec. range=6σ (Centered)

Parts per million (PPM) value corresponds to least amount of defective/scrapped parts in one million parts produced. Again, as you can see ±3σ LSL and USL width, at least %99.73 of the curve is covered and defective part counts as PPM is at least 2700. More explanations and formulations for C_{pk}, P_{pk} and C_{pm} are presented in the next section.

TABLE 12.1 Process capability constants for centered standard normal distribution

Sigma =	1	Mean =	0	Target =	0	WITHIN			
LSL	USL	C_p	C_{PL}	C_{PU}	C_{PK}	P(PL)	P(PU)	P(PU)-P(PL)	PPM
-1	1	0.33	0.33	0.33	0.33	0.15866	0.84134	0.68269	317311
-2	2	0.67	0.67	0.67	0.67	0.02275	0.97725	0.95450	45500
-3	**3**	**1.00**	**1.00**	**1.00**	**1.00**	**0.00135**	**0.99865**	**0.99730**	**2700**
-4	4	1.33	1.33	1.33	1.33	0.00003	0.99997	0.99994	63
-5	5	1.67	1.67	1.67	1.67	0.00000	1.00000	1.00000	1
-6	6	2.00	2.00	2.00	2.00	0.00	1.00000	1.00000	0.002

TABLE 12.2 Process performance constants for centered standard normal distribution

sigma =	1	Mean =	0	Target =	0	OVERALL				
LSL	**USL**	**P_p**	**P_{PL}**	**P_{PU}**	**P_{PK}**	**C_{PM}**	**P(PL)**	**P(PU)**	**P(PU)-P(PL)**	**PPM**
-1	1	0.33	0.33	0.33	0.33	0.33	0.15866	0.84134	0.68269	317311
-2	2	0.67	0.67	0.67	0.67	0.67	0.02275	0.97725	0.95450	45500
-3	**3**	**1.00**	**1.00**	**1.00**	**1.00**	**1.00**	**0.00135**	**0.99865**	**0.99730**	**2700**
-4	4	1.33	1.33	1.33	1.33	1.33	0.00003	0.99997	0.99994	63
-5	5	1.67	1.67	1.67	1.67	1.67	0.00000	1.00000	1.00000	1
-6	6	2.00	2.00	2.00	2.00	2.00	0.00	1.00000	1.00000	0.002

Similarly, for ***non-centered*** distribution, we get the standard normal distribution as depicted in Figure 12.7 where $\bar{x} =$ 1.5σ, LSL= -3σ, USL=+3σ. As you notice the black colored area in Figure 12.7 represents the number of scraps in PPM. If you compare PPM=2700 from Table 12.1 with PPM=66811 from Table 12.3, you shall quickly realize the big change in a greater amount of lost/waste! Big lost takes place at USL line violation, there is almost no lost at LSL side. See the black colored area for this argument! There is a penalty incurred for not being able to keep the process centered i.e. the more you move away from the center the more you are ending up with scraps.

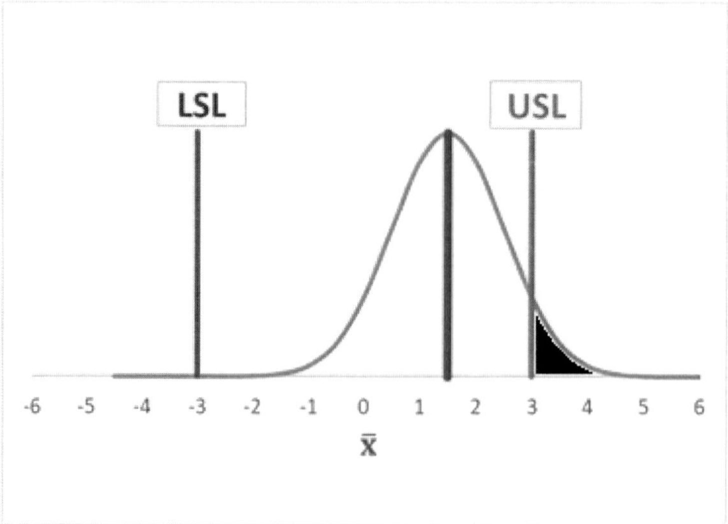

FIGURE 12.7 C_p's for spec. range = 6σ (not-centered)

TABLE 12.3 Process capability indices for not-*centered* standard normal distribution

Sigma =	1	Mean =	1.5	Target =	0	WITHIN			
LSL	USL	C_p	C_{PL}	C_{PU}	C_{PK}	P(PL)	P(PU)	P(PU)-P(PL)	PPM
-1	1	0.33	0.83	-0.17	-0.17	0.00621	0.30854	0.30233	697672
-2	2	0.67	1.17	0.17	0.17	0.00023	0.69146	0.69123	308770
-3	**3**	**1.00**	**1.50**	**0.50**	**0.50**	**0.00000**	**0.93319**	**0.93319**	**66811**
-4	4	1.33	1.83	0.83	0.83	0.00000	0.99379	0.99379	6210
-5	5	1.67	2.17	1.17	1.17	0.00000	0.99977	0.99977	233
-6	6	2.00	2.50	1.50	1.50	0.00	1.00000	1.00000	3.398

TABLE 12.4 Process performance indices for not-*centered* standard normal distribution

sigma =	1	Mean =	1.5	Target =	0	OVERALL				
LSL	USL	P_p	P_{PL}	P_{PU}	P_{PK}	C_{PM}	P(PL)	P(PU)	P(PU)-P(PL)	PPM
-1	1	0.33	0.83	-0.17	-0.17	0.18	0.00621	0.30854	0.30233	697672
-2	2	0.67	1.17	0.17	0.17	0.37	0.00023	0.69146	0.69123	308770
-3	**3**	**1.00**	**1.50**	**0.50**	**0.50**	**0.55**	**0.00000**	**0.93319**	**0.93319**	**66811**
-4	4	1.33	1.83	0.83	0.83	0.74	0.00000	0.99379	0.99379	6210
-5	5	1.67	2.17	1.17	1.17	0.92	0.00000	0.99977	0.99977	233
-6	6	2.00	2.50	1.50	1.50	1.11	0.00	1.00000	1.00000	3.398

Under the light of this introduction, many companies choose their specific process capability C_{pk}, targets above 1.33 and an expected goal of 2.0. If the process is stable and in statistical control, C_{pk} can be used to estimate the expected percentage of defective material(s) as shown in PPM. Process Capability Studies are performed to see what the process is "*capable*" of doing under controlled/ideal conditions for over a short period of time (i.e. one hour to twenty-four hours.).

Thus, process capability study lets us determine the *"short"* term *stability* and *capability* of processes. For this reason, when performing the process capability study, use your valuable experiences and as a rule of thumb,

- Try to eliminate/minimize special causes of variation. We should always
  - use the same equipment/machine
  - use the same operator and shifts
  - improve conditions of working atmosphere
  - motive the operators
  - use the same batch of material in the manufacturing, why? Different batches may mean variations in material properties!
  - stick to process and machining plan etc.

- Always try to collect a minimum of m=20-30 consecutive data readings with at least subgroups of n=3-7 samples for process capacity analysis

Calculation of C_{pk} using Z-Value

We know z-value (standard transformation) is used to normalize any generated/collected data (distributed normally) and using z-value we look up to **Z-Table** to find out corresponding *c.d.f.* probability readings. Similarly, if you have a z-value (also called z-score), then C_{pk} can be determined by dividing this z-score by *three* as presented in Equation 12.6. The advantage of using z-score is that one can visualize the data distribution in $\pm 1\sigma, 2\sigma, 3\sigma, 4\sigma, 5\sigma, 6\sigma$ scales, this format may make more sense for statisticians when analyzing/evaluating the data distribution characteristics.

How do we calculate z-value? Defects/scraps of normally distributed process curve generally fall on either side of the specification limits as small (leftover) areas as demonstrated in Figure 12.8. If you add up these both areas (remember these areas corresponds to the cumulative probabilities which are read from **Z-Table**) and virtually *assume* you put them on the right tail of the normal distribution curve (Notice to z label in Figure 12.8), z = z-value corresponding to these added areas(scraps/probabilities) can be read backward from **Z-Table** as *Z-bench* value.

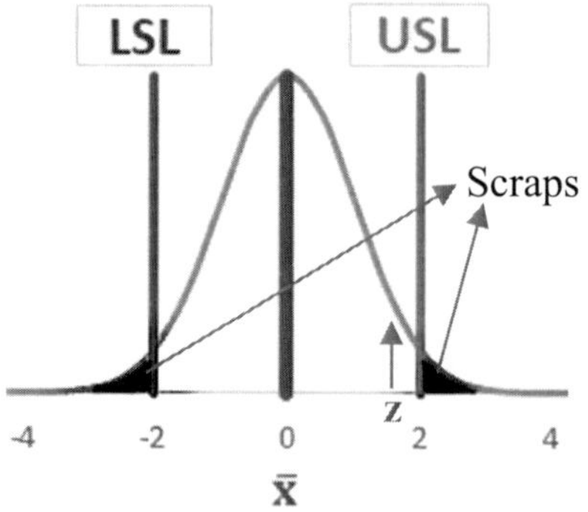

FIGURE 12.8 *Z-Bench* calculation

Z-bench value is used to describe the *sigma capability* of the process with a normal distribution. By using Z-bench statistics we can compare or have an idea about the process capability as well. Some people accept the Sigma capability as the short-term Z-bench value under potential (within) capability. Others define the Sigma capability as 1.5σ plus the long-term Z-bench value in overall capability. The difference between these two Z-bench values is called Z-shift and assumed to be 1.5σ. For example, if Z-bench=2σ, then Sigma capability=Z-bench + (Z-shift =1.5σ) = 3.5σ.

You may also examine *Z*-Bench calculation procedure as clearly presented in Equation 12.6 using **Z-Table** given in Appendix 2.

$$z = \frac{x - \mu}{\sigma_{within}} \rightarrow z - \text{value}$$

$$z_{USL} = \frac{USL - \bar{\bar{x}}}{\sigma_{within}} \rightarrow \text{above the USL} \rightarrow \textbf{Z - Table} \rightarrow P(Z \le z_{USL})$$

$$z_{LSL} = \frac{\bar{\bar{x}} - LSL}{\sigma_{within}} \rightarrow \text{below the LSL} \rightarrow \textbf{Z - Table} \rightarrow P(Z \le z_{LSL}) \qquad 12.6$$

$$|Z_{Bench}| \leftarrow \textbf{Z - Table} \leftarrow [1 - P(Z \le z_{USL})] + P(Z \le z_{LSL}) \text{ or}$$

$$Z_{Bench} \leftarrow \textbf{Z - Table} \leftarrow P(Z \le z_{USL}) - P(Z \le z_{LSL})$$

$$C_{pk} = \frac{\min(z_{USL}; z_{LSL})}{3}$$

Note above arrow directions show the forward (→) or the backward (←) **Z-Table** reading directions. In Equation 12.6, for potential(*within*) capability use $\mu = \bar{\bar{x}}$, $\sigma = \sigma_{within}$ and for *overall* capability use $\mu = \bar{\bar{x}}$, $\sigma = \sigma_{overall}$.

If you examine Table 12.5 and 12.6 given for not-centered cases, you would find all required calculations needed for process capability analysis. Note if you enter *Mean=0* in these tables, you will get centered case.

TABLE 12.5 Process Capability Constants for not-*Centered* Standard Normal Distribution

Sigma=	1	Mean=	1.5	Target =	0	WITHIN			
LSL	USL	Z_{Bench}*	Z_{USL}	Z_{LSL}	C_{pk}*	P(PL)	P(PU)	P(PL)-P(PU)	PPM
-1	1	-0.52	-0.50	2.50	-0.17	0.00621	0.30854	0.30233	697672
-2	2	0.50	0.50	3.50	0.17	0.00023	0.69146	0.69123	308770
-3	**3**	**1.50**	**1.50**	**4.50**	**0.50**	**0.000003**	**0.933193**	**0.933189**	**66811**
-4	4	2.50	2.50	5.50	0.83	0.00000	0.99379	0.99379	6210
-5	5	3.50	3.50	6.50	1.17	0.00000	0.99977	0.99977	233
-6	6	4.50	4.50	7.50	1.50	0.00000	1.00000	1.00000	3.398
					* Cpk=Zpk/3				

TABLE 12.6 Process capability constants for not-*centered* standard normal distribution

sigma=	1	Mean=	1.5	Target =	0	OVERALL				
LSL	USL	Z_{bench}	Z_{LSL}	Z_{USL}	C_{pk}^{*}	C_{PM}	P(PL)	P(PU)	P(PU)-P(PL)	PPM
-1	1	-0.52	2.50	-0.50	0.83	0.18	0.00621	0.30854	0.30233	697672
-2	2	0.50	3.50	0.50	1.17	0.37	0.00023	0.69146	0.69123	308770
-3	**3**	**1.50**	**4.500**	**1.500**	**1.50**	**0.55**	**0.000003**	**0.933193**	**0.93319**	**66811**
-4	4	2.50	5.50	2.50	1.83	0.74	0.00000	0.99379	0.99379	6210
-5	5	3.50	6.50	3.50	2.17	0.92	0.00000	0.99977	0.99977	233
-6	6	4.50	7.50	4.50	2.50	1.11	0.00000	1.00000	1.00000	3.398
					* Cpk=Zpk/3					

So far we have introduced quite much information and formulations on Process capability analysis. Now, we can go ahead and demonstrate all of this knowledge on Example 12.1.1. Process capability problem presented next!

TABLE 12.7 Gear hardness data

Days	x_1	x_2	x_3	x_4	x_5	$\bar{x}$	r
1	51.3	49.2	50.0	51.2	45.8	49.50	5.5
2	42.9	49.3	49.6	45.5	44.6	46.38	6.7
3	51.9	50.2	51.6	52.3	46.4	50.48	5.9
4	48.3	49.3	46.3	55.9	47.4	49.44	9.6
5	51.0	52.6	55.8	56.3	57.9	54.72	6.9
6	48.5	58.6	54.8	51.8	51.7	53.08	10.1
7	59.9	59.2	48.0	54.8	46.3	53.64	13.6
8	53.0	54.1	48.4	53.5	51.5	52.10	5.7
9	56.2	58.5	56.2	50.8	53.1	54.96	7.7
10	50.8	44.7	49.7	51.2	56.3	50.54	11.6
11	48.3	47.5	48.4	39.4	43.3	45.38	9.0
12	44.7	39.1	40.6	48.2	41.9	42.90	9.1
13	50.6	42.0	40.5	41.7	45.8	44.12	10.1
14	43.3	45.4	41.8	43.5	41.4	43.08	4.0
15	36.3	44.9	36.5	36.9	38.5	38.62	8.6
16	35.1	36.9	37.4	39.4	43.7	38.50	8.6
17	39.0	35.2	38.6	38.5	36.0	37.46	3.8
18	42.1	43.0	43.4	40.4	41.1	42.00	3.0
19	49.8	51.3	46.2	47.7	44.5	47.90	6.8
20	49.7	40.0	47.3	44.7	44.3	45.20	9.7
					$\bar{\bar{x}}$ =	47.0	
						$\bar{r}$ =	7.8
						σ_{within} =	3.3534
						$\sigma_{overall}$ =	6.16412

Example 12.1: Gear Hardness

Mechanical gears are widely used in many machine components in automobiles, airplanes, motorcycles, mixers, watches and many toys etc. Gear material and its carbon ratio are both important for proper heat treatment to obtain required hardness in accordance with the gear's intended operation.

Carbon steel gear materials for structural machines is one of the commonly used steel, percentage of carbon in this material is one of the major determinants in obtaining the desired hardness level. The other determinant is the heat treatment method which is assumed to be *induction hardening* and its relevant parameters need to be set properly.

After induction hardening, we may get around 40-60 HRC (Rockwell C Hardness) in helical gears for the transmission of an automobile. In *early* gear hardening trials, gear hardness data are recorded for twenty consecutive days, i.e. m=20. Each day five samples are taken with a sample size of n=5 (subgroups) and all measurements are recorded and presented in Table 12.7.

In Table 12.7, certain sum/average calculations of the statistics are also given in the *Excel* table for easier hand calculations. Now, let's start making the required calculations to draw $\bar{X}$-R control charts.

For ***Day 1***,

$$\bar{x}_1 = \frac{x_1 + x_2 + x_3 + x_4 + x_5}{n}$$

$$\bar{x}_1 = \frac{51.3 + 49.2 + 50 + 51.2 + 45.8}{5} = 49.5$$

$$r_1 = max(x_1, x_2, x_3, x_4, x_5) - min(x_1, x_2, x_3, x_4, x_5) =$$

$$max(51.3, 49.2, 50, 51.2, 45.8) - min(51.3, 49.2, 50, 51.2, 45.8) =$$

$$r_1 = 51.3 - 45.8 = 5.5$$

In Table 12.7, we calculated $\bar{\bar{x}}$=47.0 HRC using *Excel* (i.e. $\bar{\bar{x}}$ - double bar is equivalent to the average of $\bar{x}_j, j = 1, \ldots, m$. $\bar{r}$ is the average of ranges, $r_j, j = 1, \ldots, m$). Table 12.8 given below is obtained from Appendix 5, which summarizes the control chart factors used in this example.

TABLE 12.8 Control chart factors for *n=5*

Chart Factors	A_2	d_2	D_3	D_4	C_4
n=5	0.577	2.326	0	2.115	0.94

$\bar{X}$-R Control Chart

We can start with drawing $\bar{X}$-control chart using $\bar{r}$ statistics! The center line, upper and lower control limits can be calculated using the control chart factors given in Table 12.8 and using the related equations;

$$UCL = \bar{\bar{x}} + A_2\bar{r} = 47.0 + 0.577 * 7.8 = 51.501$$

$$CL = 47.0$$

$$LCL = \bar{\bar{x}} - A_2\bar{r} = 47.0 - 0.577 * 7.8 = 42.5$$

$$\sigma_{within} = \frac{\bar{r}}{d_2} = \frac{7.8}{2.326} = 3.3533$$

$$\sigma_{overall} = \sqrt{\frac{1}{mn-1}\sum_{i=1}^{mn}(x_i - \bar{\bar{x}})^2} = \sqrt{\frac{1}{99}\sum_{i=1}^{100}(51.3 - 47)^2 + \cdots + (44.3 - 47)^2}$$

$$= 6.16412$$

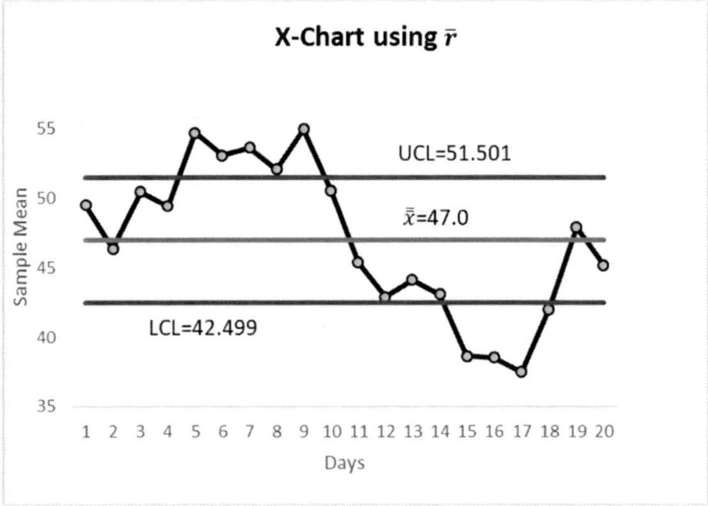

FIGURE 12.9 $\overline{\mathrm{X}}$-Control chart using $\bar{r}$

By examining $\overline{\mathrm{X}}$-Control chart given in Figure 12.9, we notice that there are many violations at both control levels for various days for which corrective measures immediately needs to be taken to put the process right back on the track!

Similarly, R-Control chart depending on the average sample range, the center line, upper and lower control limits are calculated below and plotted as shown in Figure 12.10. One thing to note that sample range distances are quite large, that's why we get the "*within*" and "*overall*" standard values quite high. They are very different from each other, which is also the indication of being the incapable/unstable process.

$$\mathrm{UCL} = D_4\bar{r} = 2.115 * 7.8 = 16.497$$
$$\mathrm{CL} = \bar{r} = 7.8$$
$$\mathrm{LCL} = D_3\bar{r} = 0$$

Where D_4 and D_3 are the control chart factors given for n sample sizes (subgroups) as presented previously in Table 12.8.

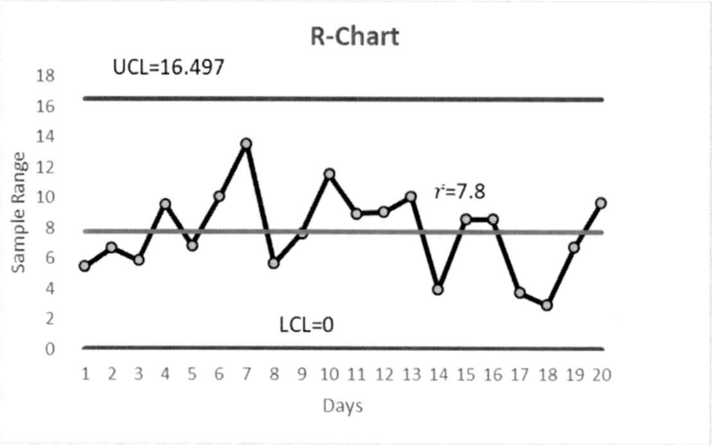

FIGURE 12.10 R-Control chart

Next calculate potential (within) capability indices,

$$C_p = \frac{USL - LSL}{6\sigma_{Within}} = \frac{60 - 40}{6 * 3.3534} = 0.99$$

$$C_{PL} = \frac{\bar{x} - LSL}{3\sigma_{within}} = C_{pL} = \frac{47 - 40}{3 * 3.3534} = 0.70$$

$$C_{PU} = \frac{USL - \bar{x}}{3\sigma_{within}} = C_{pU} = \frac{60 - 47}{3 * 3.3534} = 1.29$$

$$C_{pk} = \min(C_{PU}, C_{PL}) = 0.70$$

Then, overall capability indices,

$$\text{and } P_p = \frac{USL - LSL}{6\sigma_{Overall}} = \frac{60 - 40}{6 * 6.16432} = 0.54$$

$$C_{PL} = \frac{\bar{x} - LSL}{3\sigma_{within}} = C_{pL} = \frac{47 - 40}{3 * 6.16432} = 0.38$$

$$C_{PU} = \frac{USL - \bar{x}}{3\sigma_{within}} = C_{pU} = \frac{60 - 47}{3 * 6.16432} = 0.70$$

$$C_{pk} = \min(C_{PU}, C_{PL}) = 0.38$$

Using *"overall"* Sigma for C_{pm},

$$C_{pm} = \frac{USL - LSL}{6\sqrt{\sigma^2 + (\mu - T)^2}} = \frac{60 - 40}{6\sqrt{6.16432^2 + (47 - 50)^2}} = 0.4862$$

or using the other equation,

$$C_{pm} = \frac{USL - LSL}{6\sqrt{\frac{\sum_{i=1}^{mn}(x_i - T)^2}{(mn - 1)}}} = \frac{60 - 40}{6\sqrt{\frac{\sum_{i=1}^{100}(x_i - 50)^2}{100 - 1}}} = \frac{60 - 40}{6 * 6.862} = 0.4858$$

Using Z-Value and Calculating Scraps

C_{pk} can be calculated by dividing the *z*-score by *three* by using the equations given below; Above the USL, we read the probability as,

$$\mathbf{z_{USL}} = \frac{USL - \bar{x}}{\sigma_{within}} = \frac{60 - 47}{3.353396} = 3.876 \rightarrow \mathbf{Z - Table}\ P(Z \leq 3.876) = 0.999948$$

$$PPM > USL = 1 - 0.999948 * 1E6 = \mathbf{52\ scrap}$$

Below the LSL, we read the probability as, (note LSL and $\bar{x}$ are swapped in the formulation below to have negative *z*-value. It makes more sense in terms of physical location of Z_{LSL} on the probability graph).

$$\mathbf{z_{LSL}} = \frac{LSL - \bar{x}}{\sigma_{within}} = \frac{40 - 47}{3.353396} = -2.087 \rightarrow \mathbf{Z - Table}\ P(Z \leq -2.087) = 0.0184244$$

$$PPM < LSL = 0.0184244 * 1E6 = \mathbf{18425\ scrap}$$

$$Z_{Bench} \leftarrow \mathbf{Z - Table} \leftarrow 1 - PPM_{total}/1000000$$

$$Z_{Bench} = 2.09 \leftarrow \mathbf{Z - Table} \leftarrow 1 - 18477/1000000$$

After summing both scrap parts, we get scraps totaling 52 + 18425 = 18477 scraps, or P_{total}= 0.999948 - 0.0184244 = 0.981524 (total area under the curve, i.e. c.d.f. probability). PPM_{total}, Total scraps = (1 - 0.981524)*1,000,000 parts = 18477 ppm.

Similarly, P_{pk} can also be calculated by dividing the *z*-score by *three* as per the equation given below, keep in mind that now we are using $\sigma_{overall}$.

Above the USL, we read the probability as,

$$\mathbf{z_{USL}} = \frac{\text{USL} - \bar{x}}{\sigma_{overall}} = \frac{60 - 47}{6.16412} = 2.11 \rightarrow \mathbf{Z - Table}\ \text{P}(\text{Z} \leq 2.1089) = 0.982527$$

$$\text{PPM} > \text{USL} = (1 - 0.982527) * 1\text{E}6 = \mathbf{17473\ scrap}$$

Below the LSL, we read the probability as, (note LSL and x̄ are swapped in the formulation below to have negative *z*-value. It makes more sense in terms of physical location of Z_{LSL} on the probability graph).

$$\mathbf{z_{LSL}} = \frac{\text{LSL} - \bar{x}}{\sigma_{\text{overall}}} = \frac{40 - 47}{6.16432} = -1.14 \rightarrow \mathbf{Z - Table}\ \text{P}(\text{Z} \leq -1.1355) = 0.128061$$

$$\text{PPM} < \text{LSL} = 0.128061 * 1\text{E}6 = \mathbf{128061\ scrap}$$

$$Z_{\text{Bench}} \leftarrow \mathbf{Z - Table} \leftarrow 1 - \text{PPM}_{\text{total}}/1000000$$

$$Z_{\text{Bench}} = 1.06 \leftarrow \mathbf{Z - Table} \leftarrow 1 - 145534/1000000$$

After summing both scrap parts, we get scraps totaling 17473 + 128061 = 145534 scraps, or P_{total} = 0.982527 - 0.128061 = 0.854466 (the total area under the curve, i.e. c.d.f. probability), PPM_{total}, Total scraps = (1 - 0.854466)*1,000,000 parts = 145534 ppm.

PPM value indicates that we are getting too much waste for a production line. C_{pk}, P_{pk} and C_{pm} values are pretty low! Something has to be done to make gear hardening process better! i.e. we need to come up with much less variation in meeting target hardness values.

It is going to cost us but we need to do some intervention to the gear hardening process. We need better equipment. Let's go ahead and make some process improvements as briefly outlined below:

- New Induction hardness machine with easy to control process parameters is purchased
- Material acquisition with more stable/precise carbon amount is decided. For this reason, vendor selection is made with care
- Workers are trained and made more motivated

Now, it is time to see if our process improvements are going to be effective and make our gear hardening process any better!

Now, we go ahead and collect a new set of process hardness data and carry out the same steps that we followed. We would like to see whether there is an improvement in the process outcome after making above process improvements.

New gear hardness result data in the same format that we used are given in Table 12.9 where again certain sum/average calculations of the gear hardness statistics are also given for easier hand calculations. If we compare these data to the previous data, we quickly notice the considerable amount of drop (improvements) in both standard deviations which are now also very close to each other. The average hardness level is now around 50.931 HRC, which is a little over the target.

TABLE 12.9 Improved gear hardness data

Days	x_1	x_2	x_3	x_4	x_5	$\bar{x}$	r
1	47.7	54.3	50.5	48.6	55.0	51.22	7.3
2	49.8	51.4	55.1	48.4	53.8	51.70	6.7
3	52.3	49.2	52.1	47.6	54.5	51.14	6.9
4	47.9	54.4	56.8	48.9	51.5	51.90	8.9
5	52.5	51.3	58.4	51.2	50.6	52.80	7.8
6	51.1	50.7	49.3	49.1	49.6	49.96	2.0
7	55.9	49.3	48.5	49.9	55.4	51.80	7.4
8	55.4	48.7	48.9	48.7	51.0	50.54	6.7
9	49.4	50.7	51.8	46.2	53.0	50.22	6.8
10	48.3	49.2	50.2	53.6	53.0	50.86	5.3
11	50.9	48.8	48.9	52.9	56.6	51.62	7.8
12	52.5	50.2	49.0	53.9	53.2	51.76	4.9
13	48.1	49.8	49.9	49.9	50.0	49.54	1.9
14	51.1	50.9	52.2	49.6	47.7	50.30	4.5
15	48.5	53.4	53.9	49.6	49.0	50.88	5.4
16	49.3	51.2	52.8	50.9	55.0	51.84	5.7
17	55.1	48.7	52.0	47.0	52.6	51.08	8.1
18	49.5	50.7	48.8	48.8	49.5	49.46	1.9
19	49.3	49.2	51.7	47.4	52.7	50.06	5.3
20	50.2	49.5	50.8	50.7	48.5	49.94	2.3
					$\bar{\bar{x}}$ =	**50.931**	
						$\bar{r}$ =	**5.68**
						σ_{within} =	**2.44196**
						$\sigma_{overall}$ =	**2.46206**

Now, let's start making the required calculations to draw $\overline{X}$-R control charts. We are going to skip some detailed instruction calculations that we did earlier.

$\overline{X}$-R Control Chart

We can again start with drawing $\overline{X}$-control chart using $\bar{r}$ statistics. The center line, upper and lower control limits can be calculated using the control chart factors as given in Table 12.8 and using the following equations,

$$\text{UCL} = \bar{\bar{x}} + A_2\bar{r} = 50.931 + 0.577 * 5.68 = 54.208$$
$$\text{CL} = 50.931$$
$$\text{LCL} = \bar{\bar{x}} - A_2\bar{r} = 50.931 - 0.577 * 5.68 = 47.654$$
$$\sigma_{within} = \frac{\bar{r}}{d_2} = \frac{5.68}{2.326} = 2.44196$$

$$\sigma_{overall} = \sqrt{\frac{1}{mn-1}\sum_{i=1}^{mn}(x_i - \bar{\bar{x}})^2} = \sqrt{\frac{1}{99}\sum_{i=1}^{100}(47.7-47)^2 + \cdots + (48.5-47)^2}$$
$$= 2.46206$$

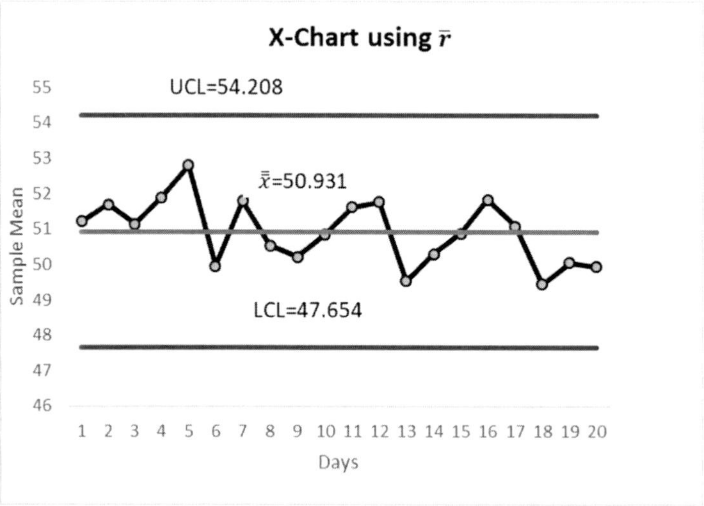

FIGURE 12.11 $\bar{X}$-Control chart using $\bar{r}$

By examining $\bar{X}$-Control Chart given in Figure 12.11, we notice that there are no violations at both control levels for all days for which corrective measures taken seems have worked well. Finally, the gear hardening process is back on the track and stable!

Similarly, R-Control chart depending on the average sample range, the center line, upper and lower control limits are calculated below and plotted as shown in Figure 12.12.

$$\text{UCL} = D_4\bar{r} = 2.115 * 7.8 = 16.497$$
$$\text{CL} = \bar{r} = 7.8$$
$$\text{LCL} = D_3\bar{r} = 0$$

Where D_4 and D_3 are the control chart factors given for n sample sizes (subgroups) as presented in Table 12.8 or in Appendix 5.

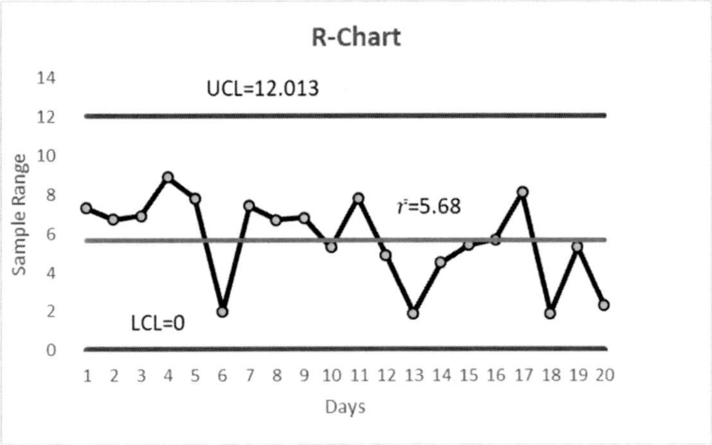

FIGURE 12.12 R-Control chart

Potential (within) capability,

$$C_p = \frac{USL - LSL}{6\sigma_{within}} = \frac{60 - 40}{6 * 2.44196} = 1.37$$

$$C_{pl} = \frac{\bar{x} - LSL}{3\sigma_{within}} = C_{pl} = \frac{50.931 - 40}{3 * 2.44196} = 1.49$$

$$C_{pu} = \frac{USL - \bar{x}}{3\sigma_{within}} = C_{pu} = \frac{60 - 50.931}{3 * 2.44196} = 1.24$$

$$C_{pk} = \min(C_{pu}, C_{pl}) = 1.24$$

Overall capability,

$$\text{and } P_p = \frac{USL - LSL}{6\sigma_{Overall}} = \frac{60 - 40}{6 * 2.46206} = 1.35$$

$$C_{pl} = \frac{\bar{x} - LSL}{3\sigma_{within}} = C_{pl} = \frac{50.931 - 40}{3 * 2.46206} = 1.48$$

$$C_{pu} = \frac{USL - \bar{x}}{3\sigma_{within}} = C_{pu} = \frac{60 - 50.931}{3 * 2.46206} = 1.23$$

$$C_{pk} = \min(C_{pu}, C_{pl}) = 1.23$$

Using *"overall"* Sigma for C_{pm},

$$C_{pm} = \frac{USL - LSL}{6\sqrt{\sigma^2 + (\mu - T)^2}} = \frac{60 - 40}{6\sqrt{2.46206^2 + (50.931 - 50)^2}} = 1.2664$$

or using the other equation,

$$C_{pm} = \frac{USL - LSL}{6\sqrt{\frac{\sum_{i=1}^{mn}(x_i - T)^2}{(mn - 1)}}} = \frac{60 - 40}{6\sqrt{\frac{\sum_{i=1}^{100}(x_i - 50)^2}{100 - 1}}} = \frac{60 - 40}{6 * 2.63387} = 1.2656$$

Using Z-Value and Calculating Scraps

Remember, C_{pk} can be calculated by dividing the z-score by *three* as per the equation given below;

Above the USL, we read the probability as,

$$\mathbf{z_{USL}} = \frac{USL - \bar{x}}{\sigma_{within}} = \frac{60 - 50.931}{2.44196} = 3.7138 \rightarrow \mathbf{Z - Table}\ P(Z \leq 3.7138) = 0.999898$$

$$PPM > USL = 1 - 0.999898 * 1E6 = \mathbf{102\ scraps}$$

Below the LSL, we read the probability as, (note LSL and $\bar{x}$ are swapped in the formulation below to have negative *z*-value. It makes more sense in terms of physical location of Z_{LSL} on the probability graph).

$$\mathbf{z_{LSL}} = \frac{LSL - \bar{x}}{\sigma_{within}} = \frac{40 - 50.931}{2.44196} = -4.4763 \rightarrow \mathbf{Z - Table}\ P(Z \leq -4.4763) = 0.000004$$

$$PPM < LSL = 0.000004 * 1E6 = \mathbf{4\ scraps}$$

$$Z_{Bench} \leftarrow \mathbf{Z - Table} \leftarrow 1 - PPM_{total}/1000000$$

$$Z_{Bench} = 3.70 \leftarrow \mathbf{Z - Table} \leftarrow 1 - 106/1000000$$

After summing both scrap parts, we get scraps totaling 4 + 102 = 106 scraps, or P_{total}= 0.999898 - 0.000004 = 0.999894 (the total area under the normal curve, i.e. c.d.f. probability). PPM_{total}, Total scraps = (1 - 0.999894)*1,000,000 parts = 106 ppm.

Similarly, P_{pk} can also be calculated by dividing the *z*-score by *three* as per the equation given below, now we are going to use $\sigma_{overall}$.

Above the USL, we read the probability as,

$$z_{USL} = \frac{USL - \bar{x}}{\sigma_{overall}} = \frac{60 - 50.931}{2.46206} = 3.6835 \rightarrow \mathbf{Z - Table}\ P(Z \leq 3.6835) = 0.999885$$

$$PPM > USL = (1 - 0.999885) * 1E6 = \mathbf{115\ scraps}$$

Below the LSL, we read the probability as, (note LSL and $\bar{x}$ are swapped in the formulation below to have negative *z*-value. It makes more sense in terms of physical location of Z_{LSL} on the probability graph).

$$z_{LSL} = \frac{LSL - \bar{x}}{\sigma_{overall}} = \frac{40 - 50.931}{2.46206} = -4.4398 \rightarrow \mathbf{Z - Table}\ P(Z \leq -4.4398) = 0.000005$$

$$PPM < LSL = 0.000005 * 1E6 = \mathbf{5\ scraps}$$

$$Z_{Bench} \leftarrow \mathbf{Z - Table} \leftarrow 1 - PPM_{total}/1000000$$

$$Z_{Bench} = 3.67 \leftarrow \mathbf{Z - Table} \leftarrow 1 - 120/1000000$$

After summing both scrap parts, we get scraps totaling 115 + 5 = 120 scraps, or P_{total} = 0.999885 - 0.000005 = 0.99988 (total area under the curve, i.e. c.d.f. probability). PPM_{total}, Total scraps = (1 - 0.854466)*1,000,000 parts = 120 ppm. Too little waste for a production line. C_{pk}, P_{pk} and C_{pm} values are now much better!

Standard deviation for "*within*" and "*overall*" are very close to each other, they exhibit almost overlapping normal distributions. C_{pk}, P_{pk} and, C_{pm} are almost identical which is a sign of "*almost*" centered distribution. Our interventions for improving the gear hardness process capability has worked out.

12.2 Supplementary Material for *Excel* Commands

Example 12.1: Gear Hardness Data

Note, in the following *Excel* sheets, **Column-L, and Column-M** list all coded *Excel commands* with required formulations for all variables used. You can easily create the same sheets on your own.

The following two *Excel* Tables are for Table 12.1, 12.2, 12.3 and 12.4.

	A	B	C	D	E	F	G	H	I	J	K	L
1	Sigma =	1	Mean =	0	Target =	0	WITHIN				C_p	=(B3-A3)/(6*B$1)
2	LSL	USL	C_p	C_{PL}	C_{PU}	C_{PK}	P(PL)	P(PU)	P(PU)-P(PL)	PPM	C_{PL}	=(D$1-A3)/(3*B$1)
3	-1	1	0.33	0.33	0.33	0.33	0.15866	0.84134	0.68269	317311	C_{PU}	=(B3-D$1)/(3*B$1)
4	-2	2	0.67	0.67	0.67	0.67	0.02275	0.97725	0.95450	45500	C_{PK}	=MIN(D3:E3)
5	**-3**	**3**	**1.00**	**1.00**	**1.00**	**1.00**	**0.00135**	**0.99865**	**0.99730**	**2700**	P(PL)	=NORM.DIST(A3,D$1,B$1,TRUE)
6	-4	4	1.33	1.33	1.33	1.33	0.00003	0.99997	0.99994	63	P(PU)	=NORM.DIST(B3,D$1,B$1,TRUE)
7	-5	5	1.67	1.67	1.67	1.67	0.00000	1.00000	1.00000	1	P(PU)-P(PL)	=H3-G3
8	-6	6	2.00	2.00	2.00	2.00	0.00	1.00000	1.00000	0.002	PPM	=(1-I3)*1000000

	A	B	C	D	E	F	G	H	I	J	K	L	M
1	sigma =	1	Mean =	0	Target =	0	OVERALL					P_p	=(B3-A3)/(6*B$1)
2	LSL	USL	P_p	P_{PL}	P_{PU}	P_{PK}	C_{PM}	P(PL)	P(PU)	P(PU)-P(PL)	PPM	P_{PL}	=(D$1-A3)/(3*B$1)
3	-1	1	0.33	0.33	0.33	0.33	0.33	0.15866	0.84134	0.68269	317311	P_{PU}	=(B3-D$1)/(3*B$1)
4	-2	2	0.67	0.67	0.67	0.67	0.67	0.02275	0.97725	0.95450	45500	P_{PK}	=MIN(D3:E3)
5	**-3**	**3**	**1.00**	**1.00**	**1.00**	**1.00**	**1.00**	**0.00135**	**0.99865**	**0.99730**	**2700**	C_{PM}	=(B3-A3)/(6*SQRT(B$1^2+(D$1-F$1)^2))
6	-4	4	1.33	1.33	1.33	1.33	1.33	0.00003	0.99997	0.99994	63	P(PL)	=NORM.DIST(A3,D$1,B$1,TRUE)
7	-5	5	1.67	1.67	1.67	1.67	1.67	0.00000	1.00000	1.00000	1	P(PU)	=NORM.DIST(B3,D$1,B$1,TRUE)
8	-6	6	2.00	2.00	2.00	2.00	2.00	0.00	1.00000	1.00000	0.002	P(PU)-P(PL)	=I3-H3
9												PPM	=(1-J3)*1000000

The following two *Excel* Tables are for Table 12.5 and 12.6.

	A	B	C	D	E	F	G	H	I	J	K	L	M
1	sigma=	1	Mean=	0	Target =	0	OVERALL					Z_{bench}	=NORM.S.INV(J3)
2	LSL	USL	Z_{bench}	Z_{LSL}	Z_{USL}	C_{pk}^{*}	C_{PM}	P(PL)	P(PU)	P(PU)-P(PL)	PPM	Z_{LSL}	=(D$1-A3)/(B$1)
3	-1	1	0.48	1.00	1.00	0.33	0.33	0.15866	0.84134	0.68269	317311	Z_{USL}	=(B3-D$1)/(B$1)
4	-2	2	1.69	2.00	2.00	0.67	0.67	0.02275	0.97725	0.95450	45500	C_{PK}	=MIN(D3:D3)/3
5	**-3**	**3**	**2.78**	**3.000**	**3.000**	**1.00**	**1.00**	**0.001350**	**0.998650**	**0.99730**	**2700**	C_{PM}	=(B3-A3)/(6*SQRT(B$1^2+(D$1-F$1)^2))
6	-4	4	3.83	4.00	4.00	1.33	1.33	0.00003	0.99997	0.99994	63	P(PL)	=NORM.DIST(A3,D$1,B$1,TRUE)
7	-5	5	4.86	5.00	5.00	1.67	1.67	0.00000	1.00000	1.00000	1	P(PU)	=NORM.DIST(B3,D$1,B$1,TRUE)
8	-6	6	5.89	6.00	6.00	2.00	2.00	0.00000	1.00000	1.00000	0.002	P(PL)-P(PU)	=I3-H3
9						* Cpk=Zpk/3						PPM	=(1-J3)*1000000

	A	B	C	D	E	F	G	H	I	J	K	L	M
1	sigma=	1	Mean=	1.5	Target =	0	OVERALL					Z_{bench}	=NORM.S.INV(J3)
2	LSL	USL	Z_{bench}	Z_{LSL}	Z_{USL}	C_{pk}^{*}	C_{PM}	P(PL)	P(PU)	P(PU)-P(PL)	PPM	Z_{LSL}	=(D$1-A3)/(B$1)
3	-1	1	-0.52	2.50	-0.50	0.83	0.18	0.00621	0.30854	0.30233	697672	Z_{USL}	=(B3-D$1)/(B$1)
4	-2	2	0.50	3.50	0.50	1.17	0.37	0.00023	0.69146	0.69123	308770	C_{PK}	=MIN(D3:D3)/3
5	**-3**	**3**	**1.50**	**4.500**	**1.500**	**1.50**	**0.55**	**0.000003**	**0.933193**	**0.93319**	**66811**	C_{PM}	=(B3-A3)/(6*SQRT(B$1^2+(D$1-F$1)^2))
6	-4	4	2.50	5.50	2.50	1.83	0.74	0.00000	0.99379	0.99379	6210	P(PL)	=NORM.DIST(A3,D$1,B$1,TRUE)
7	-5	5	3.50	6.50	3.50	2.17	0.92	0.00000	0.99977	0.99977	233	P(PU)	=NORM.DIST(B3,D$1,B$1,TRUE)
8	-6	6	4.50	7.50	4.50	2.50	1.11	0.00000	1.00000	1.00000	3.398	P(PL)-P(PU)	=I3-H3
9						* Cpk=Zpk/3						PPM	=(1-J3)*1000000

The following *Excel* table and related calculations correspond to gear hardness data and initial calculations for process capability analysis.

	A	B	C	D	E	F	G	H	I	J	K	L	M	N	O	P
1	Days	x_1	x_2	x_3	x_4	x_5	$\bar{x}$	r	$\bar{\bar{x}}$	UCL	LCL	$\bar{r}$	UCL	LCL		Excel Formula
2	1	51.3	49.2	50.0	51.2	45.8	49.50	5.50	47.0	51.501	42.499	7.8	16.497	0		
3	2	42.9	49.3	49.6	45.5	44.6	46.38	6.70	47.0	51.501	42.499	7.8	16.497	0	$\bar{x}$	= AVERAGE(B2:F2)
4	3	51.9	50.2	51.6	52.3	46.4	50.48	5.90	47.0	51.501	42.499	7.8	16.497	0	r	=MAX(B2:F2)-MIN(B2:F2)
5	4	48.3	49.3	46.3	55.9	47.4	49.44	9.60	47.0	51.501	42.499	7.8	16.497	0	$\bar{\bar{x}}$	= G$22
6	5	51.0	52.6	55.8	56.3	57.9	54.72	6.90	47.0	51.501	42.499	7.8	16.497	0	UCL	= G$22+0.577*H$23
7	6	48.5	58.6	54.8	51.8	51.7	53.08	10.10	47.0	51.501	42.499	7.8	16.497	0	LCL	= G$22-0.577*H$23
8	7	59.9	59.2	48.0	54.8	46.3	53.64	13.60	47.0	51.501	42.499	7.8	16.497	0	$\bar{r}$	= H$23
9	8	53.0	54.1	48.4	53.5	51.5	52.10	5.70	47.0	51.501	42.499	7.8	16.497	0	UCL	= 2.115*H$23
10	9	56.2	58.5	56.2	50.8	53.1	54.96	7.70	47.0	51.501	42.499	7.8	16.497	0	LCL	= 0
11	10	50.8	44.7	49.7	51.2	56.3	50.54	11.60	47.0	51.501	42.499	7.8	16.497	0		
12	11	48.3	47.5	48.4	39.4	43.3	45.38	9.00	47.0	51.501	42.499	7.8	16.497	0		
13	12	44.7	39.1	40.6	48.2	41.9	42.90	9.10	47.0	51.501	42.499	7.8	16.497	0		
14	13	50.6	42.0	40.5	41.7	45.8	44.12	10.10	47.0	51.501	42.499	7.8	16.497	0		
15	14	43.3	45.4	41.8	43.5	41.4	43.08	4.00	47.0	51.501	42.499	7.8	16.497	0		
16	15	36.3	44.9	36.5	36.9	38.5	38.62	8.60	47.0	51.501	42.499	7.8	16.497	0		
17	16	35.1	36.9	37.4	39.4	43.7	38.50	8.60	47.0	51.501	42.499	7.8	16.497	0		
18	17	39.0	35.2	38.6	38.5	36.0	37.46	3.80	47.0	51.501	42.499	7.8	16.497	0		
19	18	42.1	43.0	43.4	40.4	41.1	42.00	3.00	47.0	51.501	42.499	7.8	16.497	0		
20	19	49.8	51.3	46.2	47.7	44.5	47.90	6.80	47.0	51.501	42.499	7.8	16.497	0		
21	20	49.7	40.0	47.3	44.7	44.3	45.20	9.70	47.0	51.501	42.499	7.8	16.497	0		
22						$\bar{\bar{x}}$ =	47.0								$\bar{\bar{x}}$	= AVERAGE(G2:G21)
23							$\bar{r}$ =	7.800							$\bar{r}$	= AVERAGE(H2:H21)

The following *Excel* table is also prepared in consisted with the above table so that all calculations and their commands/formulations refer to the above table. Therefore you may place both *Excel* tables in the same *Excel* sheet for a complete view of the problem solution. Due to space limitations, we are presenting them at two separate pages. In **Column T and W**, all *Excel* commands or formulations in *Excel* codes are presented for your reference.

	Q	R	S	T	U	V	W
1							
2		**Process Data**		**Excel Formula**			**Excel Formula**
3		LSL=	40		**Potential (Within) Capability**		
4		Target	**50**		Cp =	0.99	= (S5-S3)/6/S8
5		USL=	60		CPL =	0.70	= (S6-S3)/3/S8
6		Sample Mean	47.0	= AVERAGE(G2:G21)	CPU =	1.29	= (S5-S6)/3/S8
7		Sample N	100	= A21*5	Cpk =	0.70	= MIN(V5,V4)
8		StDev (Within)	3.3534	= H23/2.326			
9		StDev (Overall)	6.16412	= STDEV(B2:F21)	**Overall Capability**		
10					Pp =	0.54	= (S5-S3)/6/S9
11					PPL=	0.38	= (S6-S3)/3/S9
12		**Exp. Within Performance**			PPU =	0.70	= (S5-S6)/3/S9
13		PPM < LSL	18424.37	=NORM.DIST(S3,S6,S8,TRUE)*1E6	Ppk =	0.38	= MIN(V11,V10)
14		PPM > USL	52.95	= (1-NORM.DIST(S5,S6,S8,TRUE))*1E6	Cpm =	0.49	= (1/6)*(S5-S3)/((SUMSQ(B2:F21)-2*S4*(SUM(B2:F21))+(20*5)*S4*S4)/(20*5-1))^0.5
15		PPM Total	18477.32	= S13+S14			
16					**Potential (Within) Capability**		
17					Z.Bench =	2.09	= NORM.S.INV(1-S15/1E6)
18		**Exp. Overall Performance**			Z.LSL =	2.09	= (S6-S3)/S8
19		PPM < LSL	128061.11	= NORM.DIST(S3,S6,S9,TRUE)*1E6	Z.USL =	3.88	= (S5-S6)/S8
20		PPM > USL	17473.18	= (1-NORM.DIST(S5,S6,S9,TRUE))*1E6	Cpk =	0.70	= MIN(V17,V18)/3
21		PPM Total	145534.29	= S19+S20			
22					**Overall Capability**		
23					Z.Bench =	1.06	= NORM.S.INV(1-S21/1E6)
24					Z.LSL =	1.14	= (S6-S3)/S9
25					Z.USL =	2.11	= (S5-S6)/S9
26					Ppk =	0.38	= MIN(V23,V24)/3
27					Cpm =	0.49	= (1/6)*(S5-S3)/((SUMSQ(B2:F21)-2*S4*(SUM(B2:F21))+(20*5)*S4*S4)/(20*5-1))^0.5

After making the process yield better gear hardness, we get the following data and corresponding process capability results.

	A	B	C	D	E	F	G	H	I	J	K	L	M	N	O	P
1	Days	X_1	X_2	X_3	X_4	X_5	$\bar{x}$	r	$\bar{\bar{x}}$	UCL	LCL	$\bar{r}$	UCL	LCL		Excel Formula
2	1	47.7	54.3	50.5	48.6	55.0	51.22	7.30	50.9	54.208	47.654	5.7	12.013	0		$\bar{x}$ = AVERAGE(B2:F2)
3	2	49.8	51.4	55.1	48.4	53.8	51.70	6.70	50.9	54.208	47.654	5.7	12.013	0		r =MAX(B2:F2)-MIN(B2:F2)
4	3	52.3	49.2	52.1	47.6	54.5	51.14	6.90	50.9	54.208	47.654	5.7	12.013	0		$\bar{\bar{x}}$ = G$22
5	4	47.9	54.4	56.8	48.9	51.5	51.90	8.90	50.9	54.208	47.654	5.7	12.013	0		UCL = G$22+0.577*H$23
6	5	52.5	51.3	58.4	51.2	50.6	52.80	7.80	50.9	54.208	47.654	5.7	12.013	0		LCL = G$22-0.577*H$23
7	6	51.1	50.7	49.3	49.1	49.6	49.96	2.00	50.9	54.208	47.654	5.7	12.013	0		$\bar{r}$ = H$23
8	7	55.9	49.3	48.5	49.9	55.4	51.80	7.40	50.9	54.208	47.654	5.7	12.013	0		
9	8	55.4	48.7	48.9	48.7	51.0	50.54	6.70	50.9	54.208	47.654	5.7	12.013	0		UCL = 2.115*H$23
10	9	49.4	50.7	51.8	46.2	53.0	50.22	6.80	50.9	54.208	47.654	5.7	12.013	0		LCL = 0
11	10	48.3	49.2	50.2	53.6	53.0	50.86	5.30	50.9	54.208	47.654	5.7	12.013	0		
12	11	50.9	48.8	48.9	52.9	56.6	51.62	7.80	50.9	54.208	47.654	5.7	12.013	0		
13	12	52.5	50.2	49.0	53.9	53.2	51.76	4.90	50.9	54.208	47.654	5.7	12.013	0		
14	13	48.1	49.8	49.9	49.9	50.0	49.54	1.90	50.9	54.208	47.654	5.7	12.013	0		
15	14	51.1	50.9	52.2	49.6	47.7	50.30	4.50	50.9	54.208	47.654	5.7	12.013	0		
16	15	48.5	53.4	53.9	49.6	49.0	50.88	5.40	50.9	54.208	47.654	5.7	12.013	0		
17	16	49.3	51.2	52.8	50.9	55.0	51.84	5.70	50.9	54.208	47.654	5.7	12.013	0		
18	17	55.1	48.7	52.0	47.0	52.6	51.08	8.10	50.9	54.208	47.654	5.7	12.013	0		
19	18	49.5	50.7	48.8	48.8	49.5	49.46	1.90	50.9	54.208	47.654	5.7	12.013	0		
20	19	49.3	49.2	51.7	47.4	52.7	50.06	5.30	50.9	54.208	47.654	5.7	12.013	0		
21	20	50.2	49.5	50.8	50.7	48.5	49.94	2.30	50.9	54.208	47.654	5.7	12.013	0		
22						$\bar{\bar{x}}$ =	50.931									$\bar{\bar{x}}$ = AVERAGE(G2:G21)
23							$\bar{r}$ =	5.68								$\bar{r}$ = AVERAGE(H2:H21)

	Q	R	S	T	U	V	W
1							
2		**Process Data**		**Excel Formula**			**Excel Formula**
3		LSL=	40		**Potential (Within) Capability**		
4		Target	50		Cp =	1.37	= (S5-S3)/6/S8
5		USL=	60		CPL =	1.49	= (S6-S3)/3/S8
6		Sample Mean	50.931	= AVERAGE(G2:G21)	CPU =	1.24	= (S5-S6)/3/S8
7		Sample N	100	= A21*5	Cpk =	1.24	= MIN(V5,V4)
8		StDev (Within)	2.44196	= H23/2.326			
9		StDev (Overall)	2.46206	= STDEV(B2:F21)	**Overall Capability**		
10					Pp =	1.35	= (S5-S3)/6/S9
11					PPL=	1.48	= (S6-S3)/3/S9
12		**Exp. Within Performance**			PPU =	1.23	= (S5-S6)/3/S9
13		PPM < LSL	3.80	=NORM.DIST(S3,S6,S8,TRUE)*1E6	Ppk =	1.23	= MIN(V11,V10)
14		PPM > USL	102.08	= (1-NORM.DIST(S5,S6,S8,TRUE))*1E6	Cpm =	1.27	= (1/6)*(S5-S3)/((SUMSQ(B2:F21)-2*S4*(SUM(B2:F21))+(20*5)*S4*S4)/(20*5-1))^0.5
15		PPM Total	105.87	= S13+S14			
16					**Potential (Within) Capability**		
17					Z.Bench =	3.70	= NORM.S.INV(1-S15/1E6)
18		**Exp. Overall Performance**			Z.LSL =	4.48	= (S6-S3)/S8
19		PPM < LSL	4.50	= NORM.DIST(S3,S6,S9,TRUE)*1E6	Z.USL =	3.71	= (S5-S6)/S8
20		PPM > USL	115.03	= (1-NORM.DIST(S5,S6,S9,TRUE))*1E6	Cpk =	1.24	= MIN(V17,V18)/3
21		PPM Total	119.53	= S19+S20			
22					**Overall Capability**		
23					Z.Bench =	3.67	= NORM.S.INV(1-S21/1000000)
24					Z.LSL =	4.44	= (S6-S3)/S9
25					Z.USL =	3.68	= (S5-S6)/S9
26					Ppk =	1.23	= MIN(V23,V24)/3
27					Cpm =	1.27	= (1/6)*(S5-S3)/((SUMSQ(B2:F21)-2*S4*(SUM(B2:F21))+(20*5)*S4*S4)/(20*5-1))^0.5

12.3 Supplementary Material for Minitab Results

Example 12.1: Gear Hardness Data Process Capability Analysis

As for to demonstrate how to perform process capability analysis in *Minitab*, please follow all the steps below.

- ✓ Start-up *Minitab* and enter all the data in **C1-C5** as x_1-x_5, corresponding to gear hardness. Then using **Stat >Quality Tools>Capability Analysis> Normal…** option, click here as shown in the following,

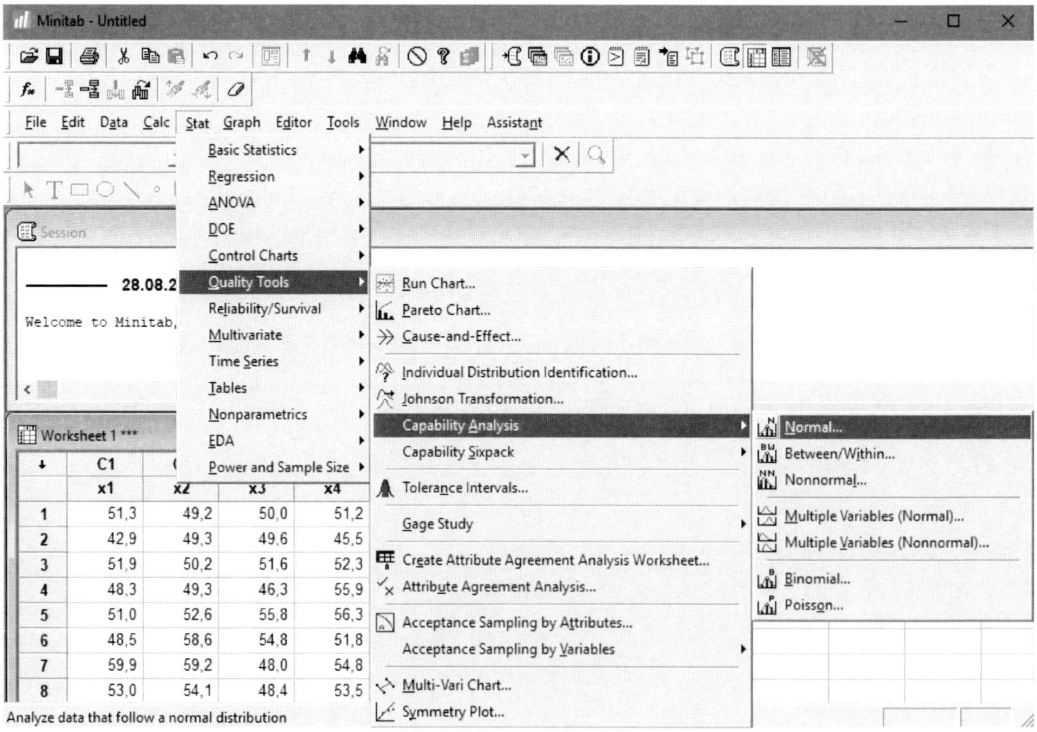

- ✓ In the pop-up **Capability Analysis (Normal Distribution)** window, first, click in the option and next in the window for **Subgroups across rows of:** (following, the variable **x** will appear, again), and select x_1-x_5 and hit **Select** then, Enter **Lower spec:40** and **Upper spec:60**

✓ Next, click on **Estimate...** tab, **Capability Analysis (Normal Distribution) – Estimation of Standard Deviation** window will appear. Click on **Rbar** option and hit OK! Window will disappear.

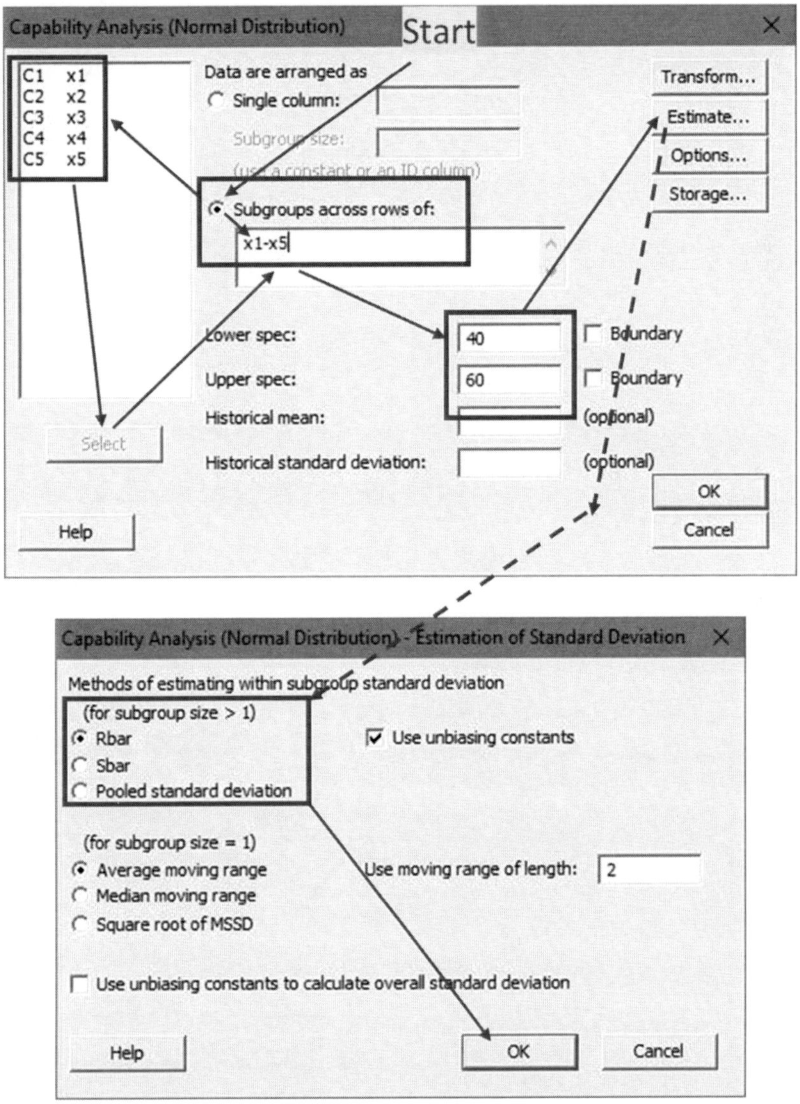

✓ Still on the open **Capability Analysis (Normal Distribution)** window, click on the **Options…** tab, **Capability Analysis (Normal Distribution) – Options** window will appear. Here input **50** for **Target (adds Cpm to table):** and **Select Capability stats (Cp,Pp)** then Hit Ok!

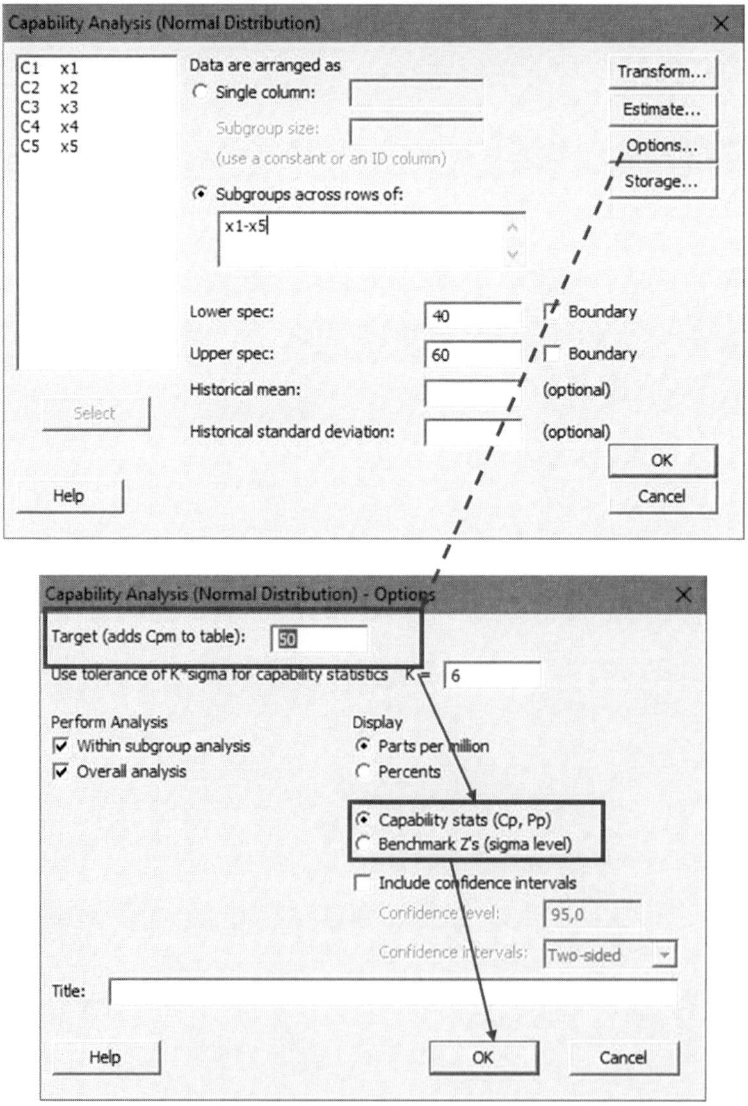

✓ You will get **Process Capability of x1; … ; x5** plot as follows;

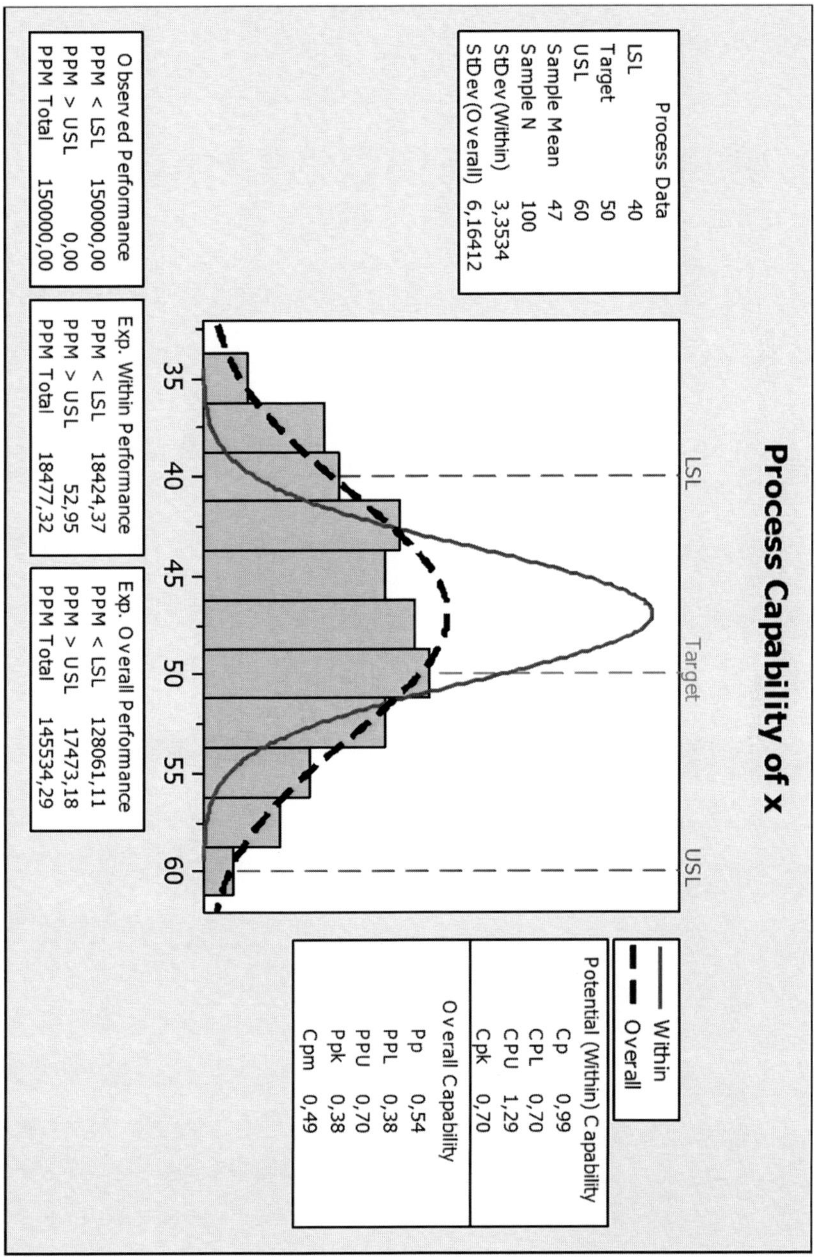

✓ Here, you can follow the same steps till the **Capability Analysis (Normal Distribution)** window, click on **Options…** tab, **Capability Analysis (Normal Distribution) – Options** window will appear, and **Select Bechmark Z's (sigma level)** then Hit Ok!

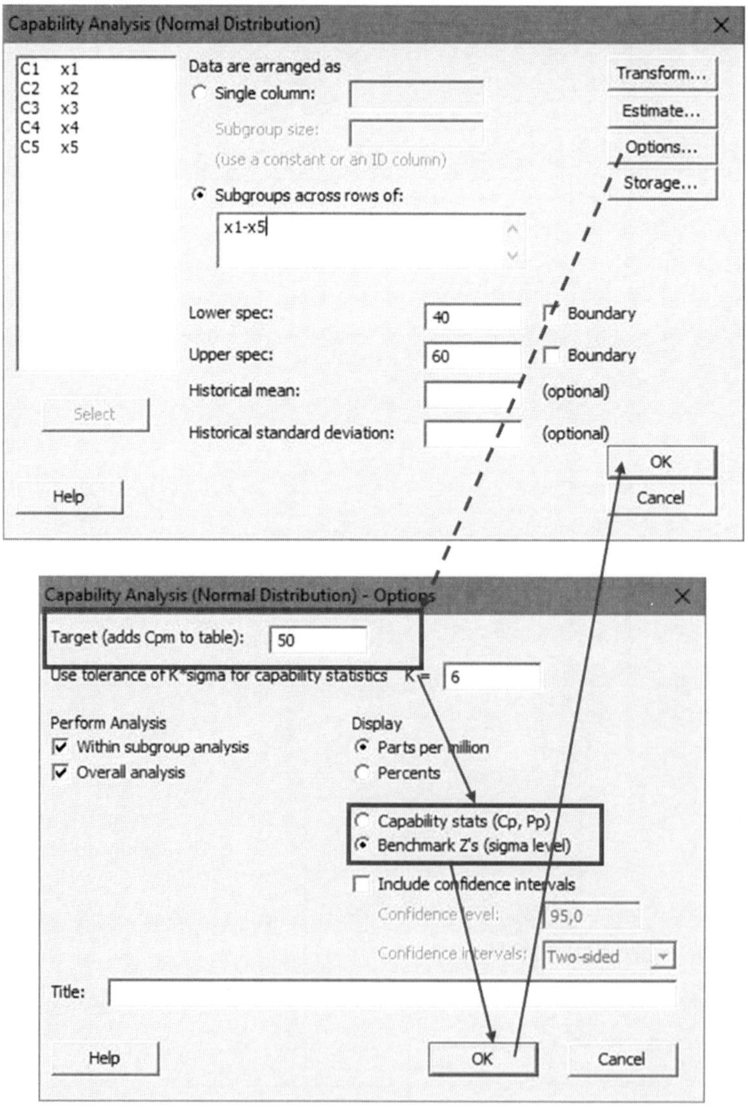

✓ You will get **Process Capability of x1; … ; x5** plot as follows;

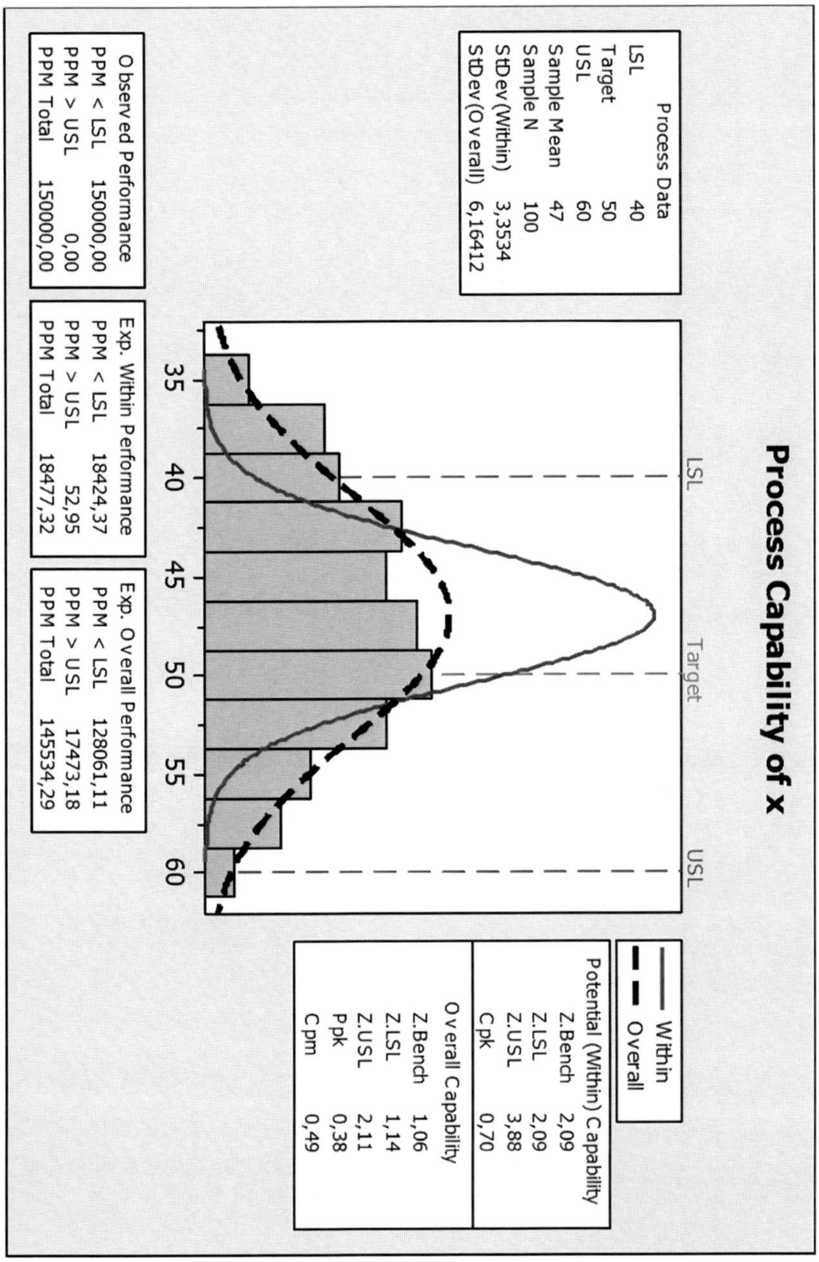

As for the continuation of our Process Capability analysis, we will perform Capability Sixpack analysis which gives us very useful control charts and their calculations to use in this context as presented in the following.

- ✓ Continuing from already open previous Minitab session, Now, using **Stat >Quality Tools>Capability Sixpack> Normal…** option, click here as shown in the following,

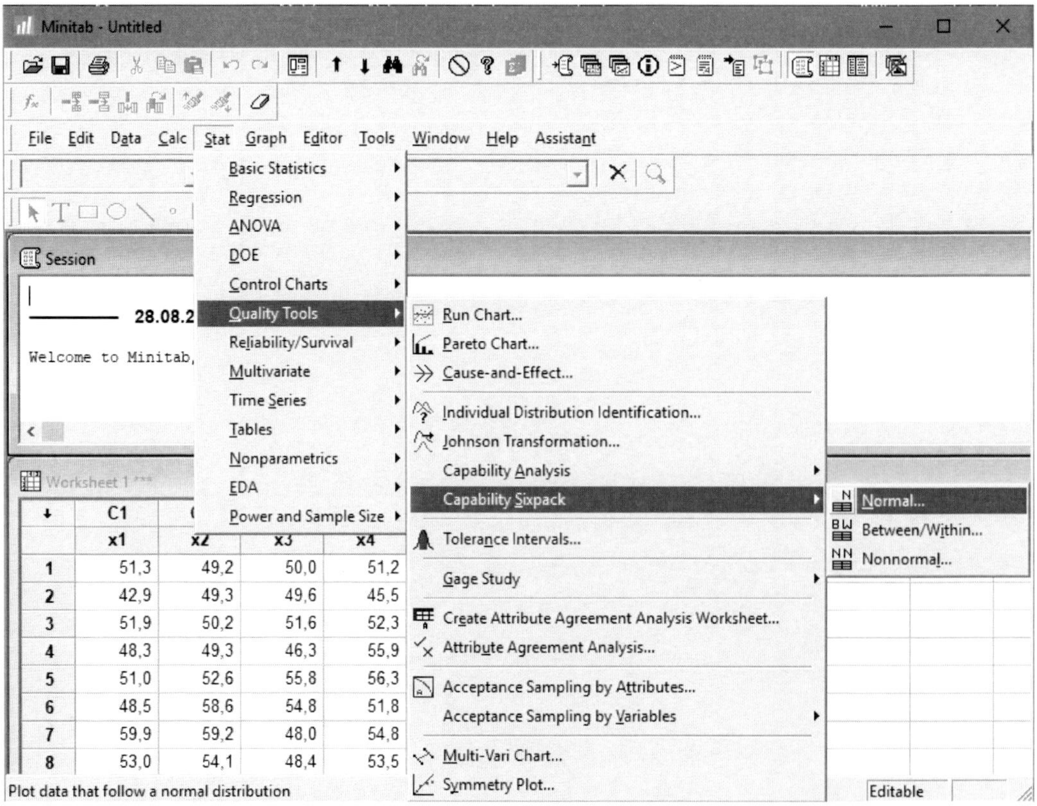

- ✓ In the pop-up **Capability Sixpack (Normal Distribution)** window, click in first the option and next the window for **Subgroups across rows of:** (following, the variable **x** will appear, again), and select x_1-x_5 and hit **Select** then, Enter **Lower spec:40** and **Upper spec:60**

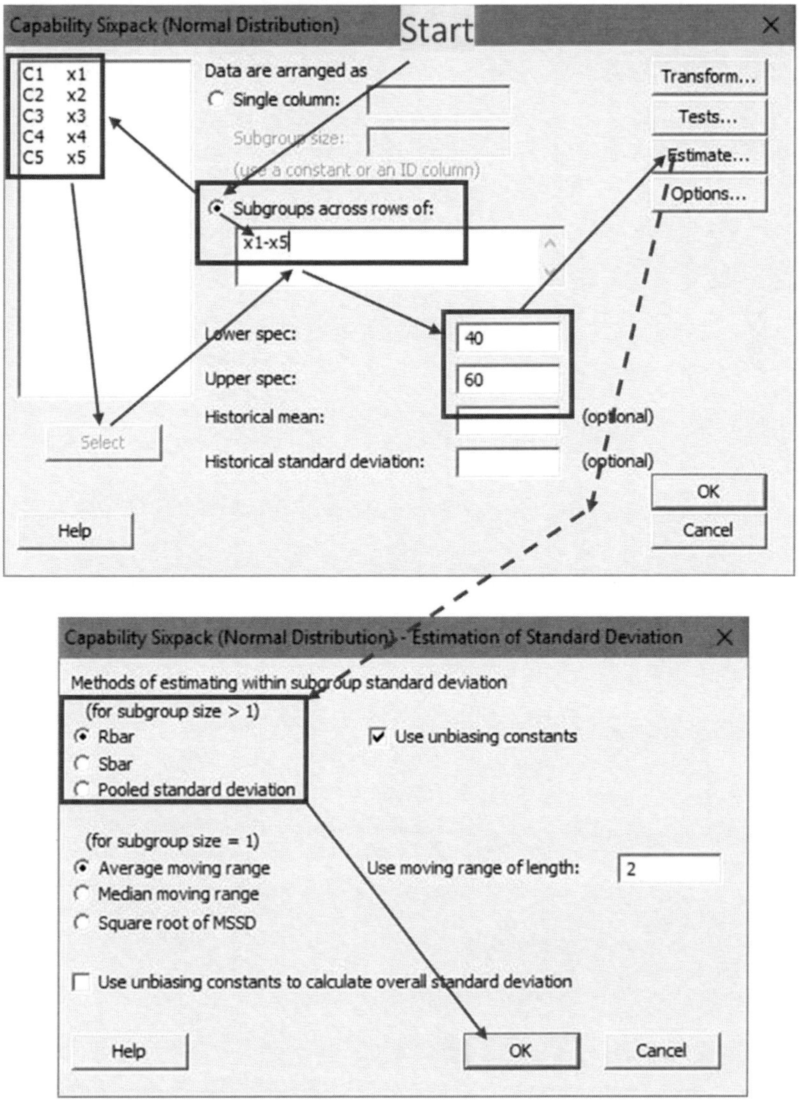

- ✓ Next, click on **Estimate…** tab, **Capability Sixpack (Normal Distribution) – Estimation of Standard Deviation** window will appear, Click on **Rbar** option and hit OK! Window will disappear.

✓ Still on the open **Capability Sixpack (Normal Distribution)** window, click on **Options…** tab, **Capability Sixpack (Normal Distribution) – Options** window will appear. Here input **50** for **Target (adds Cpm to table):** and **Select Capability stats (Cp,Pp)** then Hit Ok!

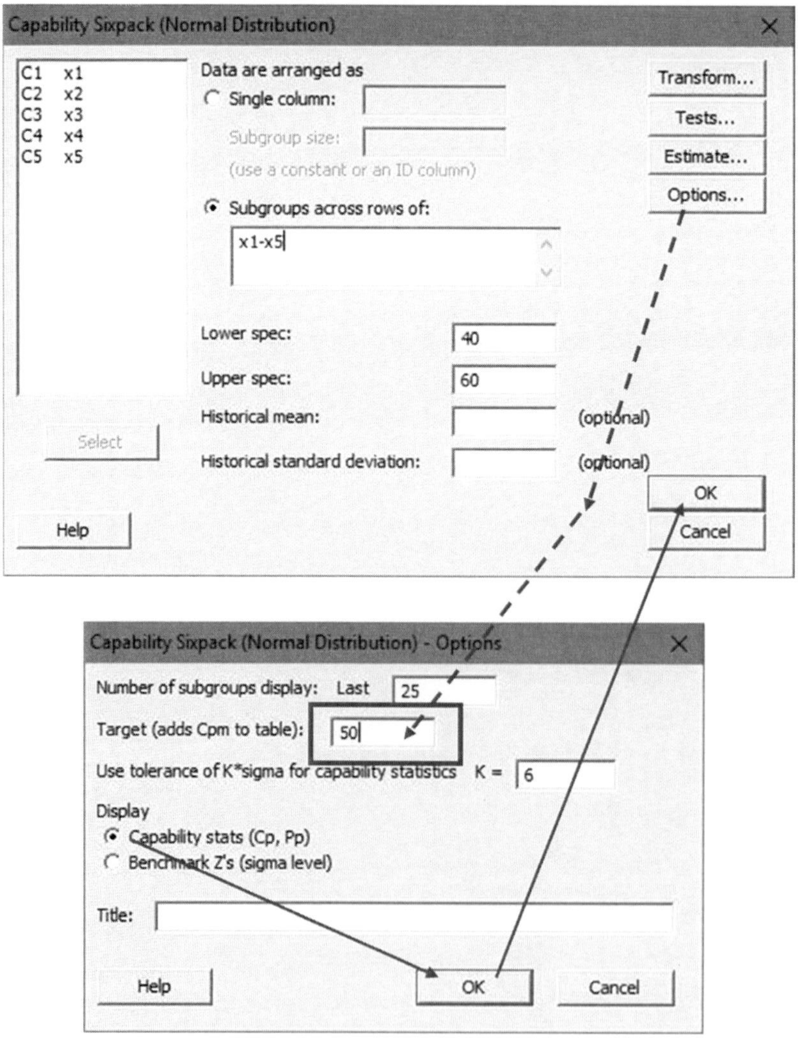

✓ You will get **Process Capability Sixpack of x1; … ; x5** plot as follows;

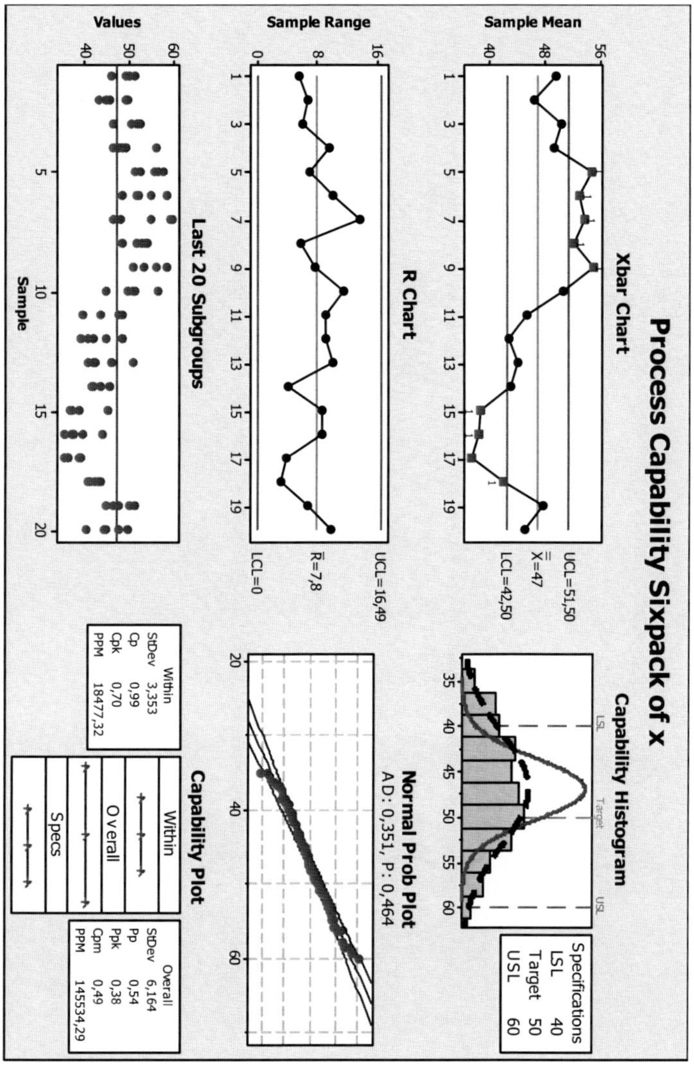

Now, we can perform the same analysis using the "*new data*" shown in the following. Since all the steps are exactly the same as presented before, only the results are presented, in the same order.

2nd Minitab results using *"new (improved) data"*

Days	x_1	x_2	x_3	x_4	x_5
1	47.7	54.3	50.5	48.6	55.0
2	49.8	51.4	55.1	48.4	53.8
3	52.3	49.2	52.1	47.6	54.5
4	47.9	54.4	56.8	48.9	51.5
5	52.5	51.3	58.4	51.2	50.6
6	51.1	50.7	49.3	49.1	49.6
7	55.9	49.3	48.5	49.9	55.4
8	55.4	48.7	48.9	48.7	51.0
9	49.4	50.7	51.8	46.2	53.0
10	48.3	49.2	50.2	53.6	53.0
11	50.9	48.8	48.9	52.9	56.6
12	52.5	50.2	49.0	53.9	53.2
13	48.1	49.8	49.9	49.9	50.0
14	51.1	50.9	52.2	49.6	47.7
15	48.5	53.4	53.9	49.6	49.0
16	49.3	51.2	52.8	50.9	55.0
17	55.1	48.7	52.0	47.0	52.6
18	49.5	50.7	48.8	48.8	49.5
19	49.3	49.2	51.7	47.4	52.7
20	50.2	49.5	50.8	50.7	48.5

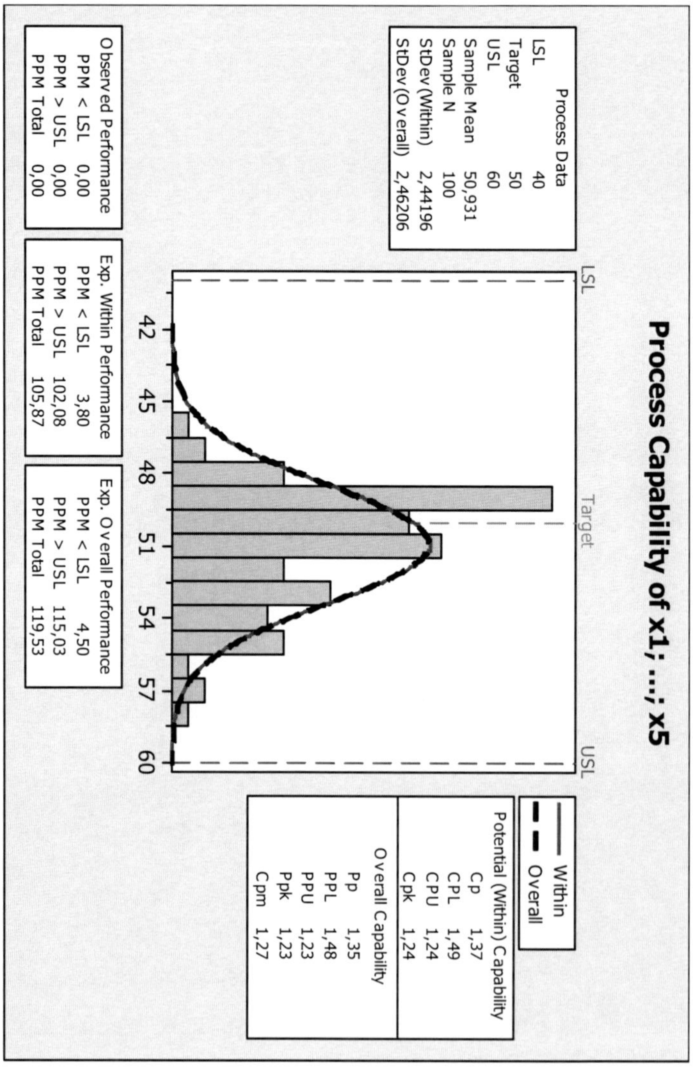

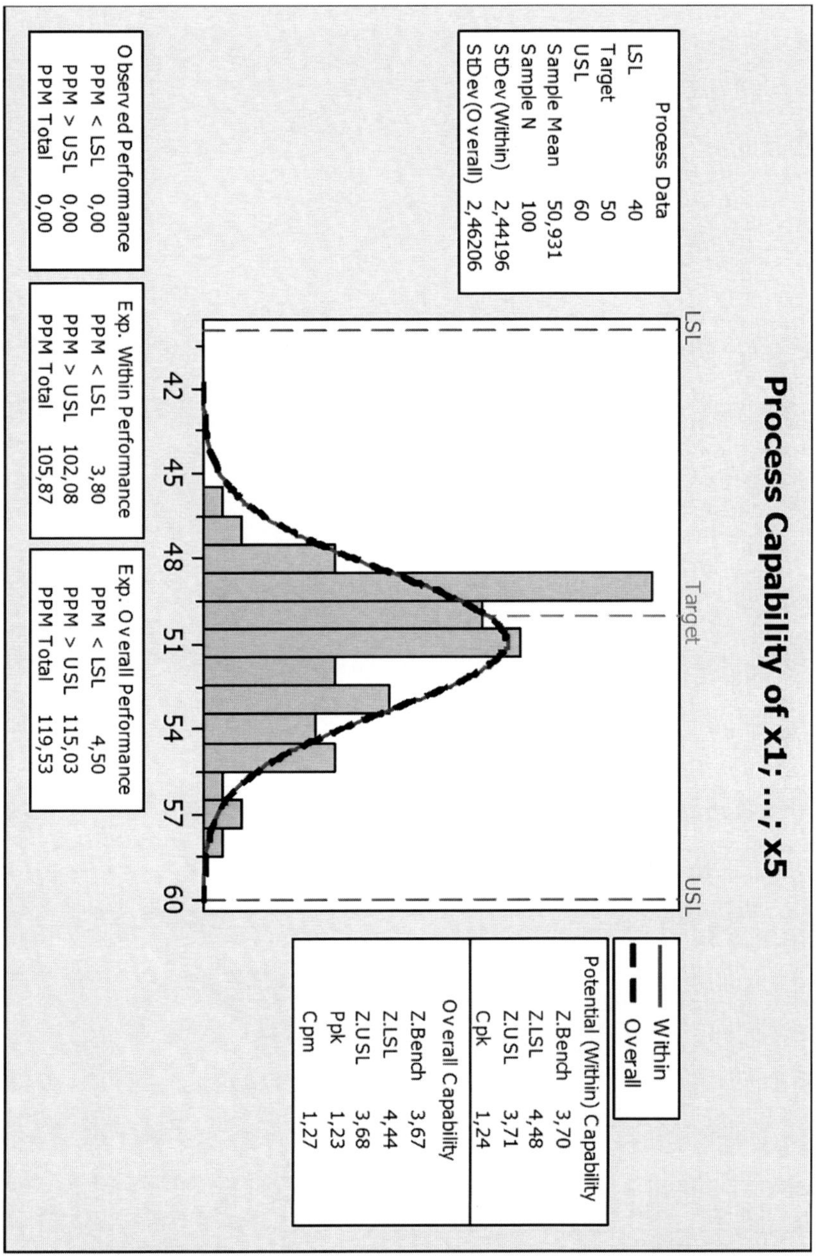
Process Capability of x1; ...; x5
Process Data
LSL 40
Target 50
USL 60
Sample Mean 50,931
Sample N 100
StDev(Within) 2,44196
StDev(Overall) 2,46206
LSL
Target
USL
Within
Overall
Potential (Within) Capability
Z.Bench 3,70
Z.LSL 4,48
Z.USL 3,71
Cpk 1,24
Overall Capability
Z.Bench 3,67
Z.LSL 4,44
Z.USL 3,68
Ppk 1,23
Cpm 1,27
42
45
48
51
54
57
60
Observed Performance
PPM < LSL 0,00
PPM > USL 0,00
PPM Total 0,00
Exp. Within Performance
PPM < LSL 3,80
PPM > USL 102,08
PPM Total 105,87
Exp. Overall Performance
PPM < LSL 4,50
PPM > USL 115,03
PPM Total 119,53

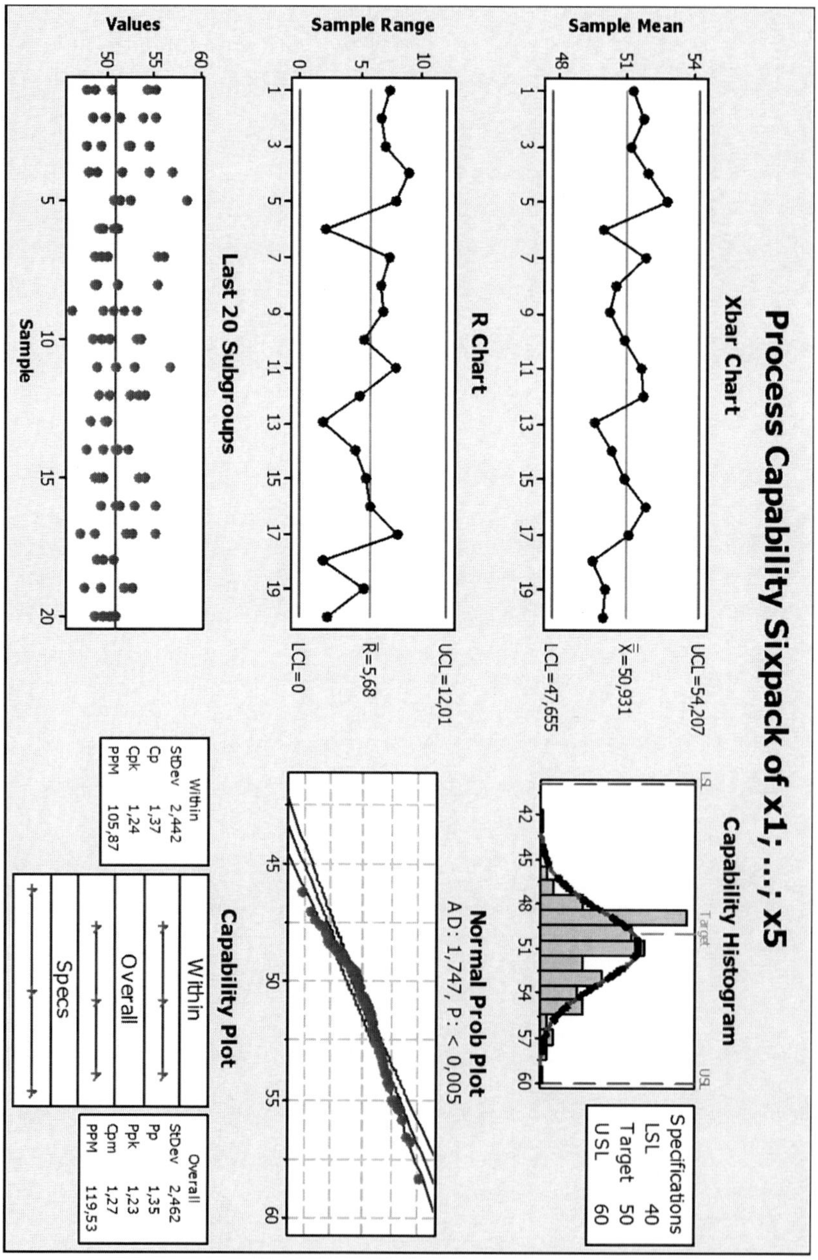
Process Capability Sixpack of x1; ...; x5
Xbar Chart
Sample Mean
UCL=54,207
X̿=50,931
LCL=47,655
R Chart
Sample Range
UCL=12,01
R̄=5,68
LCL=0
Last 20 Subgroups
Values
Sample
Capability Histogram
LSL
Target
USL
Specifications
LSL 40
Target 50
USL 60
Normal Prob Plot
AD: 1,747, P: < 0,005
Capability Plot
Within
StDev 2,442
Cp 1,37
Cpk 1,24
PPM 105,87
Overall
StDev 2,462
Pp 1,35
Ppk 1,23
Cpm 1,27
PPM 119,53
Within
Overall
Specs

12.4 Exercises

1) Describe the *process* using your own words and give a simple example comprised of several *processes.*

2) What are conformance and (in a statistical sense) control?

3) What is the difference between USL/LSL and UCL/LCL? Why should not they be mixed?

4) Process data obtained assumed to exhibit normal distribution characteristics. If the target is set at the center of the distribution, and specification limits $\pm\sigma$, $\pm 2\sigma$, $\pm 3\sigma$, $\pm 4\sigma$, $\pm 5\sigma$, $\pm 6\sigma$ make a table and calculate the corresponding C_{pk} and Z-values in a table and comment on the Process Capability.

5) What is C_{pm} index? What is the difference between C_{pm} and C_{pk}? Please explain with descriptive drawings.

6) Assume we manufacture shafts with *mean* of 30 mm with customer defined tolerances of 29.6-30.4 mm, customer lower specification limit is μ-2σ and upper specification limit is μ+2σ.

- Calculate the standard deviation
- Calculate the process capability
- For 6σ process, what is the σ should be?

6) Go back to the gear hardness problem, Assume after induction hardening we get around 45-55 HRC (Rockwell C Hardness) in helical gears for the transmission of an automobile. In early gear hardening trials, gear hardness data are recorded for twenty consecutive days, i.e. m=20. Each day five samples are taken with a sample size of n=3 (subgroups) and all measurements are as presented in the next table.

Days	X_1	X_2	X_3
1	51.3	49.2	50.0
2	42.9	49.3	49.6
3	51.9	50.2	51.6
4	48.3	49.3	46.3
5	51.0	52.6	55.8
6	48.5	58.6	54.8
7	59.9	59.2	48.0
8	53.0	54.1	48.4
9	56.2	58.5	56.2
10	50.8	44.7	49.7
11	48.3	47.5	48.4
12	44.7	39.1	40.6
13	50.6	42.0	40.5
14	43.3	45.4	41.8
15	36.3	44.9	36.5
16	35.1	36.9	37.4
17	39.0	35.2	38.6
18	42.1	43.0	43.4
19	49.8	51.3	46.2
20	49.7	40.0	47.3

- Using the data perform process capability analysis by hand
- Repeat this analysis by *Excel* and Minitab
- Comment on the process and its capability

Chapter 13

Quality Loss Function

Robust design is a complete philosophy/methodology that can be used in the design of products or processes to ensure that they perform as intended regardless of variation in a product's manufacturing process, variation resulting from deterioration, variation by the environmental effects and variation in usage by the end user. Robust design can only be achieved when all potential sources of variation are fully understood and taken into consideration in design. The remedy is to take required actions to make the product (or process) insensitive to all potential sources of variation, as much as possible. In this sense, improved quality of any product is directly linked to the application of robust design methodologies.

Further, improving quality characteristics of any product (or process) requires relating this product's targeted quality measure(s), in terms of deviation/variation in "*quality loss*" associated with a desired/targeted reference value. For this purpose, we need to use fundamental mathematics to construct a certain form of the cost function and use this cost function to evaluate the pertaining quality loss level for the product (or process). Once we formulated the *"quality loss"*, then we can go ahead and try to reach near zero or minimum quality loss goal, as our ultimate objective. So that we can make sure a high level of performance specifications are both enforced and met by/after evaluating the quality characteristics.

In the literature, we encounter with two widely-used approaches when expressing quality loss. The first approach is based on *"goal post"* i.e. very simple and easy to use, i.e. the hard form of quality loss function (i.e. step function) is used. "*goal post*" term is derived from goal-based sports where players score when they pass the ball through the goalie (i.e. frame) i.e. soccer, football goalie. Otherwise we do not score or in other words, we do not get the job done. Although it is very simple, this traditional approach is not without its drawbacks as would be introduced later.

In the second approach, a more popular (i.e. continuous) form of the cost function is used, which is in more representative (more softly translational) form, in terms of "*quadratic*" loss function usage. This loss function penalizes all deviations on the basis of not being on the specified target value (quality characteristics). Once we assess the quality loss of the product to the customer, manufacturer or briefly to the society quantitively, then we can determine/assess our success in achieving the quality characteristics (or level) of the product (or process) in question. In the following sections, we are going to focus more on the quality lost formulations and further elaborate on both approaches by their usefulness and capabilities.

13.1 Introduction

Once upon a time, the quality was thought to be as "*luck*". I remember 40 years ago, our elders were saving their money for quite a long time to purchase a black and white TV, which was not easy to get since it was expensive. Although the same brand TV was purchased say by five different families of our neighbor, only half of the families were happy with the quality of TV, the other half was having a lot of fixing problems. At that time, the common perception for this low percentage level of satisfaction with the TV or its performance had been interpreted with the term "*luck*". When did you question why you got a bad TV? "*It is a little bit luck*", that was the *"reasonable or valid"* answer you got! (as depicted in Figure13.1).

In a modern industrial age, the concept of "*luck*" is proven to be misbelief and quality was considered an integral but greatly designed/controlled aspect of the product characteristics. How? The quality can be designed, implemented and measured briefly enforced by using all modern means/tools.

FIGURE 13.1 Quality perception in *old times*

As introduced earlier, in evaluating and enforcing the quality, the first simple approach used is based upon goal post approach, where assume you have a *target, m* which characterizes the expected/planned characteristics/performance of the product *"best"*. We know that the performance/robustness means quality in the eye of the end-user. As long as you hit your target specification, *m* you are right on! as far as the product quality is concerned where the manufacturer very closely meets the key/target product specifications. Finally, end customers get good products (or service), quality term shines in our comprehension. Well, not exactly!

13.2 Goal Post Loss Function

As far as the traditional approach is involved, let's think about a dart chart as shown in Figure 13.2, where we set the rules of the game so that the best target score is at *location 0,1,2,3,4, and 5* with 0 points and the worst score is at *location 6 and 7* with 10 points. Clearly, your defined objective is to hit the lowest score at *location 0* as to hit the target, *m*! Not the highest scores at *locations 6* and *7*, which are shown as "*dotted circled-lines*". Scoring in this example is similar to the Golf scoring! The lower is the score the better is!

Similarly, traditional goal post quality approach says as long as you do not hit the *locations 6* and *7*, you get straight 0 scores! No matter what! even though you hit the *locations 5,4,3* and *2*. In other words, as long as you pass the ball through goalie it does not matter. This means all products equally satisfy the desired/targeted quality requirements as long as "goal scoring" is there. Note, you get (say) 10 penalty points only when you hit *locations 6* and *7* or this means you missed the goalie. Out of the total number of any hit trials, whoever gets the lowest total score is the winner! Clearly, in this approach, some fundamental points are either overlooked or missing. But it is simple!

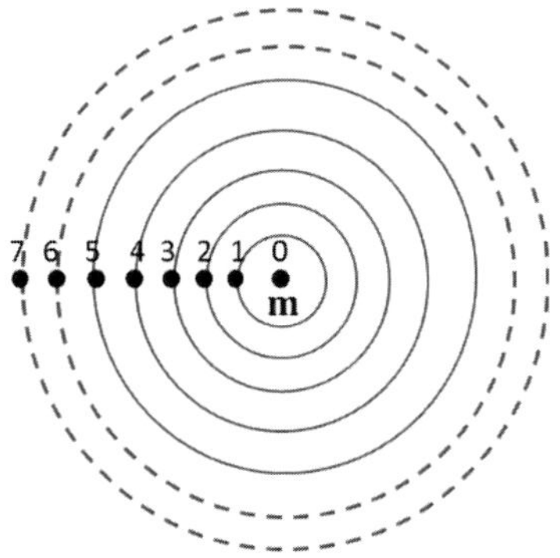

FIGURE 13.2 Dart chart for goal post "*traditional*" quality loss function

Let's continue our discussion, considering a waste/rework cost of the product that in question is A_o=\$10, when it does not meet the specifications. Also, assume our target is ***m***=\$0. If you examine Figure 13.2, the quality loss does appear right after location 5, suddenly the cost of fixing the product goes from \$0 to \$10. Thick black line (i.e. frame) looks like football goalie. Also, you may find ***L(y)*** cost in terms of ***y*** location. Notice to dramatic change i.e. even a small distance move from y= 5.98 to y=6.02, draws a very sensitive loss/cost threshold line. It is obvious that in practical life *locations 5* and *6* are *not so different* from each other. How? According to traditional approach *location 5* satisfies quality characteristics that's why it is ok, i.e. no loss case! but *locations 6* and *7* are both losses as "*waste or rework*" and not ok, i.e. a loss case! Any quality measure falling at location 6 and 7 will cost to the manufacturer in the amount of say, A_o=$10.

As you may notice this approach employs a poorly defined "*Step function*" as a quality loss measure and does not improve the quality as expected! The lower and upper specification limits draw a critical thin line between satisfaction and dissatisfaction.

Engineering specifications are invariably stated as $m \pm \Delta_o$ (by which product characteristics are controlled) and step function to quantify the quality loss in "*goal post*" approach where A_o is the \$ cost of replacement, repair or rework and Δ_0 is the enforced design tolerances as formulated in Equation 13.1 and depicted in Figure 13.3;

$$L(y) = \begin{cases} 0 & \text{if } |y - m| \leq \Delta_0 \\ A_0 & Otherwise \end{cases} \qquad 13.1$$

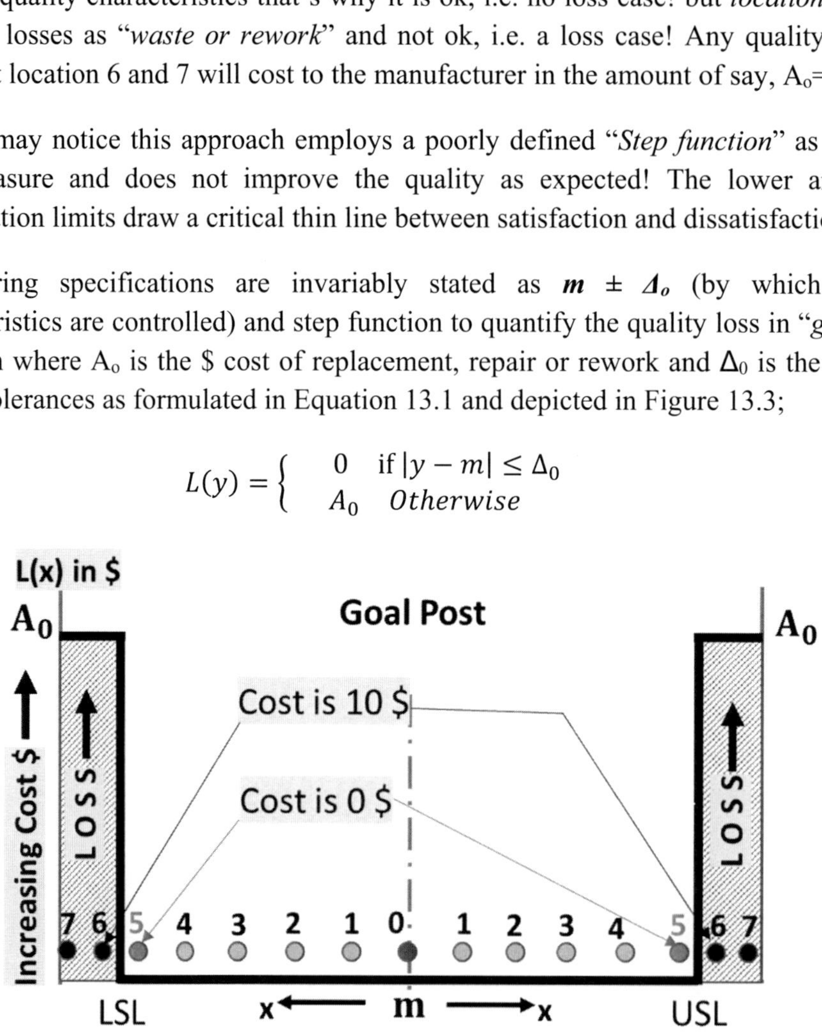

X	A_o	7	6.02	5.98	5	4	3	2	1	m=0	1	2	3	4	5	5.98	6.02	7	A_o
L(x)	$10	$10	$10	$0	$0	$0	$0	$0	$0	$0	$0	$0	$0	$0	$0	$0	$10	$10	$10

FIGURE 13.3 Goal post "*traditional*" loss function

Now, much better and widely accepted way to account for the quality loss in "*Taguchi quadratic loss function*" is introduced in the next section. This concept has caused huge developments in the Japanese industry and made their products preferred/respected all over the world in terms of quality by meeting customer requirements/expectations

13.3 Taguchi Quadratic Loss Function

By Acknowledging the corresponding weaknesses of the goal post approach, Genichi Taguchi did state that, the specification limits do not clearly identify customer satisfaction levels for the product for obtaining the desired outcome, i.e. the quality. To overcome these weaknesses, Taguchi proposed "*a quadratic loss function*" by incorporating a simple but very useful rationale (through defining a 2nd order function), which penalizes all the deviations from the target, ***m*** value even if the outcome falls in between upper and lower specifications.

Eventually, we can highlight and be aware, how each non-perfect part produced hits the manufacturer in loss as a return! In other words, these deviations reflect itself in very unwanted form as customer dissatisfaction. Unhappy customers are a nightmare for every business, eventually hunt down the business itself, sooner or later. If the variation goes beyond the specification limits, the product is marked to be defective in the form of "*loss*" or "*rework*". In product/process quality, losses initiate when marching out "*set/accepted/agreed*" target(s).

A well-known statistician Deming, in his book, states that;

"A minimal loss at the nominal value, and an ever-increasing loss with departure either way from the nominal value." - W. Edwards Deming, Out of the Crisis.

Taguchi basically states that any deviation(variation), falling away from the target ***m***, will result in poor performance. Eventually, any deviation will incur quality loss followed by the customer dissatisfaction. This is clearly depicted in Figure 13.4. The more variation occurrences will cause gradually the more dissatisfied customer as anticipated. This is certainly not good news for the companies.

For example, assume maximum pressure sealing in a hydraulic cylinder is achieved during manufacturing by controlling the piston sealing diameter which has a target value of 9.95 mm for 10 mm cylinder diameter. When you get a 9.8 mm piston diameter, you will get more loss in pressure than that of i.e. 9.9 mm. Satisfaction in obtaining right piston pressure is highly depending on how close you are to 9.95 mm. But you should also note that any diameter over 9.95 mm is not possible due to assembly and feasible functioning considerations.

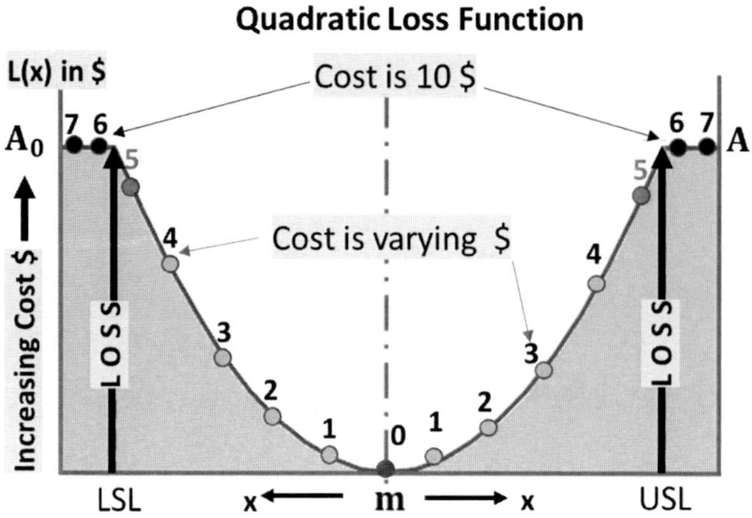

x	A_o	7	6.02	5.98	5	4	3	2	1	m=0	1	2	3	4	5	5.98	6.02	7	A_o
L(x)	$10	$10	10	$9.93	$6.94	$4.44	$2.50	$1.11	$0.28	$0	$0.28	$1.11	$2.50	$4.44	$6.94	$9.93	$10	$10	$10

FIGURE 13.4 "*Taguchi quadratic loss*" function

Taguchi clearly states that; staying in between specification limits by themselves is not the true/definite criterion for providing product/process quality and thus customer satisfaction. Instead, the continuous penalty has to be incurred gradually as the quality measure of our interest move away from the target, ***m***. Therefore companies have to minimize the variations using any tools/capabilities they have so that they can achieve the target performance, ***m*** or get minimal variation so that customer dissatisfaction from a product/process is either eliminated or minimized.

As for the quality considerations, engineering specifications are often, considered as quality characteristics(responses), targeted with an ideal/ultimate target value of, ***m***, for a lower side of $m - \Delta_o$ and an upper side of $m + \Delta_o$ tolerances, where Δ_o can be thought of either manufacturer or customer tolerance. These functional limits $m \pm \Delta_o$, are the points after which the product will fail or produce an unacceptable/unsatisfactory performance.

As illustrated in Figure 13.3, the goal post approach uses a simple step function to distinguish/decide the conformity of the engineering specification of interest with respect to the target, ***m***. This function can also be considered as *"gauge"* used to measure dimensional conformity of high precision machine parts such as shafts, discs etc.

Whereas, easy to use Taguchi's quadratic loss function is simply a second-order(quadratic) function for approximating the "*quality loss*" in *dollars* by penalizing "*amplified variability*" as quality characteristics of our interest moves away, i.e. $(y\text{-}m)^2$, from the design target, ***m***. The quality loss function ***L(y)*** is defined as a loss due to the deviation of ***y*** from ***m*** in "*dollars*" as;

$$L(y) = k(y - m)^2 \tag{13.2}$$

where ***k*** is (*economic*) constant also called quality loss coefficient, ***y*** is quality characteristic as a function of the measured response of our process/product and ***m*** is the target value for ***y*** as the product's response. Note, when ***y=m***, the loss is zero and the loss increases the further ***y*** deviates from the target, ***m***.

$$\text{Maximum cost}, L(y) = A_0 \; at \; y = m \pm \Delta_0 \tag{13.3}$$

Assuming "*maximum quality loss*" is denoted as A_o, i.e. the sum of economically measurable consequences of all failures and the value of quality loss function at $m \pm \Delta_o$, which include repair, labor, replacement, loss of use, waste, rework etc. Let the total cost of all losses at $m \pm \Delta_o$ be equal to A_o in *dollars* (note, $LSL = m - \Delta_o$ and $USL = m + \Delta_o$ where Δ_o is the symmetric tolerances. Δ_o can very well be non-equal(asymmetrical) value for lower and upper sides of the target, ***m*** as $\Delta_L \neq \Delta_U$).

As for $y = m - \Delta_0$ (or $y = m + \Delta_0$), if you insert it into Equation 13.2, you can get *k* constant as,

$$A_o = k(\cancel{m} - \Delta_0 - \cancel{m})^2 \tag{13.4}$$

$$k = \frac{A_o}{\Delta_0^{\;2}}$$

where we can obtain *k* constant in terms of A_o and Δ_o. Substituting the functional limits $m \pm \Delta_o$ and the total losses of A_o into Equation 13.2, we get Taguchi quality loss function (in dollars), i.e. note that all of the losses (i.e. part, labor, set up time, shipping time etc.) due to not meeting the target *m*, will be included in A_o.

$$L(y) = \frac{\boldsymbol{A_0}}{\boldsymbol{\Delta_0}^{\boldsymbol{2}}}(y - m)^2 \tag{13.5}$$

We obtain the general formulation of the two-sided quality loss function.

13.4 Variations of the Quadratic Loss Function

In the earlier section, the Equation 13.5 is derived for "*the nominal-the-best"* scenario. This equation has to be either directly used or modified for our "*targeted*" Taguchi's loss function of our interest.

For example, we may want to have any of the following three scenarios, which are possible and best suited for our product quality characteristics. The quality loss function, *L(y)*, as being our primary target/objective, may have the following three scenarios;

- The nominal-the-best (Symmetric/Asymmetric)
- The smaller-the-better
- The larger-the-better

As for the first scenario, we have derived the *L(y)* in Equation 13.5. For the remaining scenarios, we need to modify this loss function accordingly as per our objective requires. Now, let's go ahead and look into these scenarios with more details as presented in the following sections.

13.4.1 The Nominal is the Best

For this scenario, both "*goal post*" and "*quadratic quality loss*" functions are introduced and derived earlier. Remember, in this scenario, our objective i.e. target ***m***, can be approached either from negative or positive sides by $\pm y$ values. This mean quality loss due to variations from our target ***m***, at both sides are equal as presented in Figure 13.5 (from a broader perspective we may have also different LSL and USL). In this scenario first, symmetric quality loss characteristics are observed as given in Equation 13.2;

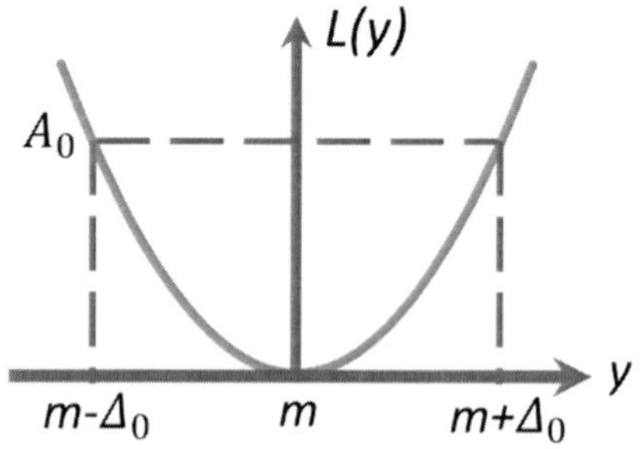

FIGURE 13.5 Symmetric "*the nominal-the-best*" quality loss function

Assume you are manufacturing a shaft on a turning machine at an outer diameter of 20 mm. Then your target ***m****=20 mm* and any deviations from ***m*** in either say, 20-0.1 = 19.9 mm or 20 + 0.1 = 20.1 directions are not tolerated when tolerances are Δ_0=*0.02 mm,* we can find **LSL**= ***m*** - Δ_0 = 20 - 0.02 =19.98 mm and **USL**= ***m*** + Δ_0 = 20 + 0.02 = 20.02 mm as follows;

$$\mathbf{m} = \mathbf{20}^{+0.02}_{-0.02} = \mathbf{20}^{\mathrm{USL}=+20.02}_{\mathrm{LSL}=-19.98}$$

Cleary, one variation to this scenario of symmetric loss function, is the "*asymmetric loss function*" which occurs when deviations(tolerances) of the quality characteristics in both directions are not the same in Δ_o , rather different as Δ_{oL} and Δ_{oU} (i.e. $\Delta_{oL} \neq \Delta_o$), which is quite expected and understandable. If you revisit shaft example assume Δ_{oL} = *0.02 mm* and Δ_{oU} = *0.01 mm,* we can find **LSL**= ***m*** - $\boldsymbol{\Delta_{oL}}$ = 20 - 0.02 =19.98 mm and **USL**= ***m*** + $\boldsymbol{\Delta_{oU}}$ = 20 + 0.01 = 20.01 mm as,

$$\mathbf{m} = \mathbf{20}^{+0.01}_{-0.02} = \mathbf{20}^{\mathrm{USL}=+20.01}_{\mathrm{LSL}=-19.98}$$

Obviously, in this case, we need to use k_L and k_U for lower and upper sides, respectively when using asymmetric quality loss functions, as presented in Equation 13.6 and shown in Figure 13.6,

$$L(y) = \begin{cases} k_U(y-m)^2 & y > m \\ k_L(y-m)^2 & y \leq m \end{cases} \qquad 13.6$$

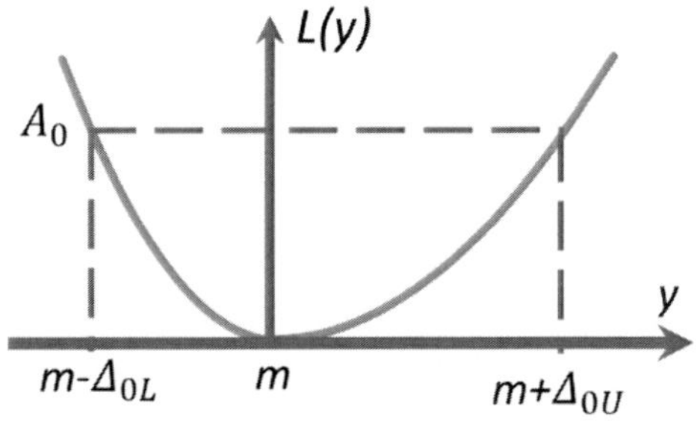

FIGURE 13.6 Asymmetric "*the nominal-the-best*" quality loss function

Example 13.1: LED TV input Voltage Fluctuations

High technology LED TV's are produced and sold all over the world by a very famous manufacturer, called ABC company. In the design of these LED TVs, power electronic engineers assumed/used 220∓30 V, i.e. voltage design range, by considering the grid electric power voltage level changes/fluctuations so that all LED TVs can operate at desirable settings when plugged into the electric outlet. This means efficiency and enjoyment of the TV watching diminishes when the input TV voltage exceeds the tolerance limits in either "*below* low" or "*above* high" voltages. Optimal or target electric voltage level is suggested as *m*=220 V.

TV Customers are given three years of very comprehensive warranty (Why? Think about competition with other TV manufacturers) for all TVs no matter what the failure is, i.e. generally unexpected extreme electric input voltage changes may damage the TVs. If anything goes wrong related to electric power fluctuations (considering both rural and urban areas), the manufacturer can repair the TVs by changing a damaged/burned down power unit at a cost of $90.

- Formulate the appropriate Taguchi quality loss (function).
- Assume the TV electric input voltage is 200V, calculate the quality loss for both goal post and Taguchi loss approaches.

By examining the problem, we see that Δ_o=30 V, m=220 V, and A_o= $80, and we want a nominal electric voltage input level for our TVs life and performance issues. Therefore we have to use Taguchi quadratic "*the-nominal-the-best*" formulation as,

$$L(y) = \frac{A_0}{{\Delta_0}^2}(\boldsymbol{y} - m)^2 = \frac{\$90}{30^2\ \mathrm{V}^2}(\boldsymbol{y} - 220\ \mathrm{V})^2 = 0.1\frac{\$}{\mathrm{V}^2}\ (\boldsymbol{y} - 220\ \mathbf{V})^2 \qquad 13.7$$

where k=0.1 $\$/V^2$, and ***y*** is LED TV operating electric voltage characteristic.

When ***y***=200 V, using Equation 13.7, we get the quality loss as, L(**200 V**)= $40 by Taguchi quality loss approach. As far as the goal post (traditional) approach is concerned, since y=200 V is very well within the 220∓30 V tolerance limits, we assume there is no quality loss according to this approach.

This example clearly shows the difference in the understanding of the quality loss methodology between the goal post and Taguchi quadratic loss functions. Note that Taguchi quadratic loss function is only zero when ***y***=220 V, any departure from this value is penalized as a quality loss in this approach. Therefore, this method can be thought of rather conservative by enforcing penalties in *any* amount of departure from, *m.*

13.4.2 Smaller is the Better

The second scenario is encountered for some characteristics; such as "*an operating noise*" of the washing machine is always desired to be at ***zero*** decibels [*db*] noise level, if possible! Of course, this is not possible in real-life! If not *zero* then we want operating noise as *minimal* as possible!

As this value increases, the human life at home progressively gets uncomfortable (worse) for people using this machine. It is clear that the "*operating noise*" value cannot be negative and always measured ***greater*** than zero! The quality loss, in this case, approximates, by substituting ***m=0*** in Equation 13.2 as,

$$L(y) = k(y)^2 \qquad 13.8$$

Note since y can take up only positive values i.e. $y \geq 0$, this loss function is one-sided. For this case, the quality lost coefficient can be obtained by substituting, y = USL= (m=0) + Δ_0 in Equation 13.8, we end up with;

$$A_o = k(\Delta_0)^2 \qquad 13.9$$

$$k = \frac{A_o}{{\Delta_0}^2}$$

The graph of the corresponding "*the-smaller-the-better*" quality loss function is shown in Figure 13.7.

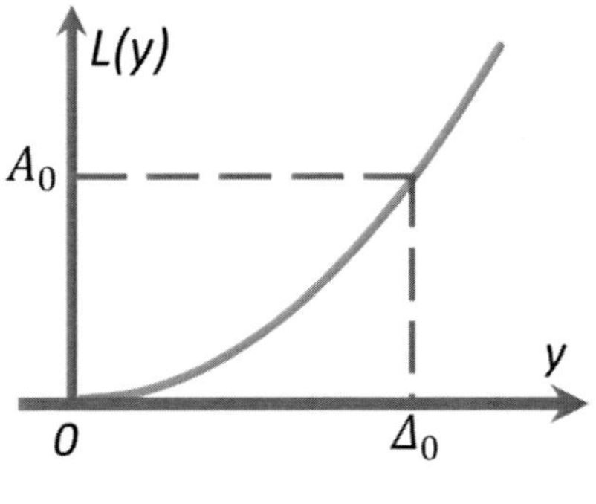

FIGURE 13.7 "*The-smaller-the-better*" quality loss function

Example 13.2: The Car Interior Noise level

A very well-known motor car company is having an increasing number of complaints about the "*excessive*" levels of interior noise for their latest brand new luxury sedan. The interior noise is attributed to the faulty insulation material and design. If customer complaints persist, each customer service visit to fix this problem is costing to the manufacturer around $490 in terms of labor, insulation material change, and test expenses. Average customer complaints are observed at the average interior noise level of 70 dB or higher at an average speed of 100 km/h speed.

In this problem, we notice that Δ_o=70 dB, m=0 dB (ideally desired but never attained noise level!) and A_o= \$490, and we can use *"the-smaller-the-better"* scenario. In this case, Taguchi quadratic loss formulation is defined as,

$$k = \frac{A_0}{{\Delta_0}^2} = \frac{\$490}{70^2\ \text{dB}^2} = 0.1\ \frac{\$}{\text{dB}^2}$$

where k=0.1 \$/dB^2, and *L(y)* is,

$$L(y) = \text{k}(\boldsymbol{y})^2 = 0.1\ \frac{\$}{\text{dB}^2}(\boldsymbol{y})^2$$

When ***y*** is car interior noise level in dB. Notice that when ***y***=70 db, L(70 dB)=\$490 and ***y***=60 db, L(60)=\$360, perhaps when ***y***=60 dB, there will not be any customer complaint, even such complaint exist it would not be accepted by automobile service since it is below the target noise level of 70 dB. Please see Figure 13.7 for better visualizing this scenario. For any noise level greater than 70 dB, i.e. y=80 dB, L(80 dB)=\$640 loss the manufacture incur, in terms of both bad quality and reputation for this luxury sedan.

13.4.3. Larger is the Better

The third scenario is encountered for some characteristics; such as "*strength*" of a connection bolt is desired at ***higher*** possible levels of MPa, if possible! As this value increases, the connection, thus the connection safety, gets better. Clearly, zero strength is the worst value and infinite strength is the best value where quality lost is zero and as the y → ∞, *L(y)* → 0. Note that, the "*strength*" value cannot be negative and always be measured greater than zero! As the quality loss, in this case, can be approximated by substituting reciprocal of *y* (i.e. $1/y$ in a smaller the better formulation for ***y*** in Equation 13.8, we get;

$$L(y) = k\frac{1}{(y)^2} \qquad 13.10$$

Note, since ***y*** can take up only positive values i.e. $\boldsymbol{y} \geq 0$, this loss function is also one-sided. The quality lost coefficient can be obtained using y = LSL= Δ_o in Equation 13.10, we get;

$$A_o = k\frac{1}{{\Delta_0}^2} \qquad 13.11$$

$$k = A_o{\Delta_\mathbf{0}}^\mathbf{2}$$

The graph of the corresponding "*the-larger-the-better*" quality loss function is also shown in Figure 13.8.

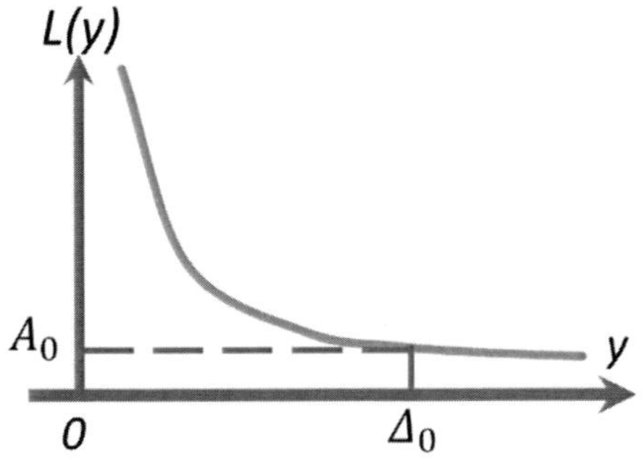

FIGURE 13.8 "*The-larger-the-better*" quality loss function

Example 13.3: The Vacuum Sealing Pressure

In an expensive industrial vacuum, an air sucking performance is critical and controlled at the gate where pressure loss is prevented with a special kind of sealing between the gate and motor (assuming the motor is fully operational) for providing higher vacuuming capability. The better the sealing achieved (i.e. in *Psi*) is the better for vacuuming performance of the industrial vacuum.

In a case this sealing is not achieved due to sealing assembly etc. failures, the fixing cost for sealing part replacement is $70 and one-hour labor cost is also charged, i.e. $130 $/per hour. According to the manufacturer's handbook, the sealing pressure, in other words, vacuum performance is lost when the sealing pressure is below 15 Psi.

- Formulate the quadratic loss function,
- Using his formulation, in case sealing pressure is down to 10 psi, calculate the quality loss in dollars.

Since we would like to have a high level of sealing force/pressure, we can use *"the-larger-the-better"* scenario with the following loss function.

$$L(y) = k\frac{1}{(y)^2}$$

In this problem, Δ_0=15 psi (a pressure limit below which the sealing starts malfunctioning), ***m***=∞ Psi (ideally desired but never attained) and A_0=\$70 (part cost) + \$130 (labor cost) = \$200. In this case, Taguchi quadratic loss function formulation can be written as,

$$k = A_0 {\Delta_0}^2 = \$200 * 15^2 \text{ Psi}^2 = 45000 \,\$ * \text{Psi}^2$$

where k=45000 \$*Psi^2, and L(15 Psi) is,

$$L(y) = 45000 \,\$ * \text{Psi}^2 \frac{1}{(y)^2} = 45000 \frac{\$ * \text{Psi}^2}{(15 \text{ Psi})^2} = \$200$$

where ***y*** is sealing pressure level in Psi. Notice that when *y*=15 Psi, L(15 Psi)=\$200, there will be not any customer complaint. Also note that when *y*=25 Psi, L(15)=\$112.5, which is much better! But never forget that there will always be some design/cost limitations or constraints.

But as for the second part of the question, when sealing pressure is low at *y*=10 Psi, we will get L(y)=\$450 in quality loss. Further note that over the years, due to vacuum usage, the vacuum will degrade beyond its allowed sealing lost limit, the additional level of losses will incur too. For example, the economic life of an automobile is given as ten years at the user manuals, beyond which you should be ready for all possible defects. Yes, the risks are always there for automobile from 0 years to over 10 years! But the probability is not the same!

13.5 Average Quality Loss – The Nominal-the-Best

So far we have considered/derived the quality loss function for a single part or product. But, in real-life, we are more concerned with the average quality loss for many parts/processes. Due to noise factors, the quality characteristics ***y*** of a product does vary from *part* to *part* and manufacturing/application *time* to *time* for a sample of products/processes under some probability distribution characteristics.

Assume ***y*** is the nominal-the-best type of quality characteristic and the target is ***m***, then the average quality loss, Q for $y_i = y_1, y_2, \ldots y_n$ for the large number of "***n sample*** *(representing the population)*" measurements, is given as the average of Q_i/n where $i=1,2\ldots,n$ as,

$$Q = \frac{1}{n}[L(y_1) + L(y_2) + \cdots + L(y_n)]$$ 13.12

$$= \frac{k}{n}[(y_1 - m)^2 + (y_2 - m)^2 + \cdots + (y_n - m)^2]$$

$$= k\left[(\mu - m)^2 + \left(\frac{n-1}{n} \cong \frac{n}{n} \cong 1\right)\sigma^2\right] = \boldsymbol{k[(\mu - m)^2 + \sigma^2]}$$

Note again, sample size, $\boldsymbol{n > 30}$, assuming it is large! Also μ and σ^2 are the mean and variance of $\boldsymbol{y}$, given by,

$$\mu = \frac{1}{n}\sum_{i=1}^{n} y_i \text{ and } \sigma^2 = \frac{1}{n-1}\sum_{i=1}^{n}(y_i - \mu)^2$$

Considering sample size i.e. $n > 30$, then we can use S^2 for σ^2 and $\bar{y}$ for μ, respectively. Then the average quality loss function for n unit is,

$$\boldsymbol{Q = k[(\bar{y} - m)^2 + S^2]}$$ 13.13

If you examine Equation 13.13, the average quality loss is comprised of two components,

i. First loss component , the deviation of the average value of $\boldsymbol{y}$ from the target $\boldsymbol{m}$, is $\boldsymbol{k(\mu - m)^2}$

ii. Second loss component due to the mean squared deviation of $\boldsymbol{y}$ around its own mean is $\boldsymbol{k\sigma^2}$

Of course it is good idea to evaluate the magnitude of these two components. In general first loss is usually easier to control than the second loss because reducing variance is much more difficult.

Example 13.4: Machine Shop Shaft Manufacturing

A small machine shop manufactures very long shafts with a diameter of *Ø =15∓0.05 mm* using high-cost titanium material. The CNC turning machine used to manufacture these shafts is very old and may need to be replaced. But the owner still thinks this machine is capable of manufacturing "*quality parts*". While most of the machined parts pass the quality control, a few parts have been rejected during assembly. A rejected part costs $30 to the manufacturer and a bad reputation as well.

- Using Taguchi quality loss function, prove to the owner of this small machine shop so that purchasing a new CNC turning machine would be of benefit of the machine shop or not.

Some of the recently manufactured shaft diameter dimensions using the old CNC machine is presented in Table 13.1.

TABLE 13.1 Shaft diameter machining data in [*mm*]

14.95	14.98	15.01	14.97	15.05	15.04
14.96	15.06	15.01	14.98	15.02	15.04
15.02	15.03	15.02	15.01	15.03	15.02
15.05	15.05	14.98	15.02	15.04	14.98
15.01	15.01	14.94	15.03	15.02	14.97

In this problem, Δ_0=0.05 mm, m =15 mm and A_o= \$30, and we want the nominal shaft diameter so we use Taguchi quadratic "*the-nominal-the-best*" formulation as,

$$L(y) = \frac{A_0}{{\Delta_0}^2}(y-m)^2 = \frac{\$30}{0.05^2\ \mathbf{mm^2}}(\boldsymbol{y}-15\ \mathbf{mm})^2 \qquad 13.14$$

$$= 12000\frac{\$}{\mathbf{mm^2}}\ (\boldsymbol{y}-15\ \mathbf{mm})^2$$

Where k=12000 \$/$\mathbf{mm^2}$, and $\boldsymbol{y}$ is outside shaft diameter. Remember an average quality loss for n sample shaft (unit) data is calculated using,

$$Q = k\left[(\mu-m)^2 + \left(\frac{n-1}{n}\right)\sigma^2\right]$$

Considering sample size i.e. $n=30$, then use S^2 for σ^2 and $\bar{y}$ for μ. Then, the quality loss function for n unit is,

$$\mu = \bar{y} = \frac{1}{n}\sum_{i=1}^{n} y_i = 15.01\text{ mm and }\sigma^2 = s^2 = \frac{1}{n-1}\sum_{i=1}^{n}(y_i-\mu)^2 = 0.0319\text{ mm}^2$$

$$Q_{ave.} = 12000\left[(15.01-15)^2 + \left(\frac{30-1}{30}\right)0.0319^2\right] = 13.04\ \$$$

Quality loss for $n=30$ shafts manufactured, we calculate, $Q_{ave}*n$=\$13.04 * 30 = \$391.2. By examining Table 13.1 we also noticed that most of the measurements are very well within the tolerances and the sample mean shaft diameter is 15.01 mm.

According to Taguchi quality loss approach, the machine shop is losing around \$391.2 per n=30 parts used in the sample. Consider manufacturing more shafts in serial manner, then the overall cost would be adding up as the part number increases.

We may assert that, the old CNC turning machine's performance is negatively effecting outputs and the machine shop is losing money, at least some of their profits goes away. Looking at closely all the sample data in Table 13.1, two shafts with 14.94 mm and 15.06 mm diameters, are not conforming the tolerance limits. Based upon that the *goal post* approach estimates the quality loss as 2*\$30=\$60, in comparison to that of quadratic loss of \$391.2.

Let's make two more quality loss evaluation using Equation 13.14, for y=15.01 and 15.04 mm (i.e. shaft diameters) as follows;

$$L(\mathbf{15.01}) = 12000 \frac{\$}{\mathbf{mm^2}} (\mathbf{15.01mm} - 15\ \mathbf{mm})^2 = \$1.2$$

$$L(\mathbf{15.04}) = 12000 \frac{\$}{\mathbf{mm^2}} (\mathbf{15.04\ mm} - 15\ \mathbf{mm})^2 = \$19.2$$

Also by looking at the above quality loss calculations (i.e. see Figure 13.4, as well). We can quickly see that, when m=15 mm, the quality loss is none. At a mean value of $\mu=y$=15.01 mm, the quality loss for this part is \$1.2. When we got the higher variation of y=15.04 mm, then the quality loss is \$19.2 where as you noticed, the change in the y value from 15.01 mm to 15.04 mm, results in almost \$19.2/\$1.2=16 times higher cost.

We can obtain the same "*results*" using the corresponding formulations in *Excel*. In the following *Excel* screen shot, the first part shows the Q_{ave} calculations using the formulation. In the second part, quality loss is obtained by summing up all individual quality loss results for obtaining an average quality loss of Q_{ave}. Thus, we can verify the Q_{ave} formulation results (i.e. left side) using "*exact calculations*" and (i.e. right side) *Excel*.

	A	B	C	D	E	F	G	H	I	J	K	L	M	N
1		**Measured Shaft Diameters**							$L(y_{i1})$	$L(y_{i2})$	$L(y_{i3})$	$L(y_{i4})$	$L(y_{i5})$	$L(y_{i6})$
2	14.95	14.98	15.01	14.97	15.05	15.04		*i=1*	30	4.8	1.2	10.8	30	19.2
3	14.96	15.06	15.01	14.98	15.02	15.04		.	19.2	43.2	1.2	4.8	4.8	19.2
4	15.02	15.03	15.02	15.01	15.03	15.02		.	4.8	10.8	4.8	1.2	10.8	4.8
5	15.05	15.05	14.98	15.02	15.04	14.98		.	30	30	4.8	4.8	19.2	4.8
6	15.01	15.01	14.94	15.03	15.02	14.97		*i=5*	1.2	1.2	43.2	10.8	4.8	10.8
7	**μ =**	**15.01**	mm		**n=**	**30**								
8	**σ =**	**0.0319**	mm							**Q_{ave}= 13.04**	$			
9										**Q_{ave}*n= 391.2**	$			
10	**Δ_0=**	**0.05**	mm											
11	**A_0 =**	**30**	$						**"Exact Calculations"**					
12	**m =**	**15**	mm											
13									$Q = \frac{1}{n}[L(y_1) + L(y_2) + \cdots + L(y_n)]$					
14	**k=**	**12000**	$/mm²	**"Overall Formulation"**										
15	**Q_{ave}=**	**13.04**	$	$Q = k\left[(\mu - m)^2 + \left(\frac{n-1}{n}\right)\sigma^2\right]$										
16	**Q_{ave}*n=**	**391.2**	$											

Following, the *Excel* commands used above are also presented in the following;

	EXCEL FORMULA
μ	=AVERAGE(A2:F6)
σ	=STDEV.S(A2:F6)
n	=COUNT(A2:F6)
k	=B11/B10^2
Q_{ave}	=B14*(((F7-1)/F7)*B8^2+(B7-B12)^2)
Q_{ave}*n	=B15*F7
$L(y_{i1})$	=B$14*(A2-B$12)^2
Q_{ave}	=SUM(I2:N6)/F7
Q_{ave}*n	=J8*F7

Example 13.5: Goal Post vs. Quadratic Loss Comparison

Y characteristic for a sample process is considered for "*the-nominal-the-best*" quality loss case. In this problem quality loss function parameters are given as target, ***m*=5**, A_0= 0.07646 and Δ_o=2, disregarding the units. Then we find; LSL= ***m*** - Δ_o =3 and USL= ***m*** + Δ_o =7, respectively.

You are asked to consider normal probability distribution (μ=5 and σ=1.15) with Taguchi quadratic formulation together. Using the Taguchi quality loss function and normal distribution characteristics,

- Obtain *Yield* and *Loss* (i.e. in percentages) for "goal post" function ,
- Obtain the same results for "Taguchi quadratic loss" function

As far as the "goal post" formulation is considered. we are required to obtain shaded area at both sides of the normal distribution divided by both LSL and USL as presented in Figure 13.9. This area can easily be calculated,

- First, by subtracting $P(Y \leq LSL=3)$ area from $P(Y \leq USL=7)$ area under the normal curve, we can find YIELD area, which is shown as "*not shaded*" under the normal curve.
- Then, since the area under normal curve is 1, subtract YIELD and find LOSS, corresponding to "shaded area".

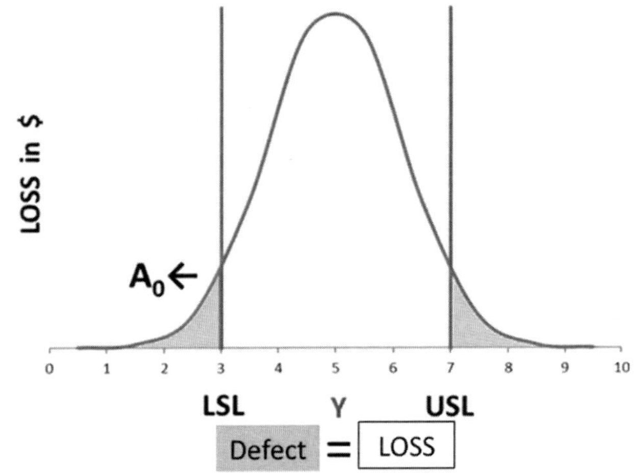

FIGURE 13.9 Calculations of goal post loss

YIELD = P(Y≤7) - P(Y≤3) = 0.958994 – 0.041006 = 0.918

Using normal probability distribution with mean, $\mu=5$ and standard deviation, $\sigma=1.15$. **YIELD = %91.8 and DEFECT = LOSS =%100 - %91.8 = %8.2, which means %8.2 of the produced parts are defective according to "goal post" approach.**

We can obtain the corresponding percentages when Taguchi quadratic loss function and normal distribution are overlapped together, where normal distribution shows the product's manufacturing distribution characteristics with respect to the this product's quadratic loss function as shown in Figure 13.10.

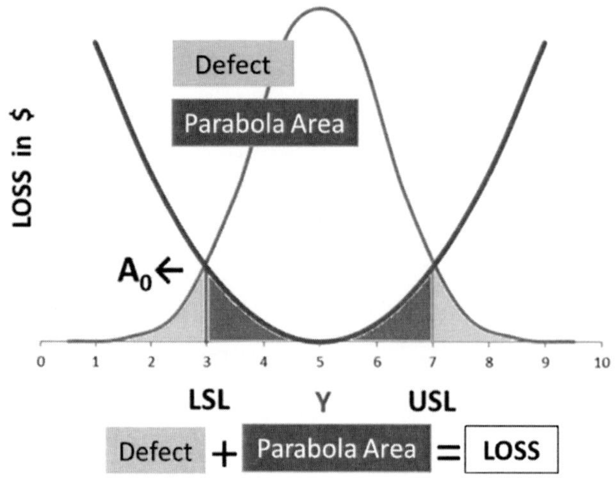

FIGURE 13.10 Calculations of quadratic loss

By examining Figure 13.10, we notice that both curves are situated symmetrically around $\mu = m = 5$. Also LSL and USL lines for this problem pass through at the both curves' intersection where maximum quality loss is given as A_o.

In Figure 13.10, the summation of shaded areas give us LOSS in terms of percentages. First, let's find out quality loss function for "*the-nominal-the-best*" case.

$$L(y) = \frac{A_0}{\Delta_0^{\ 2}}(\boldsymbol{y} - m)^2 = \frac{\mathbf{0.07646}}{2^2}(\boldsymbol{y} - 5)^2 = 0.019115\ (\boldsymbol{y} - 5)^2$$

In addition, the area (also part of the LOSS) under the quadratic loss curve between LSL and USL can be found by integrating the loss function between these limits. Let's name this area as "Parabola Area" which is obtained as,

$$Parabola\ Area = \int_{LSL=3}^{USL=7} 0.019115\ (\boldsymbol{y} - 5)^2\, dy = 0.019115 \left[\frac{(\boldsymbol{y} - 5)^3}{3}\right]_3^7$$

$$Parabola\ Area = 0.019115 \left[\frac{(\mathbf{7} - 5)^3}{3} - \frac{(\mathbf{3} - 5)^3}{3}\right] = 0.1019$$

and, the same *defect* percentage that was calculated in *goal post* based calculation using normal distribution can also be used here,

Defect = 1-YIELD = 1 - 0.918 = 0.082

Finally, for Taguchi quadratic loss approach, we obtain the LOSS=0.082+0.1019=0.1839 and in percentage; LOSS=%18.4 and YIELD=81.6, respectively.

	A	B	C
1			
2	**Y Characteristic Constants**		
3	**Mean=**	**5**	
4	**St. Deviation =**	**1.15**	**EXCEL FORMULA**
5	**Target, m=**	5	
6	A_0=	0.07646	=NORMDIST(B9,B$3,B$4,FALSE)
7	Δ_0 =	2	
8	k =	0.0191	=B6/B7^2
9	**LSL= m - Δ_0 =**	**3**	=B5-B7
10	**USL= m + Δ_0 =**	**7**	=B5+B7
11			
12	**Goal Post Loss**		
13	**LSL N. Curve Area =**	0.041006	=NORMDIST(B9,B3,B4,TRUE)
14	**USL N. Curve Area =**	0.958994	=NORMDIST(B10,B3,B4,TRUE)
15	**Goal Post YIELD =**	**91.8%**	=(B14-B13)
16	**Goal Post LOSS =**	**8.2%**	=(1-B15)
17			
18	**Taguchi Quadratic Loss**		
19	**Defect =**	0.0820	=2*B13
20	**Parabola Area =**	0.1019	=B8/3*((B10-B5)^3-(B9-B5)^3)
21	**YIELD =**	**81.6%**	=1-B22
22	**LOSS =**	**18.4%**	=(B19+B20)

All of these calculations are also performed and summarized as shown on *Excel* sheet above. As you may notice, when the goal post LOSS=8.2% vs. Taguchi quadratic LOSS=18.4% are compared, the quadratic loss is twice more than that of goal post loss. This basically says Taguchi quadratic loss is more severe in penalizing the variations, which forces engineers to take further precautions to lower the variations thus increasing the product quality and customer satisfaction.

13.6 Average Quality Loss – The Smaller-the-Better

Similar formulation (derivation) approach applies to this case as well, as demonstrated in the previous section, and we can find the average quality loss for the-smaller-the-better case for a sample size of n, as;

$$Q = \frac{1}{n}[L(y_1) + L(y_2) + \cdots + L(y_n)] = \frac{1}{n}k\sum_{i=1}^{n} y_i^2 = k[\sigma^2 + \mu^2] \qquad 13.15$$

Remember that the target ***m=0***, and S and μ are given for the population. When it comes to the sample, $\bar{y}$ is the deviation of the mean from ***zero*** and S^2 is variance. Then, using S^2 for σ^2 and $\bar{y}$ for μ. The quality loss function for *n* unit is now,

$$\boldsymbol{Q} = \boldsymbol{k}\left[\boldsymbol{S}^2 + \bar{\boldsymbol{y}}^2\right] \qquad 13.16$$

13.7 Average Quality Loss – The Larger-the-Better

Similarly, we can find the average quality loss for the-larger-the-better case for sample size of *n*, as;

$$Q = \frac{1}{n}[L(y_1) + L(y_2) + \cdots + L(y_n)] = \frac{1}{n}k\sum_{i=1}^{n}\frac{1}{y_i^2} = k\left(\frac{1}{\mu^2}\right)\left(1 + 3\frac{\sigma^2}{\mu^2}\right) \qquad 13.17$$

Remember target ***m***=∞ ***(infinite)*** for this case and S and μ are given for population. When it comes to the sample, as $\bar{y}$ is the mean and S^2 is variance. Then, using S^2 for σ^2 and $\bar{y}$ for μ. The quality loss function for *n* unit is now,

$$\boldsymbol{Q} = k\left(\frac{1}{\bar{y}^2}\right)\left(1 + 3\frac{S^2}{\bar{y}^2}\right) \qquad 13.18$$

Example 13.6: Railway Road Tie Bolts Purchase

Statewide railway company needs some large number of highly reliable railroad tie bolts to use in their new construction site. If these bolts fail, a repair cost is estimated to be \$20 including labor and part cost. Two different price quote for bolts are received from two different vendor companies i.e. A and B.

Buyer randomly selects n=30 bolt samples out of each company's products and test them internally. Ultimate tensile strength (UTS) is chosen as the testing criterion in MPa units as presented in Table 13.2. The lowest specification level for UTS is required to be at least 300 MPa and this time the total quantity of n=50,000 bolts to be purchased. The bolt unit cost per products A and B are to be proposed as \$0.60 and \$0.50, respectively. As a consulting engineer, your views/assessments are asked,

- Advise the railway company about which bolt to buy, A or B!

TABLE 13.2 Rail road tie bolts A and B - strength in [MPa]

BOLT TYPE - A						BOLT TYPE - B					
321	332	322	324	340	364	325	331	311	305	309	304
335	312	333	355	322	355	322	311	307	320	306	327
322	329	330	311	339	327	308	309	316	310	309	305
315	327	325	322	344	342	302	309	301	303	310	305
359	322	312	340	346	310	311	305	310	308	312	331

In this problem, as you will see, Δ_0=300 MPa, **m**=∞ and A_0= \$20, and we want the highest level bolt strength for our objective, so we use Taguchi quadratic "*the-larger-the-better*" formulation as,

$$L(y) = \frac{A_0{\Delta_0}^2}{\boldsymbol{y}^2} = \frac{\$20 * 300^2}{\boldsymbol{y}^2}\frac{\mathbf{MPa}^2}{\mathbf{MPa}^2} \qquad 2.19$$

Where k=18E05 \$ **MPa**^{2}, and $\boldsymbol{y}$ is bolt UTS strength. Remember an average quality loss for n bolt-sample is calculated using,

$$Q = k\left(\frac{1}{\mu^2}\right)\left(1 + 3\frac{\sigma^2}{\mu^2}\right)$$

Considering the sample size i.e. n=30, then use S^2 for σ^2 and $\bar{y}$ for μ. Then, the quality loss function for n unit is,

BOLT-A:

$$\mu = \bar{y} = \frac{1}{n}\sum_{i=1}^{n} y_i = 331.23 \text{ MPa and } \sigma^2 = s^2 = \frac{1}{n-1}\sum_{i=1}^{n}(y_i - \mu)^2 = 207.71 \text{ MPa}^2$$

BOLT-B:

$$\mu = \bar{y} = \frac{1}{n}\sum_{i=1}^{n} y_i = 311.40 \text{ MPa and } \sigma^2 = s^2 = \frac{1}{n-1}\sum_{i=1}^{n}(y_i - \mu)^2 = 66.37 \text{ MPa}^2$$

$$\mathbf{Q_{ave.-A}} = k\left(\frac{1}{\mu^2}\right)\left(1+3\frac{\sigma^2}{\mu^2}\right) = 18E05 * \left(\frac{1}{331.23^2}\right)\left(1+3\frac{14.41^2}{331.23^2}\right) = \$16.50$$

$$\mathbf{Q_{ave.-B}} = k\left(\frac{1}{\mu^2}\right)\left(1+3\frac{\sigma^2}{\mu^2}\right) = 18E05 * \left(\frac{1}{311.40^2}\right)\left(1+3\frac{8.15^2}{311.40^2}\right) = \$18.60$$

For *n= 50,000* bolts purchased, purchase costs; Cost-A=$0.6 * 50,000 = $30,000 and Cost-B=$0.5 * 50,000 = $25,000. We have also noticed that the strength measurements are within the tolerances and further the mean bolt strength of 300 MPa is obviously, the better!

According to Taguchi quality loss approach both Bolt A and B, conforms the buyer requirements in means. If Bolt-A is purchased, around $30000-$25000=$5000 per n=50,000 bolts are overpaid, but in terms of quality Taguchi loss savings 50,000*(18.60-16.50)=$105,000 is acquired in potential failure cost covered.

We can obtain the same "*above results*" using all formulations presented in *Excel*. In the following, the first part shows the same calculations presented above. In the second part quality loss is summed for all data and the formulation results for an average quality loss of Q_{ave} is also obtained/verified using directly all corresponding real calculations.

Since similar *Excel* commands are presented in the previous examples, they are not repeated in this example. Formulations will differ only for Q_{ave} and *L(y)* to be consistent with this "*the-larger-the-better*" example.

Measured Shaft Diameters

321	332	322	324	340	364
335	312	333	355	322	355
322	329	330	311	339	327
315	327	325	322	344	342
359	322	312	340	346	310

$\mu =$ **331.23** MPa   n= **30**

$\sigma =$ **14.41** MPa

$\Delta_0 =$ **300** MPa

$A_0 =$ **20** \$

m = ∞ Mpa

$$Q = k\left(\frac{1}{\mu^2}\right)\left(1 + 3\frac{\sigma^2}{\mu^2}\right)$$

k= **1800000** \$/MPa **"Overall Formulation"**

Q_{ave}= **16.50** \$

	$L(y_{i1})$	$L(y_{i2})$	$L(y_{i3})$	$L(y_{i4})$	$L(y_{i5})$	$L(y_{i6})$
i=1	17.47	16.33	17.36	17.15	15.57	13.59
.	16.04	18.49	16.23	14.28	17.36	14.28
.	17.36	16.63	16.53	18.61	15.66	16.83
.	18.14	16.83	17.04	17.36	15.21	15.39
i=5	13.97	17.36	18.49	15.57	15.04	18.73

Q_{ave}= **16.50** \$

"Exact Calculations"

$$Q = \frac{1}{n}[L(y_1) + L(y_2) + \cdots + L(y_n)]$$

Measured Shaft Diameters

325	331	311	305	309	304
322	311	307	320	306	327
308	309	316	310	309	305
302	309	301	303	310	305
311	305	310	308	312	331

$\mu =$ **311.40** MPa   n= **30**

$\sigma =$ **8.15** MPa

$\Delta_0 =$ **300** MPa

$A_0 =$ **20** \$

m = ∞ Mpa

$$Q = k\left(\frac{1}{\mu^2}\right)\left(1 + 3\frac{\sigma^2}{\mu^2}\right)$$

k= **1800000** \$/MPa **"Overall Formulation"**

Q_{ave}= **18.60** \$

	$L(y_{i1})$	$L(y_{i2})$	$L(y_{i3})$	$L(y_{i4})$	$L(y_{i5})$	$L(y_{i6})$
i=1	17.04	16.43	18.61	19.35	18.85	19.48
.	17.36	18.61	19.10	17.58	19.22	16.83
.	18.97	18.85	18.03	18.73	18.85	19.35
.	19.74	18.85	19.87	19.61	18.73	19.35
i=5	18.61	19.35	18.73	18.97	18.49	16.43

Q_{ave}= **18.60** \$

"Exact Calculations"

$$Q = \frac{1}{n}[L(y_1) + L(y_2) + \cdots + L(y_n)]$$

Before we end this chapter, we can go ahead and summarize all these there formulations and drawings as presented in a Table 13.3.

TABLE 13.3 Summary - Quadratic loss function types

The Nominal-the-best *"Loss for an individual Part"*	**The Smaller-the-better** *"Loss for an individual Part"*	**The Larger-the-better** *"Loss for an individual Part"*
$k = \frac{A_0}{\Delta^2}$	$k = \frac{A_0}{\Delta^2}$	$k = A_0\Delta^2$
$L(y) = k(\mathbf{y} - m)^2$	$L(y) = k(\mathbf{y}^2)$	$L(y) = k(\frac{1}{\mathbf{y}^2})$
Average Quality Loss(Sym.) *"Average loss per part in a distribution"*	**Average Quality Loss** *"Average loss per part in a distribution"*	**Average Quality Loss** *"Average loss per part in a distribution"*
$\boldsymbol{Q = k\left[(\bar{y} - m)^2 + S^2\right]}$	$Q = k\left[S^2 + \bar{y}^2\right]$	$\boldsymbol{Q = k\left(\frac{1}{\bar{y}^2}\right)\left(1 + 3\frac{S^2}{\bar{y}^2}\right)}$
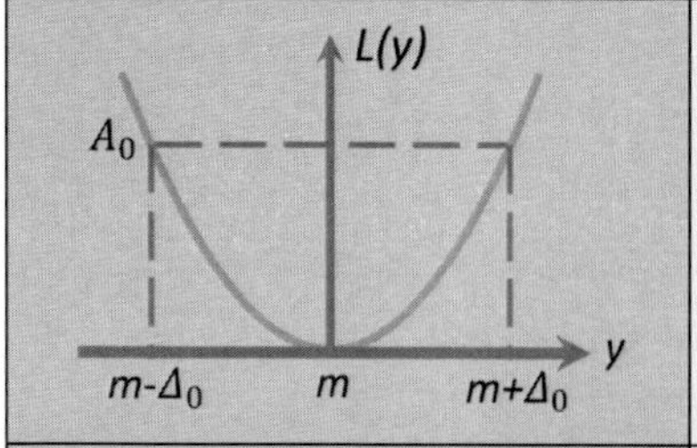	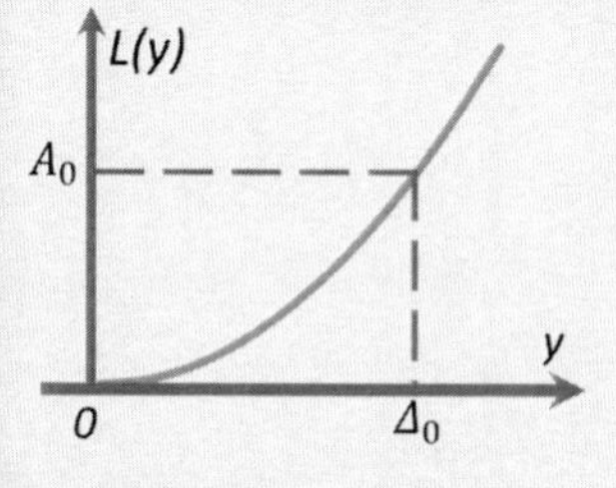	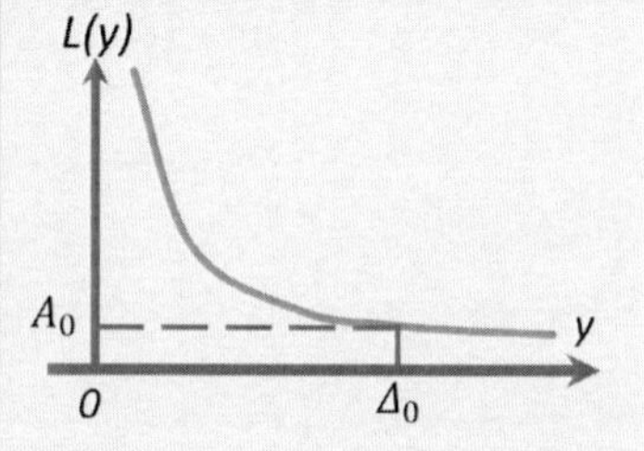
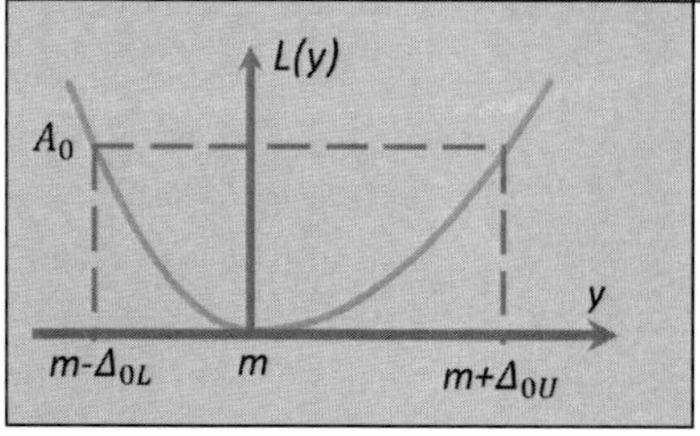		

13.8 Exercises

1) Compare "goal post" vs. "Taguchi quadratic loss function" in terms of their outcomes i.e. the capability of meeting quality in the product or process design. State their strengths and weaknesses by considering their usage and capabilities.

2) Considering sample size i.e. n>30, then use S^2 for σ^2 and $\bar{y}$ for μ. Then, the quality loss function (the-nominal-the-better case) for *n* unit is given by;

$$\boldsymbol{Q = k\left[(\bar{y} - m)^2 + s^2\right]}$$

Here, the average quality loss is comprised of two components,

i. First loss component, the deviation of the average value of ***y*** from the target ***m***, is $\boldsymbol{k(\mu - m)^2}$
ii. Second loss component due to the mean squared deviation of ***y*** around its own mean is $\boldsymbol{k\sigma^2}$

Why, in this formulation, reducing the contribution of the second loss component is much more difficult than reducing the contribution of first loss component? Please explain your answer after revising the definition of $\boldsymbol{S^2}$.

3) ABC, a very famous car battery manufacturer, produces small size batteries for mid-size cars. Using field and customer survey reports, it is observed that this type of car battery lasts on average, 4 years, with an acceptable deviation of ±1 years, depending on the battery/usage.

The battery lives are assumed to be normally distributed and warranted for 4 years. Find the quality loss associated with these batteries when the customer complaints and returns are taking place at 3 years of usage for not meeting quality characteristics (i.e. dead battery). Battery replacement cost is $50 for part and labor cost is $30 dollars for 20 minutes of battery replacement service.

4) A sphere steel balls in a ball bearing have their diameter specifications to be 10±0.01 mm. Any sphere part falling outside of these specifications will be rejected since these ball bearings require a high precision level of rotating balance.

The customer returns (not repairs, because it is not possible after assembly) happen when the ball bearing performance is low (noise, excessive heating, imbalance etc.) at a replacement cost of $100. Formulate the loss function.

5) Statewide trucking company needs to make a decision about the purchase of 10,000 truck tires which are expected to be durable and dependable in every season. If any tire wears out, a replacement cost is around $500 per tire including tire, tire change labor and service time lost cost due to time delay.

Two different price-quote for the tires are received from two different vendor companies i.e. A and B. Buyer randomly selects n=30 tires out of each company's products and test them internally for some time.

Tire (wear) life is chosen as the testing criterion in "*months*" and corresponding tire life data are as presented in Table 13.4. The higher the tire life, the better is for the company. Each tire is required to last around 12 months. The tire unit cost per products A and B are to be proposed as $160 and $200, respectively. As an consulting engineer your views/assessments are asked,

- Advise the trucking company about which brand tire to buy, A or B!

TABLE 13.4 Tire lasting data for Type A and B - Life in [*Months*]

TIRE TYPE - A						TIRE TYPE - B					
12	12	10	11	12	11	12	14	12	10	12	11
9	9	9	10	11	15	13	13	13	11	12	12
11	12	9	12	11	11	13	15	13	12	13	12
12	11	12	13	13	10	12	16	12	12	13	13
12	11	9	11	11	9	11	15	11	15	14	14

Chapter 14

Total Quality Management

"Total Quality Management, TQM is a management philosophy that seeks to integrate all organizational functions (marketing, finance, design, engineering, and production, customer service, etc.) to focus on meeting customer needs and organizational objectives." Khurram Hashmi (2007).

Management term is so popular and commonly used, yet when it comes to displaying the true spirit of management and its affections, not so many people are fully aware and can acknowledge its true meaning and commitment. Now let's try to define what is *management?*. Management is true ***leadership*** and defined with mainly three qualities as; (i.e. depicted in Figure 14.1)

- ✓ the *art* when it comes to dealing with people or situations
- ✓ the *act* when it comes to organizing, make planning and taking decisions
- ✓ the *way* of directing

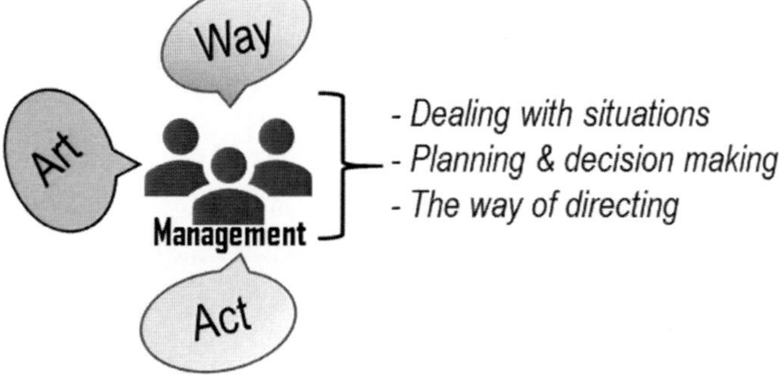

FIGURE 14.1 Dimensions of management with expected outcomes

Why? All these three items are means to achieve expected goals/objectives of the company, as far as the *white collared* management people are involved. Management people generally take up nice office spaces with a very clean office environment. But often they get bored with meetings, they mostly sit during work hours with so many phone calls and customer meetings. Yes, they are part of the TQM.

Because, one more key factor when considering quality improvement is involvement, appreciation, and commitment of the management people in this process. If these management people are not well versed on quality and did not fully understand its importance, then investment in time/men/money for quality may not find support and advocates. This is certainly a major setback for a company.

14.1 TQM Basic Concepts

Total Quality Management (TQM), constitutes organizational skills to improve companies in-house processes and products towards providing quality, thus improving ***customer satisfaction***. Modern TQM management skills are crucial in managing modern quality philosophy. Under this managemental system, quality is an overall commitment and not to be attributed to the only certain departments i.e. mostly to the inspection, to some product and process design departments.

"From cradle to the grave" concept, it is a modern culture creating awareness that ***employee involvement***, i.e. each individual of the company certainly has some say/contribution in yielding good products in meeting the customer needs. If employees at various levels in the company believe, top-level leaders have the appreciation and commitment in the company's vision for outputting high-quality products/processes, then they respect to upper-level demands for enforcing quality and take it more seriously. Eventually, rear tires do follow front tires.

Similar to *6-Sigma* philosophy and adoption/training/popularity in all GE businesses, all upper-level management is to incorporate and plan for "*quality methodology*" all in their present/future activities. The top-down hierarchy must also be feedbacked with bottom-up views, when necessary.

Quality decisions/actions should be constructed on a valid/proven systematic (statistical) approaches/fundamentals, and not to be left to the individuals or groups relative incentives. Keep in mind that for any overall evaluation/analysis quality must be measured in the form of ***performance measures***, i.e. quantitively, if not qualitatively. Measuring whatever needed has key importance to monitor the performance of interest.

Employers are companies' assets/fortune, not root-cause of the problems, so all levels of employees should be trained with quality realization and techniques. Multidisciplinary teams of employees should be brought together and encouraged towards the instituting the "*quality understanding*" in the company.

Any corrective action must concentrate on the cure of the problems for a permanent fix, not for treatment of symptoms for a temporary fix. Successful and committing employees must be appreciated and rewarded. Of course, quality requires ***continuous process improvement***, and a long-time investment and culture, and not to be earned overnight.

Further, ***Supplier partnership*** provides continuing/uninterrupted support for TQM, having loyal suppliers and long term partnerships are always preferred in controlling either outsourcing some jobs or in supplier support. Long run mutual trust is very valuable in hostile market environments. Figure 14.2 depicts all TQM basic concepts presented above together, we can list ***TQM basic concepts*** as itemized below.

1. Leadership
2. Customer satisfaction
3. Employee involvement
4. Performance measures
5. Continuous process improvement
6. Supplier partnership

FIGURE 14.2 TQM concepts

14.2 Underdoing/Overdoing of TQM

However, quality has to be fed to into the organization naturally in spirit and material manner. Not unnaturally, as show-off/dictated, i.e. "*do today, give up tomorrow*" or as "*come and go desires*". Nevertheless, overdoing in training/meetings should also be avoided; eventually, TQM systems must be applied to be humanly, i.e.

- Long meeting hours, with many meaningless talks and greed.
- Sheer eagerness for being in the front row/statue/salary raise!
- Unnecessary social events or training, like child play
- Favoritism among workers, taking care of some people over the remaining in an unjust manner
- Unfair responsibility-sharing etc.

Such things may reduce the belief/usefulness in the eye of the employers, once they are caught up bored/overloaded/overburdened in the whole total quality management process. If done properly, an optimization can prevent *waste*, do the things just as needed! no less! no more, at the right time, at the right place! Of course, these remarks are easy to say but difficult to do and apply!

Think of medicine, it is always prescribed according to age, gender, need, health, and condition of the patient for good intention and cause. Measuring and prescribing the exact amount is the way to go. If given little amount, you may not recover or it takes longer to recover. If given more than it should be, it damages your body.

14.3 Limitations of TQM

TQM has its limitations as *cost in time and money* associated with risks involved. Risks, most of the management people' nightmare. Therefore, many of them generally do not want to take risks. Each new thing comes with its unknowns and cross-influential effects. These unknowns maybe better or may be not! But risk takers are not much! *Fear of change* always occupy people' mind. Traditional people, old people, short-sided people do not like changes. Because change may bring risks with it. Risk can cause lost! Loss of seat, loss of revenue, loss of thrust, loss of profit, loss of the company. At the same time, risk can bring along good new things with them. To acquire success, management people should be willing to facing the risk to perform better than their competitors.

In mid-'90s, outsourcing for big companies like GE was very promising, since it is very cheap and clock hours around the world were being utilized. China and India were looking promising for outsourcing. GE was one of the pioneers in this sense i.e. was willing to try this concept first. The other big companies were watching GE, and wondering how GE would perform? Since they cannot take the risk of outsourcing by themselves in the first place. Once GE was successful in outsourcing in these countries, the other players, of course, by-delay, did established outsourcing. Call centers, manufacturing units, design centers etc. were all outsourced with the huge cost of savings. But of course, outsourcing came with its problems to deal with. But since mid-90's it is so popular/advantageous, many small/mid/big scale companies keep using outsourcing today.

14.4 Exercises

1) Describe TQM using your own terms.

2) Assume, you are one member of upper-level management in a company, how would you interact with employees? What are your criterions? Redline ? Please discuss in a few paragraphs.

3) Assume you had a company of 100 people and your company is manufacturing door locks. How do you plan TQM actions?

Appendix

Appendix 1. *Excel* Commands for Statistics

Excel Command	Description
=COUNT(A2:A7)	Counts the number of cells A2 through A7
=SUM(A2:A7)	Adds/sums values of cells
=AVERAGE(A2:A7)	Averages the numbers of cells
=MEDIAN(A2:A7)	Returns the median of the given numbers
=MODE.SNGL(A2:A7)	Returns the mode of the given numbers
=MAX(A2:A7)	Returns the maximum of the given numbers
=MIN(A2:A7)	Returns the minimum of the given numbers
=VAR.S(A2:A7)	Calculates sample variance in given cells
=STDEVA(A2:A7)	Estimates standard deviation of a sample in given cells
=KURT(A2:A7)	Returns the kurtosis of a data
=SKEW(A2:A7)	Returns the skew of a data
=CONFIDENCE.T(α,Standard_dev, Size)	Returns the confidence interval for a population mean (Standard_dev), for α (level of confidence) and n (size/count of data) using a Student's t distribution.
=CONFIDENCE.NORM(α,Standard_dev, Size)	Returns the confidence interval for a population mean (Standard_dev), for α (level of confidence) and n (size/count of data) using a normal distribution

=NORMDIST(x,mean,standard_dev, cumulative)	Returns the normal distribution for the specified mean and standard deviation for x value where cumulative term is either TRUE or FALSE – If cumulative is TRUE, NORMDIST returns the cumulative distribution function; i.e. the normal distribution formula is the integral from negative infinity to x of the given formula – if FALSE, it returns the probability mass/density function i.e. $f(x,\mu,\sigma)$
=NORM.INV(probability,mean,standard_dev)	Returns the inverse of the normal cumulative distribution for the specified mean and standard deviation at probability value
=NORM.S.DIST(z,cumulative)	Returns the standard normal distribution (has a mean of zero and a standard deviation of one). Where cumulative term is either TRUE or FALSE – If cumulative is TRUE, NORMDIST returns the Normal cumulative distribution function – if FALSE, it returns Normal probability distribution function, i.e. $f(z,\mu=0,\sigma=1)$
=TINV(probability,deg_freedom)	Returns the two-tailed inverse of the Student's t-distribution using α and dof. For example: =TINV(α,n_1+n_2-2)
=TDIST(x,deg_freedom,tails)	Returns the p-probability for the Student t-distribution for dof and tails at x(i.e. T_o) value. For example: =TDIST(T_o,n_1+n_2-2,2)

$$P(Z \leq z) = \Phi(z) = \int_{-\infty}^{z} \frac{1}{\sqrt{2\pi}} e^{-\frac{1}{2}y^2} dy$$

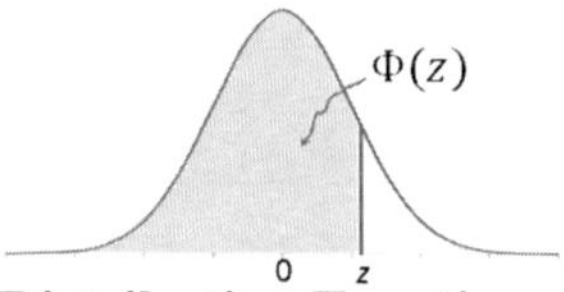

Appendix 2: Z-Table for Cumulative Standard Normal Distribution Function

z	0	0.01	0.02	0.03	0.04	0.05	0.06	0.07	0.08	0.09
0	0.5	0.503989	0.507978	0.511966	0.515953	0.519939	0.523922	0.527903	0.531881	0.535856
0.1	0.539828	0.543795	0.547758	0.551717	0.55567	0.559618	0.563559	0.567495	0.571424	0.575345
0.2	0.57926	0.583166	0.587064	0.590954	0.594835	0.598706	0.602568	0.60642	0.610261	0.614092
0.3	0.617911	0.62172	0.625516	0.6293	0.633072	0.636831	0.640576	0.644309	0.648027	0.651732
0.4	0.655422	0.659097	0.662757	0.666402	0.670031	0.673645	0.677242	0.680822	0.684386	0.687933
0.5	0.691462	0.694974	0.698468	0.701944	0.705401	0.70884	0.71226	0.715661	0.719043	0.722405
0.6	0.725747	0.729069	0.732371	0.735653	0.738914	0.742154	0.745373	0.748571	0.751748	0.754903
0.7	0.758036	0.761148	0.764238	0.767305	0.77035	0.773373	0.776373	0.77935	0.782305	0.785236
0.8	0.788145	0.79103	0.793892	0.796731	0.799546	0.802337	0.805105	0.80785	0.81057	0.813267
0.9	0.81594	0.818589	0.821214	0.823814	0.826391	0.828944	0.831472	0.833977	0.836457	0.838913
1	0.841345	0.843752	0.846136	0.848495	0.85083	0.853141	0.855428	0.85769	0.859929	0.862143
1.1	0.864334	0.8665	0.868643	0.870762	0.872857	0.874928	0.876976	0.879	0.881	0.882977
1.2	0.88493	0.886861	0.888768	0.890651	0.892512	0.89435	0.896165	0.897958	0.899727	0.901475
1.3	0.9032	0.904902	0.906582	0.908241	0.909877	0.911492	0.913085	0.914657	0.916207	0.917736
1.4	0.919243	0.92073	0.922196	0.923641	0.925066	0.926471	0.927855	0.929219	0.930563	0.931888
1.5	0.933193	0.934478	0.935745	0.936992	0.93822	0.939429	0.94062	0.941792	0.942947	0.944083
1.6	0.945201	0.946301	0.947384	0.948449	0.949497	0.950529	0.951543	0.95254	0.953521	0.954486
1.7	0.955435	0.956367	0.957284	0.958185	0.95907	0.959941	0.960796	0.961636	0.962462	0.963273
1.8	0.96407	0.964852	0.96562	0.966375	0.967116	0.967843	0.968557	0.969258	0.969946	0.970621
1.9	0.971283	0.971933	0.972571	0.973197	0.97381	0.974412	0.975002	0.975581	0.976148	0.976705
2	0.97725	0.977784	0.978308	0.978822	0.979325	0.979818	0.980301	0.980774	0.981237	0.981691
2.1	0.982136	0.982571	0.982997	0.983414	0.983823	0.984222	0.984614	0.984997	0.985371	0.985738
2.2	0.986097	0.986447	0.986791	0.987126	0.987455	0.987776	0.988089	0.988396	0.988696	0.988989
2.3	0.989276	0.989556	0.98983	0.990097	0.990358	0.990613	0.990863	0.991106	0.991344	0.991576
2.4	0.991802	0.992024	0.99224	0.992451	0.992656	0.992857	0.993053	0.993244	0.993431	0.993613
2.5	0.99379	0.993963	0.994132	0.994297	0.994457	0.994614	0.994766	0.994915	0.99506	0.995201
2.6	0.995339	0.995473	0.995604	0.995731	0.995855	0.995975	0.996093	0.996207	0.996319	0.996427
2.7	0.996533	0.996636	0.996736	0.996833	0.996928	0.99702	0.99711	0.997197	0.997282	0.997365
2.8	0.997445	0.997523	0.997599	0.997673	0.997744	0.997814	0.997882	0.997948	0.998012	0.998074
2.9	0.998134	0.998193	0.99825	0.998305	0.998359	0.998411	0.998462	0.998511	0.998559	0.998605
3	0.99865	0.998694	0.998736	0.998777	0.998817	0.998856	0.998893	0.99893	0.998965	0.998999
3.1	0.999032	0.999065	0.999096	0.999126	0.999155	0.999184	0.999211	0.999238	0.999264	0.999289
3.2	0.999313	0.999336	0.999359	0.999381	0.999402	0.999423	0.999443	0.999462	0.999481	0.999499
3.3	0.999517	0.999534	0.99955	0.999566	0.999581	0.999596	0.99961	0.999624	0.999638	0.999651
3.4	0.999663	0.999675	0.999687	0.999698	0.999709	0.99972	0.99973	0.99974	0.999749	0.999758
3.5	0.999767	0.999776	0.999784	0.999792	0.9998	0.999807	0.999815	0.999822	0.999828	0.999835
3.6	0.999841	0.999847	0.999853	0.999858	0.999864	0.999869	0.999874	0.999879	0.999883	0.999888
3.7	0.999892	0.999896	0.9999	0.999904	0.999908	0.999912	0.999915	0.999918	0.999922	0.999925
3.8	0.999928	0.999931	0.999933	0.999936	0.999938	0.999941	0.999943	0.999946	0.999948	0.99995
3.9	0.999952	0.999954	0.999956	0.999958	0.999959	0.999961	0.999963	0.999964	0.999966	0.999967

$$P(Z \leq z) = \Phi(z) = \int_{-\infty}^{z} \frac{1}{\sqrt{2\pi}} e^{-\frac{1}{2}y^2} dy$$

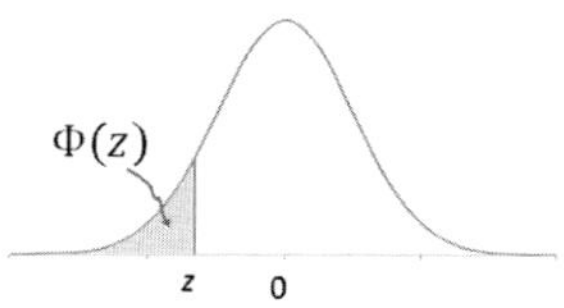

Appendix 2: Z-Table for Cumulative Standard Normal Distribution Function

z	-0.09	-0.08	-0.07	-0.06	-0.05	-0.04	-0.03	-0.02	-0.01	0
-3.9	0.000033	0.000034	0.000036	0.000037	0.000039	0.000041	0.000042	0.000044	0.000046	0.000048
-3.8	0.000050	0.000052	0.000054	0.000057	0.000059	0.000062	0.000064	0.000067	0.000069	0.000072
-3.7	7.53E-05	7.84E-05	8.16E-05	8.5E-05	8.84E-05	9.2E-05	9.57E-05	9.96E-05	0.000104	0.000108
-3.6	0.000112	0.000117	0.000121	0.000126	0.000131	0.000136	0.000142	0.000147	0.000153	0.000159
-3.5	0.000165	0.000172	0.000178	0.000185	0.000193	0.0002	0.000208	0.000216	0.000224	0.000233
-3.4	0.000242	0.000251	0.00026	0.00027	0.00028	0.000291	0.000302	0.000313	0.000325	0.000337
-3.3	0.000349	0.000362	0.000376	0.00039	0.000404	0.000419	0.000434	0.00045	0.000466	0.000483
-3.2	0.000501	0.000519	0.000538	0.000557	0.000577	0.000598	0.000619	0.000641	0.000664	0.000687
-3.1	0.000711	0.000736	0.000762	0.000789	0.000816	0.000845	0.000874	0.000904	0.000935	0.000968
-3	0.001001	0.001035	0.00107	0.001107	0.001144	0.001183	0.001223	0.001264	0.001306	0.00135
-2.9	0.001395	0.001441	0.001489	0.001538	0.001589	0.001641	0.001695	0.00175	0.001807	0.001866
-2.8	0.001926	0.001988	0.002052	0.002118	0.002186	0.002256	0.002327	0.002401	0.002477	0.002555
-2.7	0.002635	0.002718	0.002803	0.00289	0.00298	0.003072	0.003167	0.003264	0.003364	0.003467
-2.6	0.003573	0.003681	0.003793	0.003907	0.004025	0.004145	0.004269	0.004396	0.004527	0.004661
-2.5	0.004799	0.00494	0.005085	0.005234	0.005386	0.005543	0.005703	0.005868	0.006037	0.00621
-2.4	0.006387	0.006569	0.006756	0.006947	0.007143	0.007344	0.007549	0.00776	0.007976	0.008198
-2.3	0.008424	0.008656	0.008894	0.009137	0.009387	0.009642	0.009903	0.01017	0.010444	0.010724
-2.2	0.011011	0.011304	0.011604	0.011911	0.012224	0.012545	0.012874	0.013209	0.013553	0.013903
-2.1	0.014262	0.014629	0.015003	0.015386	0.015778	0.016177	0.016586	0.017003	0.017429	0.017864
-2	0.018309	0.018763	0.019226	0.019699	0.020182	0.020675	0.021178	0.021692	0.022216	0.02275
-1.9	0.023295	0.023852	0.024419	0.024998	0.025588	0.02619	0.026803	0.027429	0.028067	0.028717
-1.8	0.029379	0.030054	0.030742	0.031443	0.032157	0.032884	0.033625	0.03438	0.035148	0.03593
-1.7	0.036727	0.037538	0.038364	0.039204	0.040059	0.04093	0.041815	0.042716	0.043633	0.044565
-1.6	0.045514	0.046479	0.04746	0.048457	0.049471	0.050503	0.051551	0.052616	0.053699	0.054799
-1.5	0.055917	0.057053	0.058208	0.05938	0.060571	0.06178	0.063008	0.064255	0.065522	0.066807
-1.4	0.068112	0.069437	0.070781	0.072145	0.073529	0.074934	0.076359	0.077804	0.07927	0.080757
-1.3	0.082264	0.083793	0.085343	0.086915	0.088508	0.090123	0.091759	0.093418	0.095098	0.0968
-1.2	0.098525	0.100273	0.102042	0.103835	0.10565	0.107488	0.109349	0.111232	0.113139	0.11507
-1.1	0.117023	0.119	0.121	0.123024	0.125072	0.127143	0.129238	0.131357	0.1335	0.135666
-1	0.137857	0.140071	0.14231	0.144572	0.146859	0.14917	0.151505	0.153864	0.156248	0.158655
-0.9	0.161087	0.163543	0.166023	0.168528	0.171056	0.173609	0.176186	0.178786	0.181411	0.18406
-0.8	0.186733	0.18943	0.19215	0.194895	0.197663	0.200454	0.203269	0.206108	0.20897	0.211855
-0.7	0.214764	0.217695	0.22065	0.223627	0.226627	0.22965	0.232695	0.235762	0.238852	0.241964
-0.6	0.245097	0.248252	0.251429	0.254627	0.257846	0.261086	0.264347	0.267629	0.270931	0.274253
-0.5	0.277595	0.280957	0.284339	0.28774	0.29116	0.294599	0.298056	0.301532	0.305026	0.308538
-0.4	0.312067	0.315614	0.319178	0.322758	0.326355	0.329969	0.333598	0.337243	0.340903	0.344578
-0.3	0.348268	0.351973	0.355691	0.359424	0.363169	0.366928	0.3707	0.374484	0.37828	0.382089
-0.2	0.385908	0.389739	0.39358	0.397432	0.401294	0.405165	0.409046	0.412936	0.416834	0.42074
-0.1	0.424655	0.428576	0.432505	0.436441	0.440382	0.44433	0.448283	0.452242	0.456205	0.460172
0	0.464144	0.468119	0.472097	0.476078	0.480061	0.484047	0.488034	0.492022	0.496011	0.5

Appendix 3: T-Table for $t_{\alpha,dof}$ Distribution

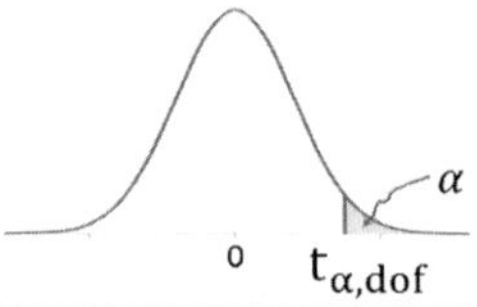

dof\α	0,4	0,25	0,1	0,05	0,025	0,01	0,005	0,0025	0,001	0,0005
1	0,3249	1,0000	3,0777	6,3138	12,71	31,82	63,66	127,32	318,31	636,62
2	0,2887	0,8165	1,8856	2,9200	4,3027	6,9646	9,9248	14,089	22,327	31,599
3	0,2767	0,7649	1,6377	2,3534	3,1824	4,5407	5,8409	7,4533	10,215	12,924
4	0,2707	0,7407	1,5332	2,1318	2,7764	3,7469	4,6041	5,5976	7,1732	8,6103
5	0,2672	0,7267	1,4759	2,0150	2,5706	3,3649	4,0321	4,7733	5,8934	6,8688
6	0,2648	0,7176	1,4398	1,9432	2,4469	3,1427	3,7074	4,3168	5,2076	5,9588
7	0,2632	0,7111	1,4149	1,8946	2,3646	2,9980	3,4995	4,0293	4,7853	5,4079
8	0,2619	0,7064	1,3968	1,8595	2,3060	2,8965	3,3554	3,8325	4,5008	5,0413
9	0,2610	0,7027	1,3830	1,8331	2,2622	2,8214	3,2498	3,6897	4,2968	4,7809
10	0,2602	0,6998	1,3722	1,8125	2,2281	2,7638	3,1693	3,5814	4,1437	4,5869
11	0,2596	0,6974	1,3634	1,7959	2,2010	2,7181	3,1058	3,4966	4,0247	4,4370
12	0,2590	0,6955	1,3562	1,7823	2,1788	2,6810	3,0545	3,4284	3,9296	4,3178
13	0,2586	0,6938	1,3502	1,7709	2,1604	2,6503	3,0123	3,3725	3,8520	4,2208
14	0,2582	0,6924	1,3450	1,7613	2,1448	2,6245	2,9768	3,3257	3,7874	4,1405
15	0,2579	0,6912	1,3406	1,7531	2,1314	2,6025	2,9467	3,2860	3,7328	4,0728
16	0,2576	0,6901	1,3368	1,7459	2,1199	2,5835	2,9208	3,2520	3,6862	4,0150
17	0,2573	0,6892	1,3334	1,7396	2,1098	2,5669	2,8982	3,2224	3,6458	3,9651
18	0,2571	0,6884	1,3304	1,7341	2,1009	2,5524	2,8784	3,1966	3,6105	3,9216
19	0,2569	0,6876	1,3277	1,7291	2,0930	2,5395	2,8609	3,1737	3,5794	3,8834
20	0,2567	0,6870	1,3253	1,7247	2,0860	2,5280	2,8453	3,1534	3,5518	3,8495
21	0,2566	0,6864	1,3232	1,7207	2,0796	2,5176	2,8314	3,1352	3,5272	3,8193
22	0,2564	0,6858	1,3212	1,7171	2,0739	2,5083	2,8188	3,1188	3,5050	3,7921
23	0,2563	0,6853	1,3195	1,7139	2,0687	2,4999	2,8073	3,1040	3,4850	3,7676
24	0,2562	0,6848	1,3178	1,7109	2,0639	2,4922	2,7969	3,0905	3,4668	3,7454
25	0,2561	0,6844	1,3163	1,7081	2,0595	2,4851	2,7874	3,0782	3,4502	3,7251
26	0,2560	0,6840	1,3150	1,7056	2,0555	2,4786	2,7787	3,0669	3,4350	3,7066
27	0,2559	0,6837	1,3137	1,7033	2,0518	2,4727	2,7707	3,0565	3,4210	3,6896
28	0,2558	0,6834	1,3125	1,7011	2,0484	2,4671	2,7633	3,0469	3,4082	3,6739
29	0,2557	0,6830	1,3114	1,6991	2,0452	2,4620	2,7564	3,0380	3,3962	3,6594
30	0,2556	0,6828	1,3104	1,6973	2,0423	2,4573	2,7500	3,0298	3,3852	3,6460
31	0,2555	0,6825	1,3095	1,6955	2,0395	2,4528	2,7440	3,0221	3,3749	3,6335
32	0,2555	0,6822	1,3086	1,6939	2,0369	2,4487	2,7385	3,0149	3,3653	3,6218
33	0,2554	0,6820	1,3077	1,6924	2,0345	2,4448	2,7333	3,0082	3,3563	3,6109
34	0,2553	0,6818	1,3070	1,6909	2,0322	2,4411	2,7284	3,0020	3,3479	3,6007
35	0,2553	0,6816	1,3062	1,6896	2,0301	2,4377	2,7238	2,9960	3,3400	3,5911
36	0,2552	0,6814	1,3055	1,6883	2,0281	2,4345	2,7195	2,9905	3,3326	3,5821
37	0,2552	0,6812	1,3049	1,6871	2,0262	2,4314	2,7154	2,9852	3,3256	3,5737
38	0,2551	0,6810	1,3042	1,6860	2,0244	2,4286	2,7116	2,9803	3,3190	3,5657
39	0,2551	0,6808	1,3036	1,6849	2,0227	2,4258	2,7079	2,9756	3,3128	3,5581
40	0,2550	0,6807	1,3031	1,6839	2,0211	2,4233	2,7045	2,9712	3,3069	3,5510
50	0,2547	0,6794	1,2987	1,6759	2,0086	2,4033	2,6778	2,9370	3,2614	3,4960
100	0,2540	0,6770	1,2901	1,6602	1,9840	2,3642	2,6259	2,8707	3,1737	3,3905

Appendix 4. Taguchi Array Selector (i.e. L4,L8,L9 and L12 Taguchi Designs)

LEVELS	NUMBER of FACTORS/INPUTS: 2	3	4	5	6	7	8	9	10	11	12	13	14	15	16	17	18	19	20	21	22	23	24	25	26	27	28	29	30	31	32
2	L4		L8				L12				L16				L32																
3	L9			L18				L27					L36																		
4	L'16				L'32																										
5	L25					L50																									

L4	A	B	C
	1	1	1
	1	2	2
	2	1	2
	2	2	1

L8	A	B	C	D	E	F	G
	1	1	1	1	1	1	1
	1	1	1	2	2	2	2
	1	2	2	1	1	2	2
	1	2	2	2	2	1	1
	2	1	2	1	2	1	2
	2	1	2	2	1	2	1
	2	2	1	1	2	2	1
	2	2	1	2	1	1	2

L9	A	B	C	D
	1	1	1	1
	1	2	2	2
	1	3	3	3
	2	1	2	3
	2	2	3	1
	2	3	1	2
	3	1	3	2
	3	2	1	3
	3	3	2	1

L12	A	B	C	D	E	F	G	H	J	K	L
	1	1	1	1	1	1	1	1	1	1	1
	1	1	1	1	1	2	2	2	2	2	2
	1	1	2	2	2	1	1	1	2	2	2
	1	2	1	2	2	1	2	2	1	1	2
	1	2	2	1	2	2	1	2	1	2	1
	1	2	2	2	1	2	2	1	2	1	1
	2	1	2	2	1	1	2	2	1	2	1
	2	1	2	1	2	2	2	1	1	1	2
	2	1	1	2	2	2	1	2	2	1	1
	2	2	2	1	1	1	1	2	2	1	2
	2	2	1	2	1	2	1	1	1	2	2
	2	2	1	1	2	1	2	1	2	2	1

Appendix 5. Table of Control Chart Factors

n	A_2	c_4	d_2	D_3	D_4
2	1.880	0.7979	1.128	0	3.267
3	1.023	0.8862	1.693	0	2.575
4	0.729	0.9213	2.059	0	2.282
5	0.577	0.9400	2.326	0	2.114
6	0.483	0.9515	2.534	0	2.004
7	0.419	0.9594	2.704	0.076	1.924
8	0.373	0.9650	2.847	0.136	1.864
9	0.337	0.9693	2.970	0.184	1.816
10	0.308	0.9727	3.078	0.223	1.777
11	0.285	0.9754	3.173	0.256	1.744
12	0.266	0.9776	3.258	0.283	1.717
13	0.249	0.9794	3.336	0.307	1.693
14	0.235	0.9810	3.407	0.328	1.672
15	0.223	0.9823	3.472	0.347	1.653
16	0.212	0.9835	3.532	0.363	1.637
17	0.203	0.9845	3.588	0.378	1.622
18	0.194	0.9854	3.640	0.391	1.609
19	0.187	0.9862	3.689	0.404	1.596
20	0.180	0.9869	3.735	0.415	1.585
21	0.173	0.9876	3.778	0.425	1.575
22	0.167	0.9882	3.819	0.435	1.565
23	0.162	0.9887	3.858	0.443	1.557
24	0.157	0.9892	3.895	0.452	1.548
25	0.153	0.9896	3.931	0.459	1.541

n=sample size

References

1. A. Field, Discovering Statistics Using SPSS, Third Edition, Sage, London, UK, 2009, ISBN 9781847879073

2. A. W. Bowman, D. R. Robinson, Introduction to Statistics, Adam Hilger, Bristol, UK, 1987, ISBN 0852744080

3. Anderson, Sweeney, Williams, Freeman, Shoesmith, Statistics for Business and Economics, Thomson Learning, London, 2007, ISBN 978-1-84480-313-2

4. B. G. Dennis, Cost of Quality, Quality Planning and the Bottom Line, 2000, G. Dennis Beecroft Ltd., Canada.

5. Chao-Ton Su, Quality Engineering - Off-Line Methods and Applications, 2013, CRC Press

5. D. A. Garvin, Competing on the eight dimensions of quality,1987, Harvard Business Review, 87(6):101-109

6. D. B. Relyea, The practical application of the process capability study, 2011, CRC Press, ISBN-13 978-1-4398-8464-5

7. D. Kannan, V. Lakshmikantham, Edited by, Handbook of Stochastic Analysis and Applications, Taylor & Francis, NY, 2001, ISBN 9780824706609

8. D. C. Montgomery, Introduction to Statistical Quality Control, John Wiley & Sons, 1991, New York

9. D. C. Montgomery, Design and Analysis of Experiments, Sixth Edition, Wiley, NY, 2004, ISBN 978-0-471-66159-7

10. D. C. Montgomery, G. C. Runger, Applied Statistics and Probability for Engineers, Wiley, NJ, 2006, ISBN 978-0-471-74589-1

11. D. C. Wood, The Executive Guide To Understanding and Implementing Quality Cost Programs, ASQ Published, 2007, United States of America.

12. E. P. Tobin, Poke-yoke for Engineers: An Introduction to Poke-Yoke for Engineers, 1st Edition, CRC Press, 2016, ISBN-13: 978-1534704268

13. G. Taguchi, E. A. Elsayed, and T. C. Hsiang, Quality Engineering in Production Systems, 1989, New York, McGraw-Hill

14. G. Taguchi , S. Chowdhury, Y. Wu, Taguchi's Quality Engineering Handbook, 2005, John Wiley & Sons, ISBN 0-471-41334-8

15. J. M. Juran, Quality Control Handbook, 1951, McGraw Hill

16. J. M. Juran, A. B. Godfrey, Juran's Quality Handbook, 1999, Fifth Edition, McGraw Hill Professional, , ISBN-13 978-0071386876

17. J. Neter, W. Wasserman, G.A. Whitmore, Applied Statistics, Third Edition, 1987, Allyn & Bacon, ISBN 0205103286

18. J. T. McClave, T. Sincich, Statistics, Prentice-Hall Inc., NJ, 2003, ISBN 978-0-130-65598-1

19. K. Hashmi, Introduction and Implementation of Total Quality Management (TQM), 2007, https://www.isixsigma.com/

20. M. A. Arslan, Introduction to Applied Statistics, 2017, S. Publishing, MAA, NY

21. Minitab16 Help, 2013

22. M. L. Berenson, D. M. Levine, Basic Business Statistics Concept and Applications, Fifth Edition, Prentice Hall, NJ, 1992, ISBN 0130995576

23. Microsoft Office *Excel* Help, 2016

24. M. S. Phadke, Quality Engineering Using Robust Design, 1989, Prentice Hall, ISBN 0-13-745167-9

25. P. B. Crosby, Quality is free, 1980, Mentor, McGraw-Hill, ISBN 0-451-62585-4

26. R. D. Moen, T. W. Nolan, L. P. Provost, Improving Quality Through Planned Experimentation, 1991, McGraw-Hill Book Company, New York

27. R. V. Hogg, J. Ledolter, Engineering Statistics, Macmillan Publishing Company, NY, Collier Macmillan Publishers Canada, 1987, ISBN 0-02-355790-7

28. S. Kotz, C. R. Lovelace, Process capability indices in theory and practice, 1998, Hodder Arnold Publication , ISBN-13 978-0-3406-9177-9

29. S. Shingo, Zero Quality Control: Source Inspection and the Poka-Yoke System, 1986, Productivity Press, ISBN: 0-915299-07-0

30. N. W. Polhemus, Process capability analysis – estimating quality, 2017, CRC Press, ISBN-13 978-1-138-03015-2

31. N. Singh, Systems Approach to Computer-Integrated Design and Manufacturing, John Wiley & Sons, 1996, ISBN 0-471-58517-3

32. N. R. Tague, *The Quality Toolbox, Second Edition*, 2005, ASQ Quality Press

33. P. J. Huber, Robust Statistics, Wiley-Inter Science, Wiley Series in Probability and Statistics, NJ, 2005, ISBN 0471650722

34. W. E. Deming, Out of the Crisis, 2000, First MIT Press, ISBN 978-0-2622-5056-6

35. W. Y. Fowlkes, C. M. Creveling, Engineering Methods for Robust Product Design, Addison Wesley, MA, 1995, ISBN 9780201633672

Index